A HI$

O⊦

THE FIRM OF

CHANCE BROTHERS & CO.

GLASS AND ALKALI MANUFACTURERS

ROBERT LUCAS CHANCE THE ELDER

A HISTORY
OF
THE FIRM OF

CHANCE BROTHERS & CO.

GLASS AND ALKALI MANUFACTURERS

BY

JAMES FREDERICK CHANCE

A History of the Firm of Chance Brothers & Co., Glass and Alkali Manufacturers. by J. F. Chance

Edited by Michael Cable, 2016

The eighth in a series of books written by authorities of their times showing how understanding of Glass Technology has developed since the seventeenth century:

Volume 1. The Art of Glass by Christopher Merrett (1662)
Volume 2. Bosc D'Antic on Glass Making (1758–80)
Volume 3. Early 19th Century Glass Technology in Austria and Germany (1820–37)
Volume 4. Apsley Pellatt on glass making (1807–49)
Volume 5. Georges Bontemps on glass making (1868)
Volume 6. Zschimmer's Chemical Technology of Glass (1913)
Volume 7. Glass Manufacture by Walter Rosenhain (1919)

© Published by the Society of Glass Technology, 2018

The objects of the Society are to encourage and advance the study of the history, art, science, design, manufacture, after treatment, distribution and use of glass of any and every kind. These aims are furthered by meetings, publications, the maintenance of a library and the promotion of association with other interested persons and organisations.

Society of Glass Technology
9, Churchill Way
Chapeltown
Sheffield S32 2PY
UK

Registered Charity No. 237438

Web site: http://www.sgt.org

ISBN 978 0 900682 80 3

CONTENTS

LIST OF ILLUSTRATIONS

A REMARKABLE SERIES OF BOOKS CREATED BY A REMARKABLE MAN

As the Third Millennium was approaching, there were some very productive discussions at the Society of Glass Technology about the historical and heritage-related aspects of glass. Professor W. E. S. Turner, founder of the Society, was throughout his life interested in these aspects of glass studies as well as his major mission to bring the light of science to bear on this most enigmatic state of matter. We felt that it was time to reinvigorate this aspect of the Society's life and work.

The Annual Conference began to devote significant amounts of programme time to the history and heritage of glass, a thread which soon attracted authors and delegates and which enriched the discussion and debate, proving a fertile ground for creative thinking and cross disciplinary collaborations. But it was Professor Michael Cable who perceived a huge lacuna in our knowledge and appreciation. Books about glassmaking and the mysteries involved in its manufacture had been produced in a number of languages during the second millennium, but most of these were no longer readily available for study.

Professor Cable recognised that these works not only gave insight into the development of glass science and technology, but also the evolution of the thought pattern and techniques which underlie today's scientific endeavour. He identified a number of major works which were clear milestones charting the path towards the technological understanding of glass which plays such a crucial role in shaping our lives and lifestyles.

Beginning with Neri's work of 1611, Professor Cable laboured to bring each of these formative texts once more to the light of day and make them accessible in the English language. Each book required painstaking research to establish the most authentic text and to place that text in its historical context. Many of the books required translation, and for such material as this, translation is not obvious - it requires great depth of understanding if the result is to be faithful to the intentions of the original authors.

Those of us who benefited as students from the mentoring of Professor Cable remember well his exacting standards. He required our work to be diligent and precise, and underpinned by understanding and creative thought. These things were the foundations of his own work, and he drove himself harder than he drove us. So these eight books provide a valuable and reliable resource for study, relevant to all researchers whatever their discipline or interest.

Sadly, Professor Cable was not to see his final volume in published form. He died on 20th August 2016 at the age of 82. We owe Michael Cable a considerable debt of gratitude for bringing into being these volumes which so enrich our appreciation of the glassmaking heritage bequeathed to us by previous generations.

David Martlew, HonFSGT

CHANCE HISTORY: FOREWORD

Few glass making firms have been the subject of detailed published historical studies. The one exception is, as might be expected Saint Gobain, the first successful maker's of plate glass, in France. Over the years several studies of the company's development have been published. Those include a notable study of the life of the village of Saint Gobain and the intimate links between the village and the company in the eighteenth century.

The best-known study of an English Glass-maker is the detailed account of Pilkington Brothers, Britain's most important manufacturers of plate glass the history of which was thoroughly investigated and reported in two notable volumes by T. C. Barker *Pilkington Brothers and the Glass Industry* (1960) and *The Glassmakers; Pilkington 1826 – 1976* (1977). The former volume covered the period up to 1918 and the latter covered the years from the establishment of the company early in the nineteenth century to the first decade of Float glass manufacture.

Corning Glass Works has been the outstanding firm in the United States for more than a century. Having established a research laboratory in the early years of the twentieth century Corning, under the leadership of the Houghton family, the company was from that time a leader in meeting new challenges for the use of glass. Some of its most famous developments were the production of Pyrex borosilicate glass and the manufacture of the Ribbon Machine for making the bulbs needed for incandescent lamps. The history of Corning published in 2001 as *Corning and the Craft of Innovation* discusses the ways in which the Company responded to important technological challenges over the years but does not discuss technical matters.

A detailed study of Chance Brothers was first published in 1919 but only intended for private circulation. This volume, *A History of Chance Brothers, Glass and Alkali Manufacturers* is, as a result, little known and rarely referred to.

This is the eighth and last in a series of books, published chronologically, which cover the period from 1612 to 1926. These were written by acknowledged European authorities and allow modern readers to gain some insight into understanding of glass technology at the times when these books were written. Knowledge of chemistry and physics was very primitive in 1612, it was still believed that all substances were composed of four elements, air, earth, fire and water. By 1926 both disciplines had reached sufficiently modern understanding to deal with almost all matters that usually concern glass although studies of glass structure were a decade away from the pioneering x-ray studies of Warren and his successors. The pioneering work on strength of glass by Griffith had been done but relations between glass composition and properties, of great practical importance, remained a mystery and are still a major field of researches today.

Christopher Merrett

The first, Christopher Merrett's *Art of Glass* (1662) a translation of Antonio Neri's *L'Arte Vetraria* published in Florence in 1612, is the most famous of all books on glass making but, although it gives dozens of recipes for coloured glasses, it provides little insight, there being little understanding of physics or chemistry at that time. Merrett was a physician and an early Fellow of the Royal Society who knew something of glass making and added notes of his own to Neri's text.

Bosc d'Antic

The next book shows a considerable advance; *Bosc d'Antic on Glass Making* (1758–80) shows a knowledgeable French author struggling to understand glass-making when the classic four element theory that all materials were made from four fundamental chemical elements air, earth, fire, and water which might be equivalent to solid, liquid, gas and heat, was just about to collapse. Bosc was a colourful character, a Protestant who qualified as a physician by making a surprisingly brief trip to Holland. Quite early in his career, having won a prize for an essay on how to improve French glass making, he was briefly involved with Saint-Gobain at a time when the Company was having serious problems. His activities there were not appreciated by the older more experienced staff who were expected to defer to him and he was soon dismissed. Towards the end of his life he attempted to set up a school to train glass makers but he was refused a licence and it failed. His introduction to his collected works contains some bitter heartfelt comments about financiers who join industrial organizations chiefly to abuse the funds.

B. Scholz & W. E. Kirn

The third book, *Early Nineteenth Century Glass Technology in Austria and Germany* (1820–1837), has works by two sadly neglected authors writing in German. Benjamin Scholz was a physician who became the first Professor of Applied Chemistry in Vienna. Amongst his other duties he was official observer of attempts to develop an effective way of using saltcake instead of potash to make colourless glass. The Emperor had offered a prize for a successful process but the main contender was a charlatan and Scholz obviously took pleasure revealing that. Scholz's understanding of some aspects of glass (such as solarization) was surprisingly advanced for his time.

Wilhelm Emerich Kirn was the second of four brothers two of whom became army officers and one a magistrate. He became an industrialist but we do not know where he gained his experience of glass making. However, he was employed to find ways of improving glass making in Württemberg. So, most unusually he was free to experiment and to publish the results of his investigations. His detailed accounts of glass making had been totally neglected for more than a century. By that time the modern basis of chemistry was fairly well established but all kinds of glass making except plate glass relied on gathering glass on a blowpipe.

Apsley Pellatt, father and son

The fourth, *Apsley Pellatt on Glass Making*, reprints works by two Apsley Pellatts, father and son, who were prestigious London glass makers, the son taking over the business when the father died in 1826. The father took out a patent for glass lights to illuminate the interiors of ships in 1807. Apsley junior, who was a Member of Parliament for some years, had seven publications, the most important was the last *Curiosities of Glass Making* (1848) which he wrote for interested members of the public, no doubt hoping to increase his firm's business. It describes the procedures used to make several different types of glassware but has no detailed science or technology.

Georges Bontemps

Georges Bontemps was the outstanding French glass works manager of his time. His *Guide du Verrier* (1868) is the most detailed critical account of making the main types of glass to be written by a knowledgeable author. The section on each major product ends with a detailed analysis of costs of manufacture; Bontemps hoped that glass makers of subsequent times would benefit from comparing their practices with those of his time. Bontemps was manager of the works at Choisy-le-Roi for about 25 years and was responsible for several important advances, for example rediscovering how to make copper ruby glass and how to reproduce old Venetian filigree work. He left there in a time of particularly bad political turbulence in France and moved to Chance Brothers in Smethwick where he had for about 15 years been assisting them in various ways especially over making sheet glass and optical glass. He was particularly engaged in setting up their manufacture of optical glass. He died a very rich man at an advanced age.

Eberhard Zschimmer

The sixth book *The Chemical Technology of Glass* (1913) by Eberhard Zschimmer is another important book with a surprising history. Zschimmer had trained as a mineralogist and joined Otto Schott's Jena Glass works in 1898 soon after completing his studies. He was a man of outstanding intellect who was interested in everything to do with glasses and soon became Schott's senior scientist. In 1913 he completed that long book, *The Chemical Technology of Glass* in which he critically reviewed understanding of most important aspects of glasses and various important problems. Schott had become a major supplier of optical glasses and one the questions considered in great detail was whether it might be possible to develop good models for predicting properties from composition (or vice versa). His main hope evidently was to develop accurate models for predicting properties from composition or vice versa: a problem that still occupies many workers. Another topic, then very important but now rarely considered, was the ageing of mercury in glass thermometers: his own investigations in that subject included studying one group of thermometers over a period of four years. The book was printed privately and dedicated to Otto Schott but, unfortunately, Schott was dismayed to see how much information it contained that he did not wish to be known outside the Company. He then bought the whole printing and destroyed all but a few copies. Consequently the book was soon forgotten until two copies came to light a few years ago. It contains much information worth studying as it shows clearly the state of knowledge about glasses and their most important properties only a few years before x-ray diffraction made it possible to begin study of glass structures and gain better insight. When Zschimmer left Schott's company he joined the Technical University at Karlsruhe where he soon became Germany's first Professor of Glass Technology. He was a serious student of philosophy, becoming a committed socialist. He published several books on philosophy. He was also very interested in the fine arts.

Walter Rosenhain

Walter Rosenhain's parents emigrated from Germany to Australia when Walter was a child. He proved to be an outstanding student graduating at Queen's College, Melbourne University in 1896. He was awarded a scholarship at Cambridge where he studied at St Johns College. There he did pioneering work together with A. Ewing on the deformation of metals.

His first position after completing researches at Cambridge was as scientific adviser to Chance Brothers where he investigated possibilities of improving the range and quality of optical glasses that the firm made. Although he soon left to take up an important post in metallurgical researches at the National Physical Laboratory, he retained a lifelong interest on glasses. He thought that too little science was employed in the glass industry and wrote *Glass Manufacture* as an outline of the main types of glass and the industry's main branches in 1908, hoping to stimulate interest. The industry had, of necessity, made considerable advances during the First World War and he thought that those advances deserved a revision of *Glass Manufacture* which was published in 1919 [the seventh in this series of volumes] and was an important review by an important materials scientist. Walter Rosenhain was a founding member of the Society of Glass Technology, the fifth Honorary FSGT, and President of The Institute of Metals.

Chance Brothers

The present book, volume 8, was published (privately) in 1919 and is of a different type, it may be described as commercial rather than technological.

Members of the Chance family first had contact with glass making through a family connection with the Nailsea glass works at Bristol. Robert Lucas Chance bought the British Crown Glass Company's works in Spon Lane, Smethwick in 1824. The company specialised in making crown window glass. The company ran into difficulty and its survival was guaranteed in 1832 by investment from Lucas's brother, William who owned an iron merchants in Birmingham. After the partnership with James and John Hartley was dissolved in 1836, Lucas and William Chance became partners in the business, which was renamed *Chance Brothers and Company*. In 1838 James Timmins Chance (1814–1902) joined and became a partner in 1839.

Chance Brothers was amongst the earliest glass works to carry out the cylinder process in Europe making the first British cylinder blown sheet glass using French and Belgian workers. In 1848, under the supervision of Georges Bontemps, a new plant was set up to manufacture crown and flint glass for lighthouse optics, telescopes and cameras. The optical glass work led to the supply of 2300 lighthouse lanterns around the world. Chance Brothers provided the glazing for the Crystal Palace to house the Great Exhibition of 1851, Smethwick neighbour Fox, Henderson and Company having won the contract to supply the ironwork and build Joseph Paxton's design. There are accounts of trials of new technology offered by Henry Bessemer, not a success, and the Siemens brothers, rapidly adopted. In 1889, the firm was incorporated as Chance Brothers and Co. Limited.

After the publication of the *History*, the company continued to work on the many new ways of making glass evolved at Chance Brothers such as the innovative welding of a cathode ray tube John Logie Baird used for television; 1929–1930 began producing pressed domestic glassware; in 1935 the company was renamed Chance Brothers Limited; in 1947 Chance set up plant in Malvern for the manufacture of syringes and precision tubing, producing interchangeable barrels and plungers for syringes.

Pilkington Brothers acquired a 50% shareholding in 1945 and by the end of 1952 Pilkington had assumed full financial control, but were not actively involved in its management until the mid- to late-1960s. The production of flat glass ceased at Smethwick in 1976. The remainder of the works closed in 1981 ending more than 150 years of glass production at Smethwick and all flat glass production was absorbed by Pilkington's St Helens factories.

In 1992, during a period of rationalisation at Pilkington, a management buy-out reverted the Chance plant in Malvern to private ownership and it became an independent company, changing its registered name to Chance Glass Limited and retaining the historical Chance logo. To this day the company has continued to develop its range of glass tubular components.

The Spon Lane site is the focus of the Chance Glass Works Heritage Trust's work in rejuvenating the area, conserving the listed buildings, restoring economic activity, promoting cultural and heritage values.

Michael Cable
Department of Engineering Materials
University of Sheffield

WILLIAM CHANCE
(died 1856)
from an oil painting by Thomas Philips R.A.
(To face page 3)

A HISTORY
OF
CHANCE BROTHERS & CO.

CHAPTER I

1824 TO 1836

THE Chances of Birmingham have been concerned with the manufacture of glass for five generations, from the time when, in 1793 or earlier, William Chance and Edward Homer, hardware merchants of Birmingham, joined in partnership with their brother-in-law John Robert Lucas, glass manufacturer of Bristol, in respect of new undertakings of his at Nailsea and Stanton Drew, or Wick, near that city. As a consequence, Edward Homer transferred his residence thither about 1794. William Chance, remaining in Birmingham, continued his partnership until 1821, when he retired from it and the direct concern of his family with Nailsea ceased, to be renewed after the lapse of half a century. But from his interest in these works resulted the establishment of his descendants at Spon Lane.

In 1811 his eldest son, Robert Lucas Chance, a young man of exceptional activity and capacity, said to have managed the Birmingham business from the age of fourteen, joined his uncles at Nailsea, marrying his cousin Louisa Homer. He did not stay there long, leaving in 1815 to set up as a glass merchant in London. But in the four years he learnt the manufacture, and was able, when it seemed good, to venture upon it on his own account. On May 18, 1824, he bought from Joseph Stock and others the works of the British Crown Glass Company and other property at Smethwick, near Birmingham, altogether some fifteen acres of what had been "Blakeley's Farm." He wrote of his prospects to his brother Henry, a barrister of Lincoln's Inn: "I have every reason for thinking that the concern will realize the most sanguine expectations I have form'd, and it presents a scope for the exercise of my acquirements as a man of business."

The property extended from the old high level Birmingham Canal to the Oldbury Road.[1] On the canal the length of its frontage was about 330 yards, starting from a point about 40 yards west of Spon Lane. That on the Oldbury Road ran for about 150 yards eastward from the gate afterwards known as the "Britannia" by No. 12 house. By that gate was entered the "Fordrove" occupation road, which formed the western boundary of the property there. Fifty yards or so to the south of the canal ran in the same direction a road, originally intended for a street with houses. South of this the property narrowed from the east by 50 yards, and here adjoined it a square piece of land known as "Phillips's," where now stand the offices. Fifty yards farther south the boundary again receded westwards, the property continuing as a strip about 150 yards wide as far as the Oldbury Road. The original glass-house, No. 1, and the two built by Robert Lucas Chance, No. 2 in 1824 and No. 3 in 1828, were situated from east to west between the street mentioned and the canal. All the rest, in 1824, was open country. The house that still stands embedded in the works was then "The Hall," a new mansion with kitchen and flower gardens, approached from the said street by a footpath to the east and by a carriage road to the west, on which road, further afield, were cottages known then, or later, as "Scotch Row." When the lower canal was cut, this road was carried over it by "Hartley Bridge."

Lucas Chance, as he was always called, having built, as said, a second glass-house, rapidly extended his business. It was largely export; in July 1827, for instance, he wrote to his brother Henry: "I have £20,000 in foreign hands, very secure, but coming home slowly." But it was not possible for him properly to conduct at once the sale of glass in London and its manufacture at Smethwick, although he thought little of travelling to Birmingham by one night coach and returning by the next. When at Nailsea he had brought thither from Dumbarton John Hartley, reputed to be the leading crown glass expert in the kingdom[2] and now he cast eyes on him again. A partnership was arranged between them as from April 1, 1828, the date of expiry of Hartley's last engagement with the Nailsea firm. He was to have a salary of £400 a year, with a house, coals, &c, and a fifteenth share of the profits, finding a capital of £4,000. Profits to be drawn were limited to £50 per share, a total of £3,000; any surplus to be

---

1   When the low level canal was cut, a few years later, the line of the high level was carried more to the north-west and over the other by the present aqueduct.

2   The story goes that Lucas Chance travelled to Dumbarton, roused Hartley out of bed, and brought him straight away.

put to capital. With him, or soon afterwards, came his elder son James to assist in the management. The style of the "British Crown Glass Company" was retained.

Lucas Chance wrote to his brother Henry: "I could not resist the temptation of giving you information that I am sure you will rejoice at." He did not expect increase of capital to be required, rather the contrary.[1] He saw great promise from "one year's good management at Smethwick under Hartley" and his own credit "vastly encreased ... I have so often been disappointed, that I dare not calculate on any thing, but the probabilities are, that I shall establish on a solid basis a manufactory that will be a credit to the family, and perhaps to the neighbourhood, with an income sufficient to make all my sons glass manufacturers." Hartley was "perfectly satisfied, and as he places the most implicit confidence in me, the engagement is in every respect every thing I wish." And of the feelings of his relatives at Nailsea: I really know of no reason for their being angry, excepting that they have lent me their pot maker, but then any other house would have done the same."[2]

However, events did not turn out as he anticipated. In 1831 he found himself in difficulties: his liabilities for capital raised on mortgage had been increased by the erection of a third glass-house and other extensions; he had incurred heavy loss by a venture with an inventor named Badams;[3] and there was great depression in the glass trade, witness memorials of the time by an association of the manufacturers for a change in the mode of levying the window tax, calculated greatly to increase the demand for glass. The high standing of Lucas Chance in the trade is shown by the fact that he was chosen for their chairman, and headed a deputation to the Chancellor of the Exchequer.

The crisis was surmounted by the aid of his brother William, a leading

---

1   "But I deem it to be prudent, with a prospect of an income as large as I anticipate, to provide as much capital as possible, to secure myself against all those contingencies to which at present all money transactions seem so exceedingly liable."

2   Letters to Henry Chance, July 1827.

3   "In regard to Badams I may lose money by him or I may gain a fortune, and as I have no fancy for things that appear to hinge on chance I am sorry to have had anything to do with him. Still, on review of all the circumstances, I have nothing to blame myself for." Liabilities in the matter were considerably reduced and repayments expected to go on steadily (To the same, July 7, 1827). British patent No. 5174, of May 16, 1823, is an application by John Badams, of Ashted, near Birmingham, chemist, concerning a new method of extracting certain metals from their ores and purifying them. No specification was enrolled. This may, or may not, have been the invention in question.

merchant of Birmingham[1] engaged with a third brother, George, in trade with America. He found capital, took over for the time the freehold of the property, and in the following year joined the partnership. Lucas Chance now removed his residence from London to Handsworth, in order to be able to give the works his regular supervision. This was the more necessary, in that the firm was now undertaking the establishment in England of the French mode of making what was known—and for many years continued to be known—as "German"or"Bohemian" sheet glass.[2]

This, and polished cast plate, are now the only kinds of glass in use for ordinary windows. In 1832, on the contrary, what was in all but exclusive use in England for the purpose was crown glass. Plate, both cast and blown, was made at two works[3] but still was chiefly used for mirrors. Sheet glass, already victorious over crown on the Continent and well established as a manufacture in the United States,[4] was barely known in England; it was made there only in the very rough form known as "spread " or "broad" glass, and in face of the heavy duties very little was imported. Isaac Cookson & Co. had attempted the manufacture, but had failed to produce an article that could compete even with the coarsest qualities of British crown. It is clear from statements made on their behalf to the Commissioners appointed to inquire into the working of the Excise Laws that they knew nothing of the methods of the French and Belgians.[5]

---

1   In the previous year he had served the office of chief magistrate, High Bailiff, of the town. It fell to him to receive in this capacity, firstly the Princess Victoria, and afterwards the Duke of Wellington and Sir Robert Peel, whose visit was partly arranged to convince them that Birmingham must have representation in Parliament. Such was their unpopularity, that when William Chance entertained them and some seven hundred others at the Royal Hotel, the windows were broken by the mob. The glass works at Spon Lane were among the industrial establishments visited both by them and by the Princess.

2   So called from the fact that its manufacture had been reintroduced into France, early in the eighteenth century, by the aid of workmen from Germany, or rather from Bohemia.

3   The British Plate Glass Company, of Ravenhead in Lancashire, the pioneers of 1773, and Isaac Cookson & Co., of Newcastle and South Shields, who entered on the manufacture some forty years later.

4   W.F. Reuss, a London merchant, states in his *Calculations and Statements relative to the Trade between Great Britain and the United States of America* (London, 1833), p. 276, that twenty-three sheet glass factories were working there. He estimated the annual output at 5,625,000 square feet, worth $850,000. His statement is borne out by that of Reynell, p. 11, note 3.

5   Isaac Cookson and William Cuthbert, giving evidence in November 1833, stated that their attempt had been of the nature of an experiment and that the glass, made from coarser metal than crown, was very inferior and would not sell. " It is what we call cockled ; it is more like horn; . . . There is no consumption in this country. . . . What we made we made for exportation. We found no demand for it in England, and did not go on with it. . . . Crown glass is not much manufactured on the Continent; they use glass manufactured in a different way, the German sheet and the inferior

Plate, crown, and sheet glass each have their advantages and their disadvantages. Made from the same materials melted together in very nearly the same proportions, they derive their differences from the processes of their manufacture.[1]

Cast plate is made by pouring the molten "metal" upon an iron table and rolling it out. Its surfaces being rough, for use of the glass in windows or for mirrors they must be ground and polished. The plates, of perfect transparency and often of immense size, are now in use wherever weight and expense are no objection.

Crown glass excels in brilliance and transparency, but yields in other respects to plate and sheet. The gathered metal is blown into the ultimate form of a large disc, known as a "table" having at its centre a lump, the "bullion" or "bull's eye." This, and the circular form of the tables, prevent the cutting of large rectangular panes from them, and there is much waste. Panes have been made of as great an area as six square feet, but the largest usual sizes have hardly exceeded four. Thickness, also, is limited to 18 oz. per square foot, the "usual" substance being 13 oz. and other, for special purposes, 9 oz. And as the tables have a slight convexity the panes, unless flattened, show a distinct curvature and distort the vision. Owing to the manufacture having been brought to great perfection in England, and the British public habituated to the small bright panes, it survived there for many years longer than on the Continent. Now, save for small quantities made at Spon Lane occasionally, it is extinct.

Sheet glass is made by blowing the metal into the form of a hollow cylinder, whose dimensions and the thickness of whose walls the workman judges with a skill that can be acquired in perfection only by years of practice. When finished, the cylinder is cut longitudinally and reheated in a flattening kiln, or "lear," where it is opened out into a flat sheet, limited in size and thickness only by the weight of glass that the blower can wield,

---

glass, I do not know exactly the mode of making it, but it is like the spread glass, a very low-priced glass." Lucas Chance, who did know all about the manufacture as carried on in France and Belgium, testified on this point in language diametrically opposite (Thirteenth Report of the Commissioners, on Glass, 1835, pp. 27, 28, 95, 133-4).

1   The processes are described in full detail by Georges Bontemps in his Guide du Verrier (Paris, 1868), by Henry Chance and H. G. Harris in The principles of Glass Making, one of the Technological Handbooks of the Society of Arts (London, 1883), and in the Cambridge edition, of the *Encyclopædia Britannica*. Henry Chance's contribution to the Principles— on the manufacture of crown and sheet glass—is, as he states, for the most part a reproduction of his paper read before the Society of Arts in 1856.

and available for use, if required, as a single pane. These may have an area of as much as twenty square feet, entailing yet larger dimensions for the other sheets, the "lagres," on which they are flattened in the lears. Waste is avoided, and thickness may be greatly varied up to a weight of 42 oz. per square foot. Disadvantages are unevenness of surface and comparative lack of brilliance.[1]

In the introduction of this manufacture into England Lucas Chance was associated with the eminent French manufacturer, Georges Bontemps. The latter had paid his first visit to England in 1828, in connexion, it may be, with his work of producing large lenses for telescopes,[2] and Lucas Chance went with one of the Hartleys to see his works at Choisy-le-Roi in 1830. As the result, in spite, as he stated at a later time,[3] of strongest opposition by his partners, he made his decision to embark on the new venture. Intermediary in the affair was a mutual friend, A. Claudet, a glass merchant in London, through whom Bontemps had for some years past imported into England sheet glass and shades.

At the same time that Bontemps, in one of his works, sets forth the advantage of the large sheets, he calls attention to the difficulties attendant on the enterprise.[4] The chief of these difficulties was to get the necessary workmen from abroad—partly from their unwillingness to leave home, partly from their self-imposed restrictions on their handicraft. A law that they had long since made unto themselves forbade instruction in the art to any but members of their own families, and it was only now beginning to be broken through.[5] However, with Bontemps' aid, this and other obstacles were overcome, and the manufacture was begun in No. 2 house at Spon Lane, to be known now as the "French house" in August 1832. The furnace had eight small pots, very soon to be changed for the somewhat larger ones for which it had been designed.

`There was one peculiar advantage in the undertaking, duly taken into

---

1   "It is evident that sheet glass will never acquire so fine a surface as crown glass, unless some method can be discovered of spreading it at a high temperature, and without contact with any area" (Bontemps, *Report on the Paris Universal Exhibition*, 1855, ii, 390).

2   See his *Guide du Verrier*, pp. 653-4.

3   To James Chance, November 22, 1848.

4   *Examen historique et critique des verres, vitraux, et cristaux, composant la XXIVe classe de l'Exposition Universelle de 1851*, pp. 20-3. Of the Spon Lane undertaking he writes : "Il a fallu toute leur persistence et leur energie pour résister a toutes les difficultés que cette nouvelle fabrication a rencontrées dans le principe."

5   For particulars on this subject, see the *Guide du Verrier* pp. 179, 180, 232-3.

account by Lucas Chance and Bontemps, which arose from an eccentricity of the excise laws. For their purpose sheet glass counted as crown, paying a duty of £3 13s. 6d per cwt.[1] Since the duty was levied on the whole weight made, and since, for export, the crown tables were mostly cut up into panes, to compensate for the waste in cutting them a "drawback" of £4 18s. per cwt. on such export was allowed. For sheet glass the terms were the same, although the waste in trimming the sheets was very small. So that a large portion of the drawback was a clear gift to the sheet glass manufacturer. It was shown to the Commissioners above-named by the rivals of the Spon Lane firm that this gift had amounted in 1833 to £1,267 2s. 8d, and it was asserted that the manufacture had been undertaken for the sake of the bounty, and that the duty on sheet glass ought to be £4 10s.[2] As a fact, the drawback was reduced in 1836 to £4 4s. per cwt., and again in 1838 to £4.

It was one thing to get the foreign workmen over, another to control them when arrived. In spite of their high wages they were dissatisfied with their surroundings, came to work late and left early, blew their cylinders of short weight, had an excessive proportion of breakage, and failed to produce more than about two-thirds of their agreed quantity of 120 per "journey."[3] And they set themselves steadily, in accordance with their customary law, against instructing Englishmen. Two flatteners, for instance, declined to do this for a less sum than 10,000 francs, while not objecting to Englishmen working by themselves at the other lear. They were offered 1,500 francs to teach two; in case of refusal, Bontemps' firm to be written to about their insubordination and desired to replace them.[4]

It must be noted that at this time each blower worked out his pot by himself, helped only by a boy, his *gamin*. Gatherers and separate blowing holes were yet of the future. This was the main reason for the small size of the pots as compared with those in use in the crown houses; each might

---

1   Much more than double the cost of manufacture. Aggregate Spon Lane figures for the sheet glass for the five years 1834-1838 show cost of manufacture £58,326, excise duty £133,422 10s.; and for the crown glass, cost £180,335, duty £457,224 12s.

2   Thirteenth report, cited, pp. 27, 95, 158. For the same five years the drawback exceeded the duty by 15½ per cent.

3   *Journée*, a day's work.

4   From the Board minutes. Reynell reported on February 21, 1834, from Charleroi that the best of the blowers at Spon Lane, Gaspard Andre, had the reputation of making his cylinders thin in order to increase their number. A minute of August 2, 1833, states that it was the men's interest to produce as little glass as possible.

only contain so much metal as the blower could deal with in the journey by himself. The usual size of the finished sheet was 36 inches by 24, and it is related that but one man could be found able to make the longer ones required about this time by Joseph Paxton for the Chatsworth conservatories, the forerunners of his "Crystal Palace."[1]

For wages, we have a definite statement by Lucas Chance that he engaged the first men to work by the piece, at 50 per cent, above the Belgian tariff.[2] In August 1833 it was resolved to engage one, recommended by Bontemps, at not exceeding 12 centimes per cylinder of ordinary dimensions, he paying his *gamin*. Before long this arrangement was altered; the men were engaged by the month, and at a higher rate. For instance, in January 1836, a contract with "le grand" Meyer was for 400 francs per month. But he, no doubt, was a man who could work large sizes and heavy weights. In 1845, when the removal of the duties on glass gave rise to an extraordinary expansion of the manufacture and heated competition for the services of foreign workmen, such best men had to be paid from 500 francs a month for 30 hours' work per week to 600 for 40 hours, while ordinary ones were engaged at 400 and 500 respectively.[3]

Among the blowers were two Germans, who differed from the French and Belgians in their methods. For one thing, they made their cylinders short and wide, instead of having the greater dimension in the length. This has the advantage that the bubbles in the glass are round instead of elongated, and so less noticeable; on the other hand the cap of the cylinder, that has to be cut off, is wider, whence more waste. But the important difference was in the men's habits of work. They had been accustomed at home to eighteen hours or more at a stretch three times a week. They paid great attention to quality, says Reynell (of whom below), "continually skimming, which I believe the Frenchmen never do," but they were slow, taking ten hours to work out the pots as against the seven or eight of the others. That meant keeping up the heat of the furnace specially for them, and made it difficult to bring in the five journeys in the week. It was discussed whether a second sheet glass furnace should not be built, to be worked by Germans only on their plan. Reynell thought that he could procure a sufficient number. The decision was for maximum production;

---

1   *The Crystal Palace and the Great Exhibition* (London, 1851) p. 34
2   Letter to William Chance, August 8, 1841.
3   Cf. on the above the *Guide du Verrier* pp. 116-17, 245-7, 273-5.

the Germans were informed that if they could not conform to the French methods they must go. Early in 1834 they left.[1]

One great improvement, however, was due to them, the use of clay rings in the pots. These, floating on the molten glass, reduce the area to be skimmed and limit the gathering to the better metal in the centre of the pot. Their employment by the Germans was followed by trial in a shade pot by the French. A few months later, when Bontemps had written of their advantages, they were tried again, and by the end of 1834 were in partial use in all the houses.[2]

Recognition of the need of full production by no means excluded that of the importance of quality. This is not surprising, in view of the vastly increased value of superior glass.[3] Bontemps in his work cited states that Chance Brothers & Co. must have abandoned the manufacture of sheet glass, had they not made glass of quality superior to that usual in France and Belgium. For there cheap glass was in principal demand, but in England the contrary. Moreover, good glass could best bear the burden of the excise duties and best prevail for export. To obtain quality, he says, methods of founding and of work had to be modified—for instance, the cylinders were blown wider, since then glass of better surface was obtained.[4] In accordance with these views runs a Board minute of August 1833:

> It would greatly increase the profit of the French house, were we to make such an article of glass as would command a sale in this country, and provided we kept to our present weight of 14 or 15 oz. per foot the price will prevent the crown manufacturers troubling themselves about it; it was therefore Resolved, that the utmost attention should be paid to every thing calculated to increase the quality.

---

1   Particulars of the above in Reynell's letter of September 24, 1833, a report of a full discussion with the foreign workmen at Spon Lane. In an undated letter to William Chance, no doubt of the same time, he wrote: "The Frenchmen have all worked to-day at their newly appointed places. A shew of resistance was again made, but not of a character to demand coercive measures."

2   The dates are interesting, since Bontemps in the Guide, p. 118, says that the rings, owed to Germany, had only lately—that is, not long before 1868—been adopted in England.

3   Henry Chance observed in his paper of 1836 that a table of crown glass of the highest quality was worth three times one of the lowest.

4   *Examen historique et critique*, p. 23. In regard to competition, Bontemps notes a saying of Lucas Chance, that his firm did not really begin to make a profit until faced with the competition of other English manufacturers. He commends to legislators not to lose sight of the statement of so experienced a man, but to grasp the fact that competition is more favourable to development and profit than monopoly.

And again on November 22, on the question of adopting the French or German mode of working, it was determined "to make all the good glass we possibly can, and find out the best market for the bad."[1]

Two things are required for fine quality of glass, good colour, obtainable by the use of pure and clean materials, and freedom from blemish, whether such result from imperfect founding or from bad workmanship. To attain the former end the sand was washed and the other materials sought as free from impurity as possible. Kelp was discarded in favour of carbonate or sulphate of soda, the latter with or without admixture of charcoal." To obtain freedom from blemish various mixtures were tried, with systematic record of the results, and the workmen were ordered to be carefully watched. As the result, it could be recorded that "our light colour has evidently given our glass a character in the market which it never had before" and, "as Cooksons and others are imitating us" it was resolved that no expense should be spared, at least in the crown houses and for shades, to get the best colour possible.[2] But in regard to workmanship, especially of the crown glass, results remained subject to constant and most vexing variation.

The "best market for the bad" was found in British possessions, under protection against foreign competition, and especially in Ireland. It is significant that the quality of crown glass, next above the "coarse" was known as "Irish." What was sent there was so bad that Reynell, travelling for the firm in August 1833, scarcely could obtain an order. But it had to be got rid of somewhere. Occasionally there was a demand for it; in 1835 old materials were ordered to be used up in No. 3 crown house to provide the stuff for Canada and the British Isles, but this was exceptional. It could not be sent to foreign countries; New York, with its native supply of inferior glass in abundance, and South America, from preference for good, would have none imported but the best, and it was hopeless to compete with the Belgians anywhere abroad, excepting in quality, when they could deliver their "demi-blanc" glass at Antwerp at 12s. 6d to 14s. 6d per 100 feet, 5s. to 7s. less than the cost of the like glass at Spon Lane.[3] Nor would London take the worst qualities willingly, in spite of an assertion by a merchant named Huth that the Londoners did not look at quality.[4] Earnestly desir-

---

1   Notice of trying sulphate occurs in a Board minute of November 20, 1832.
2   Board minute of March 26, 1834.
3   Board minute of November 22, 1833.
4   Reynell, January 8, 1833.

ous of extending his foreign trade, Lucas Chance made it his steady aim to produce the highest quality compatible with full production.

To investigate the opportunities of foreign markets, John Reynell undertook an adventurous journey for the firm through Europe in the first months of 1833. James Hartley accompanied him as far as Charleroi, returning thence by way of Paris and visiting on his way factories in the north of France. At Charleroi they established intimate relations with M. Houtard of Mariemont,[1] "the acting partner of probably the largest glass-works in Belgium" and with M. Drion of Jumet, "chief proprietor" also of the principal glass-works near Valenciennes. "We have found the people" says Reynell, "exceedingly communicative of everything relating to their works." He noticed the duty on Belgian imports into Holland raised, since the late revolution, from 5 to 25 per cent., the same as for other nations, and anticipated a new outlet for British manufactures there, as well as to the Dutch East Indies and through them Siam and other countries of the East: "not, I imagine, an insignificant market, altho' I am ignorant of its extent." Concerning production in Belgium, he found at Jumet 22 pots, producing in five journeys per week 66,000 square feet, and he estimated the whole yield from eight factories at 17,472,000 feet a year. Drion took the number of works in France to be three or four times as great.

After parting with Hartley, Reynell went on to examine what was being done in Bavaria and Bohemia. He made inquiry at all places of importance on his way, and was able to write from Frankfort: "By the time I shall have completed the tour your house will, I think, be pretty well known in this part of Europe"—his letters of introduction increasing in number "in the ratio almost of the square." Near Nuremberg he was able, under difficulties, to see the principal plate glass-works in Bavaria, Messrs. Fischer's, and then accomplished on horseback or by sledge "an adventurous excursion of upwards of 100 miles through mountains entirely covered with snow and forest" in western Bohemia. He had to hunt out the various scattered works,

---

1   "We have become very intimate there and have induced M. Houtard to pay a visit to Birmingham" especially in connexion with a device of his for flattening sheet glass. When the visit had taken place Reynell reported him to have been much interested in the crown manufacture and to have changed his opinion, expressed at Mariemont, that England could not compete with Belgium in that of sheet. "Mr. James Hartley overheard him remark to his London partner that his apprehensions were much increased in finding that you obtain your alkalies and coal so much cheaper than in Belgium." Houtard was found to be very jealous of and to distrust Bontemps. He would not furnish models of his new invention until "convinced that Mr. Bontemps is not a partner at West Bromwich." In the Guide du Verrier (p. 287) he is named Houtard Cosse.

found only German spoken, and at one place was all but arrested as a spy, but nevertheless could term the journey "amusing." He was satisfied that if Belgian competition could be met, certainly the Germans need not be feared, nothing being cheap in Bohemia but fuel; wages high through the men's "freemasonry" and materials having to be brought from a distance. He found that all the good glass was sent to Spain and Portugal, confirming what he had previously learnt in London and at Charleroi, that consumers of those nations, whether in Europe or in South America, were most particular about quality. Altogether he found about fifteen seven-pot furnaces of circular form; the pots not much more than half the size of the French, and the cylinders made wider and shorter than in Belgium. Among other details he mentions that very good shades were being made for the Dutch market, and that strong competition had arisen of late years in Austria, whence glass was shipped from Trieste for Constantinople and the Levant.[1] Next, he described a visit to the little town of Heide, "the residence of all the German glass merchants who have their establishments in foreign countries," and among other things, discussed the possibility and risks of export to Mexico.

From Bohemia he went on by Dresden and Leipzig to Berlin,[2] finding in Prussia more encouragement than he had expected, but at the same time hearing of other English travellers in glass, offering at considerably lower prices than his own. He was told that the only practical way of doing business was to attend the Leipzig and Frankfort fairs with samples. At Lübeck he found no demand for quality, and at Copenhagen little to be done. Thence he took ship for Petrograd, only to discover that no glass was permitted entry into Russia, excepting watch-glasses. He visited the "exceedingly fine" imperial factory and others near the capital, but found the chief seat of the glass industry to be the neighbourhood of Moscow.[3] The general result of his inquiries may be taken to be that no export trade could be done excepting to British possessions, saving in glass of superior

---

1   He had been informed in London by a cousin, Charles Joyce, arrived with large orders from Alexandria, that glass of inferior colour was being made there by an Italian firm.

2   "An exceedingly fine city of 200,000 inhabitants, with innumerable palaces, really good glass is appreciated."

3   Interesting particulars about the Russian manufacture are as follows. It had been established some thirty years before and formerly, as from Germany, much glass had been sent to the United States, but not since factories had become numerous there. The workmen were chiefly from Bohemia and adhered to their system of excluding foreigners; no Russians had learnt the art, unless in one factory near Moscow. The glass was very good both in quality and workmanship. Mirrors were a government monopoly; the largest size seen about 105 by 47 inches, inferior in colour to the French.

quality. For another tour he proposed to attend the fairs at Frankfort and Leipzig, to proceed thence by way of Italy to Constantinople, and to return through Spain and Portugal. But this did not come off. He was again in Prussia in January 1834, and described the working with covered pots at Gernheim, near Minden, and other devices of the owner, Herr Schrader, to obviate the injurious effects of his bad coal.[1] Thence he was recalled to Belgium to seek for workmen, and it was proposed to establish him as an agent at Sydney in Australia. But in October he left the service of the firm.

In December 1833 John Hartley died, a severe loss, although Lucas Chance wrote: "I don't know that his death will occasion any unfavourable change in my prospects, especially as the manufactory is in a very efficient state, as compared with any former period."[2] Hartley's sons, James and John, were now taken into partnership, the firm adopting the style of "Chances & Hartleys." Young John Hartley was desired to begin to qualify himself for general management of the manufacturing by taking special charge of the work in the crown houses; to be always with the men, when working, was impressed upon him as a matter of extreme importance.[3] He very soon distinguished himself by his invention of the "bullion cup," a device which occasioned "such a vast improvement in the workmanship" that it was decided to take out a patent for it. It came at once into general and remained in permanent use. In 1837 and 1838 it brought in royalties amounting to £75 a year.[4]

Before Hartley's death it had been decided to build a fourth furnace. Discussion whether it should be for sheet or crown glass had resulted in favour of the latter, on the ground that inferior crown could be made as cheaply as inferior sheet. The pots were to be large enough to yield habitually 200 tables each, and to be worked, if required, by two sets of men.[5]

---

1   He found the workmen's wages greatly reduced of late and negotiated with them privately about coming to England. Their mode of blowing, he says, was "precisely that of your two Germans." But he took the reduction of wages to show the superiority of the French mode, "which it is clear can make far cheaper glass and really, I suspect, as good. In fact, I am more and more inclined to the French manner of making glass" Schrader himself thought it likely to displace the German in Germany (letter of January 21, 1834).

2   To his brother Henry, December 19, 1833.

3   Board minute of July 1, 1834.

4   The patent, No. 6702, of October 22, 1834, was taken out in the names of both the Hartleys, but in the minutes John has the credit of it.

5   Board minutes of October 22, 1833. It may be noted that the "usual" weight of crown tables at this period was something under 9 lb. A few were blown "large" weighing 13 lb. or more, and some "thick" 15 to 16 lb.

The furnace was to be "double"—that is to say, to have eight pots, four on either side. Lucas Chance, who had planned the new arrangement, anticipated from it decided economy, more time both for founding and for settling, the possibility of increasing the proportion of lime in the mixture, with consequent reduction of cost and improvement in quality, and "that if it be true that rings are only advantageous in the two first pots worked, then we should thereby be able to use them in four pots instead of two." The metal, he expected,

would be worked at a lower temperature, the glass would be better in quality, the workmanship would be better, crizzling would in a great degree be avoided, and the waste of metal would be less. . . . Mr. Chance submitted also, and it was Resolved, that in No. 4 house at present, and in the best furnace in general, it will be expedient to use the best materials we have on the premises, using only the cullet itself produces and carefully excluding the use of the moils and everything calculated to injure the colour.[1]

The furnace was started on December 11, 1834, and soon it was "manifest that all the advantages anticipated . . . are more than realised."[2] Similar reconstruction and working of No. 1 was considered, but resolution on the subject was deferred. In February, 1836, the increased production obtained was pronounced to be clearly advantageous.

It had been resolved also to rebuild the sheet-glass furnace (No. 2) for more and larger pots, for ten, namely, of a size to furnish each 120 cylinders of 36 inches by 24, two being used for making shades. But the change had to await the engagement of more foreign workmen. The two Germans were leaving, as said, and others were unwilling to renew their contracts. Bontemps was reminded of "the urgency of our having additional workmen. "Reynell, in Belgium, was desired to stay until he had obtained a full set "for the next campaign." On his report that he had secured three excellent blowers and a *gamin* and had four others in view, it was resolved to complete the number of eleven as early as possible, and to put the old furnace out as soon as that number was obtained. Reynell was able to engage two more at Strasburg at an advance of 20 per cent, on the Belgian tariff, but on proceeding to Rive-de-Gier in the Lyonnais his quest brought him into collision with the law, and he barely escaped

---

1   Board minute of December 2, 1834.
2   Board minute of January 9, 1835.

imprisonment.[1]

The enlarged furnace was started in the summer of 1834. In October a notable change was made, foreshadowing the later introduction of separate blowing holes—the erection of a separate furnace for finishing the larger shades. The reason was that the size of the working holes required for them had been found unduly to reduce the furnace-heat.[2] The innovation was carried out when information was had of the practice of the flint glass makers.[3] In March 1835 working holes were ordered to vary from 10½ to 16 inches in diameter.

Still the blowers made their cylinders of short weight. To remedy the evil, it was resolved to pay them only for what they actually produced.[4] But the decision was ineffective; in February 1835, while fully working out the pots, they were found to be making glass of but 70 to 75 lb. weight per 100 feet, instead of 88 to 92, as proper;[5] that is to say, of 11 to 12 oz. per foot, instead of the regulation 14 to 15.

With the new furnace at work, production of sheet glass rose from some 14,000 to over 18,000 lb. per week. This was not kept up; for the years 1837 and 1838 the figures were but 15,400 and 14,800 respectively. There were made besides a small quantity of fluted and a little coloured glass, and shades in varying quantity.

Recognised as a weak point in the sheet glass manufacture was the flattening and annealing. It was the custom at this date to conduct the two operations together, with manifest disadvantage.[6] Manufacturers abroad

---

1   Board minutes of January 31 to April 16, 1834; letter from Reynell to M. Demazeau at Rive-de-Gier, from Lyons, May 2.

2   "As vast quantities of metal have been injured by the two large holes in the present French furnace, Mr. James Hartley's attention was called to the importance of having a flashing furnace for making large shades combined with the power of making large quantities of small shades at the same time " (Board minute of March 4, 1834. A resolution in accordance was adopted on June 21).

3   "Mr. Chance having incidentally learnt from Mr. Bacchus in London, that they use a sort of bottoming hole for making large shades, &c, which the flint makers call a "Glory hole" it was resolved that one be immediately erected in the French house, but to be called a "Shade Hole" (Board minute of September 16, 1834).

4   Board minute of December 20, 1834. The blowers at work at the beginning of the month were Gaspard and Joseph Andre, Desguines, Stengre, Felix Bournique, Meyer, Reppert, Villard, Zeller, Rapper, Cresset; the five first named having been in employment from the beginning. Stengre and Cresset were making sheets of 40 inches by 30, Gaspard Andre various, including lagres, the others of 36 inches by from 23 to 28. The shade-makers were the Andres, Desguines and Meyer. In January 1836 Zeller was dismissed as an example, and "le grand" Meyer, above-mentioned, engaged in his place.

5   Board minute of February 6, 1835.

6   To quote James Chance's patent of 1842, noticed in the next chapter: "the flattening and annealing kilns being connected, the former has to be cooled down together with the latter, and the

were busy about methods for removing the flattened sheets continuously from the lear and annealing them separately, to the fore among them Hutter & Co. of Rive-de-Gier and Houtard of Mariemont, who has been mentioned. In 1833 was tried at Spon Lane the "Lyons kiln" no doubt the Hutter rotatory arrangement described by Bontemps as improved by himself.[1] This was soon abandoned as a failure, even for coarse glass. Next came under consideration an invention by Houtard, to which Reynell called the attention of the board, but the excise regulations laid a difficulty in the way. A boy had to be stationed inside the kiln to unload the "chariots" while the annealing chamber was kept by the excise officers under lock and seal, until they should attend to weigh the finished glass. It was resolved, on counsel's opinion, that the kiln "could not be used without a new law or order, either from the Board of Excise or the Treasury, and that it was not probable that a boy would in any case be allowed to remain locked up in the kiln for the purpose of unloading the chariots"; wherefore it was inexpedient to apply for a licence until a change in the law were contemplated, and further information should be awaited.[2] The matter was kept in view, Lucas Chance recording his opinion

> that the increasing sales of sheet glass for exportation and the better intelligence we have as to the various markets of the world will, he believes, take off the whole of our manufacture without overloading any market, but that as we may expect the North Tyne glass to be fully equal to ours, and we shall find it necessary to adopt every improvement, and also to economise the cost, he therefore thinks we should adopt the Mariemont plan of flattening, and expressed his wish that Messrs. Hartleys should consider the subject well.

But James Hartley set himself against the Houtard kiln, and trial of it was not made until he and his brother were gone.[3]

The above-mentioned is but a minor instance of the vexatious interference of the excise laws. The "Thirteenth Report of the Commission of Excise Inquiry" on Glass, presented in 1835, exposes their harassing

---

operation of the flattening is thus suspended during every interval required for annealing and drawing the glass, as well as for reheating the kiln to the temperature required for the process of flattening."

1   For descriptions of this and of other new flattening and annealing kilns, see the *Guide du Verrier*, pp.279 *et seq.*

2   Board minute of April 9, 1834.

3   3 Board minute of March 20, 1835. The reference to "North Tyne" glass would seem to show that Messrs. Cookson, or other enterprising firm in that district, were embarking on the manufacture of sheet glass.

restrictions, their contradictions, their powerlessness against wilful fraud, the obstacles that they placed in the way of progress. It is worth digression to take note of its curious and often entertaining revelations. An infinity of particulars had to be registered by the officers, and everything to be conducted under their eyes. Four were in constant attendance at Spon Lane, besides a special inspector of glass packed for export and a supervisor to check collusion on the part of his subordinates. If pots were to be arched for heating up, if they were to be filled, if the annealing kilns were to be packed or emptied, hours of notice had to be given in advance. Each pot was numbered and registered, and the weight of materials and the kind of glass to be made had to be declared. The annealing kilns, when filled, were locked up by the officers and sealed, and the weight of glass taken from them checked by previous gauging of the metal in the pots. The last safeguard was to lock up the produce in a special room, until the supervisor could attend to see that all was right.

These are but samples of the burdensome routine. The Commissioners condemned the interference utterly. The window-glass manufacturers, however, considered the regulations, their anomalies expunged, necessary for proper collection of the revenue. Afraid of foreign competition, they did not want the duties repealed. Lucas Chance thought "the present regulations for the charge and collection of these duties unexceptionable" saving that gauging the pots was useless. He found no hindrance to experiments, and to increase consumption would have repeal or reduction of the window tax rather than of the duties.

Fraud, not infrequently connived at by the excise men, was rampant. If discovered it was only compromised, and the fraudulent manufacturer could soon make the fine good. The honest ones, who suffered by his lowering of prices, complained bitterly of the inadequacy of the punishment. In the window-glass trade, indeed, fraud was difficult, but there were ways. One device of an unnamed crown glass manufacturer in Lancashire, known to evade the duties regularly, was to have a whole side of his annealing kiln removable. Thomas Hawkes said: "There are various ways of extracting goods from the lear; I have heard of immense long forks, with a lever which is used for that purpose, and I take it upon myself to know that such practices have been carried on." It was possible also to export glass as valuable, with benefit of the drawback, and to reimport it as worthless, free of duty. For small articles of flint glass, on the other hand, hundreds of illicit furnaces, hidden away in outhouses and cellars, escaped detection. The loss to the revenue by them, in London alone, was reckoned at £65,000 a year. The open manufacturers deposed that they were forced either to close their works or to carry them on at a loss. They declared it impossible to make a profit but by fraud.[1] The worst was in the domain of art and science. Arthur Aikin, secretary to the Society of Arts, showed how the degradation of painting and enamelling on glass in England was due to the illegality of necessary experiment, and how lenses for telescopes could not be made at all. "I built a small furnace ... for the purpose of investigating the action of some of the causes that affect the quality of optical glasses. On mentioning the circumstance to the late Mr. Carr, then solicitor to the Excise, and with whom I was personally acquainted, I received such an answer as determined me to give up my intention." Dollond, the optician, was quoted to have said: "It is totally impossible, under the present regulations of the Excise, to make many experiments and to make that description of glass which he requires, achromatic glass, for which a premium has been offered by the Board of Longitude formerly." Dollond himself obtained leave to get crown glass made at Spon Lane of sufficient thickness to cut up for lenses. But here intervened the law that crown glass might not exceed in thickness one-ninth of an inch. The reason for this was that at one time the glass had been exported as plate, with advantage in the drawback. Although that advantage had been removed by subsequent legislation, the limitation was convenient for distinguishing crown glass from plate by gauge, and was retained. Dollond, therefore, though he saw promise of success, and had permission from the Treasury to continue the experiments, was forbidden to do so by the excise supervisor.[2]

And the same with the new French apparatus for lighthouses. Cookson & Co., requested by the Northern Lighthouse Board to try whether they

---

1   Lucas Chance gave evidence that attendance of the supervisor was "essential to the security of the revenue." Thomas Hawkes, M.P., of Dudley, stated about flint glass that "the nature of the trade is such, that the officers are compelled to sit, night and day, over the hot furnaces," hence inattention and connivance. The Board of Excise he knew to be aware thereof, and he did not doubt that great pains were taken to prevent the frauds, "but there is no remuneration to be had for the seizing officer, and therefore, probably, there is not that vigilance used by the officer that there might be." Thomas Badger, another flint glass manufacturer of Dudley, deposed: "with glass, when it is once made, it is the act of only two minutes to defraud of a thousand pounds in weight, which may be carried out of the room; and if the officer comes in, he cannot find it out."

2   Letter from G. Dollond, March 17, 1833, *Thirteenth Report* p. 52.

could construct a Fresnel "polyzonal" lens, similar to one obtained from France by Sir David Brewster, had succeeded in the attempt, using plate glass. But that might neither be less than one-eighth of an inch in thickness, nor more than five-eighths. Thicker, the plates were charged with a duty of £4 18*s*. per cwt. instead of £3, and that was found prohibitive. Leave to make under the lower duty required an Act of Parliament.[1] Another instance is supplied from the bottle trade. Frederick Fincham found out how to render his green bottle glass fit for acid-resisting flasks and phials required by chemists. His experiments were peremptorily stopped. Say the Commissioners:

> He was informed that his work could not be allowed to continue, because he produced an article so good, that it could not be sufficiently distinguished from flint glass, the danger being that this article, which for a great variety of purposes was admitted to be in all respects as good as the comparatively highly taxed, and therefore high-priced, article of flint glass, would be substituted for that description of glass to the detriment of the revenue, however much the substitution might conduce to the convenience of the public.

Besides which, nothing might be made by a bottle manufacturer of less capacity than six ounces.

Chances & Hartleys made at Spon Lane the alkali for their glass, as well as the glass itself, using at first bought sulphate of soda. Yields being very poor, only 10 cwt. of white ash from a ton of sulphate, established manufacturers — Adkins & Co. and Clay & Muspratt — were consulted, and in April 1834 the firm resolved to put up a vitriol chamber and salt cake furnaces for themselves. These "acid works" as they were called, occupied ground from No. 2 house southwards, on part of which No. 7 was built at a later time.[2] At the end of the year named William Neale Clay, engaged as manager, could reckon on obtaining from 6 tons of sulphur and 11 cwt. 3 lb. of nitre used in the vitriol chamber, 16 tons of acid of

---

1   Evidence of Isaac Cookson and William Cuthbert, p. 104. The reason for the restriction was stated to be that the plate glass makers had occasionally used the refuse glass left in their pots for "bull's eyes" for shipping, thus coming into collision with the flint glass people, the duty for both at the time having been £4 18*s*. per cwt. Although subsequently the duty on flint glass had been reduced to £2 16*s*., and that on plate glass, within the prescribed limits, to £3, yet for plates beyond the higher limit the old charge was still imposed.

2   In connexion with an enlargement of No. 5 house, in 1842, a road was ordered to be made through the "acid yard" towards No. 4, to replace the old one taken in. Presumably this was the way that now passes No. 7 house on the north.

1.750, 17 of sulphate of soda of 98 per cent., and 12 of white ash of 45 per cent. It is interesting to find him with ideas for recovering the lime and sulphur from the vat waste, a problem finally solved at Oldbury only after the lapse of fifty years.

In January 1835 an analyst named Richard Phillips was engaged, who at once developed a process for making sulphate of soda by furnacing salt with sulphate of iron (copperas), obtained by atmospheric oxidation of iron pyrites.[1] This, chiefly supplied by the "coal-brasses" of the collieries, was eagerly sought throughout the Midland counties, in South Wales, and wherever else it could be found. To work the process, and to provide room for further extension, land was taken on lease at Oldbury, as hereafter told. Clay was given the management there, and allowed £45 a year to provide himself with a house in the immediate neighbourhood. Leonard and Robert Potts took over the manufacture of chemicals at Spon Lane on contract, with most satisfactory results.[2] When, however, the Hartleys claimed and were allowed relief from any share in the undertaking at Oldbury, Clay was desired to give his whole attention at Spon Lane. It was agreed on April 18, 1836, that Lucas and William Chance should take over the Hartleys' interest, and credit the firm with the amount expended.

It was in these years that the Smethwick property was severed into two parts by the cutting of the low level canal. A first scheme appears to have been to carry it for portions of the distance through tunnels, but it was agreed instead to give Lucas Chance lands in exchange for what was taken from him. He wrote to his brother Henry on February 2, 1828: "In the place of the tunnels I get Moilliet's land, No. 1 (including the piece they reserve it cost them £1,000), Phillips's, No. 2, with 6 cottages, which cost them £800, No. 3, of which about 2/3rds of an acre is not to be spoil'd and which will much improve the house ultimately."[3] But his determination to have a proper marketable title for his acquisitions kept the lawyers employed for several years, the canal conveyance being dated May 15, 1838, six weeks after the canal had been opened for traffic.

Phillips's land has been mentioned. Moilliet's, afterwards known as the "Island," lay beyond the upper canal on the north-west. No. 3 was a triangle of an acre and three-quarters, whose base extended along the

1   English patent No. 6846, of June 4, 1835. Phillips is described as a lecturer in chemistry at St. Thomas' Hospital, London.
2   Board minutes, April to September 1835.
3   Although most of the kitchen garden was cut away.

south bank of the new cut nearly to Spon Lane. (The figures refer to a plan.) Besides these parcels was obtained the site of the old line of the upper canal, diverted for the purpose of carrying it over the lower level. The balance of gain amounted to about an acre and a half.

In 1835 and 1836 other plots of land were bought, aggregating in extent two-thirds of an acre, and extending the property to or towards Spon Lane. These were: from the two Archibald Kenricks 1475 square yards in the corner between Spon Lane and the upper canal for £221 5s.; from George Neale 853 square yards adjoining Phillips's land, with five cottages, for £530; from Darnel Pearsall 580 square yards for £85 19s; and from John Callaway and others 330 square yards, with four cottages, for £370. There is mention in December 1838 of a tunnel to be made under No. 1 furnace, to terminate in Kenrick's land.

Among additions to the plant were gasworks and a sawmill, while the offices were enlarged, and a coach-house and stable provided. Novelties included the introduction of covered pots for coloured glass and shades, the former practically limited before to green pot-metal made in open pots. Also was made in open pots a special glass for staining, for which "kelp frit" was brought into use again.

Towards the end of 1836 the Chance and Hartley partnership was dissolved. From the beginning of it the seniors on either side had failed to agree. Lucas Chance was alert and sanguine, all for enterprise and progress and not afraid of risks, James Hartley conservative, difficult to move, disposed to accept conditions as they stood. An instance is his recorded opinion about the sheet glass, "that we are never likely to get a much larger proportion of good glass than we have hitherto done." In spite of the evident advantage of the rings, he refused to be persuaded that they were not injurious. He set himself against the Houtard kiln, as said, and even when it was reported a success still declared that it had never answered, and never would.[1] Lucas Chance, on the other hand, was resolute in belief (for instance), "that any plan which would habitually secure good work and well annealed glass would also secure an unprecedented demand, and, if combined with good metal, would render us in the glass manufacture what Wedgwood was in that of earthenware."[2] Especially galling to him was the bad work in the crown houses, which Hartley failed entirely to remedy.

---

1   Board minutes of April 1 and December 19, 1834, and July 15, 1836.
2   Board minute of June 21, 1834.

Good founding had been obtained, after it had been determined that they should settle the mixtures for each journey jointly,[1] and Lucas Chance was able to set down that "our great redeeming qualities are our small bullions and good metal when free from the wave" but the faults of the gatherers and blowers persisted. This, he pronounced, "at a period when we have steadily kept to one mixture and the metal has been uniformly excellent, is very singular and is a sacrifice of profit we have no reason to calculate on." He "could not believe that bad work to the extent prevailing in this manufactory was unavoidable" and declared "that he should not be satisfied till the bad work was almost entirely got rid of"[2] The principal offenders were the men under charge of two Hartley cousins, James and Joseph, their results contrasting sadly with those of a third set under another foreman, Stamp. The two Hartleys were constantly cautioned and reprimanded, but the bad work went on, and in course of time they were dismissed. As for the sheet glass, it came in April 1835, when a long minute sets forth the causes of the faults and their remedies, to Lucas Chance taking over entire control of that house himself.

Three months later there was open breach. James Hartley refused to attend a Board meeting, and it was written down in anger that he had "greatly lessened the profits of the concern by the negligent manner in which he has conducted his department," and resolved that only No. 1 house should be left in his charge. The particular cause of the quarrel seems to have been contention for credit of the successful arrangement of working in No. 4. At the next board Hartley's thorough approval of the new mode, he being present, was registered, with an exposition of the reasons for it and its advantages, and further, that Lucas Chance considered himself to have been very lightly treated.[3] Then, both Hartleys absenting themselves from the works for the occasion and James having sent in a long answer to the above, it was resolved "that Mr. Chance having in those observations denied that the mode of working, which Mr. Hartley formerly proposed, had been adopted, it behoves Mr. Hartley to bring forward his proofs instead of reiterating his assertions."[4]

The quarrel was composed, it being agreed on September 2 that "all

---

1   Board minute of July 25, 1834.
2   Board minutes of October 7 and 15, 1834.
3   "It certainly never entered into Mr. Chance's mind that Mr. Hartley would claim to be the author of that resolution."
4   Board minutes of July 1835.

past differences were entirely and satisfactorily arranged," that there should be no reference to them for the future, that the "French house" should be placed under control of the two jointly, and that Hartley should have entire management of the crown houses, with his brother, Withers and Stamp each in charge of one of them under him. Yet the bad work and consequent complaints of customers continued, and soon recrimination was renewed. In March 1836 Hartley again refused to appear at a Board meeting, although on the works, or even to reply to a letter summoning him to come. On a second occasion he absented himself in Birmingham. On July 16, when he delivered his opinion about the Houtard kiln as above said, Robert Lucas Chance was authorised to take such measures in the matter as he should think fit. After this James Hartley seems to have ceased to attend the Board meetings altogether, although John came to them. On November 18 the partnership was formally dissolved, the Hartleys departing to establish themselves at Sunderland.

In conclusion of this chapter, notice must be taken of proposals of April 1836 for amalgamation of Chances & Hartleys with the Birmingham Plate Glass Company, newly formed.[1] They came to nothing, for the demands of the former, justified by the fact that theirs was a well-established and flourishing business, did not prove acceptable to the other side. There would be little occasion to introduce the subject, but that the papers supply valuable information about conditions at Spon Lane.

Negotiations having been opened after (it would seem) the issue of the Company's prospectus, Chances & Hartleys submitted new drafts in substitution. One of these proposed formation of a joint concern with a capital of £500,000, to be styled "The Birmingham Plate Glass, Crown Glass, German Sheet Glass, French Shade, and Alkali Company." William Chance was to be a director, Lucas Chance and the two Hartleys managing partners. The advantageous situation of the Spon Lane works, affording ready access to all parts at lowest rates of carriage and cheap supply of fuel, was dwelt upon; it was recalled that the manufacture of sheet glass and shades had been introduced by the present proprietors at great expense, and was carried on nowhere else in the United Kingdom; conviction was expressed that to add that of plate glass would increase the trade in alkalis; and emphasis

---

1   Its prospectus was issued under date March 18, 1836. Forecasts included anticipation that the expected removal of the duties on glass would bring "a very considerable additional profit by the consequent increase of the consumption, and the adoption of the Article in every respectable House and Shop in the Empire."

was laid on the saving of outlay in taking over established works instead of erecting new ones. In the allotment of shares preference was to be given to "respectable dealers" in glass, such as it was intended only to supply.

Another draft added to the above the statement that "unusually large" profits had been made, and proceeded in the usual style of this class of document:

> The Proprietors feel it due to themselves to state, that their business is in a highly flourishing state, the returns and profits having greatly increased from year to year, and at no former period has the trade afforded such a prospect of advantage as at present; in fact, with the aid of additional capital, and under the present system of management, it would be difficult to fix a limit, to which the profits may be extended. The Proprietors therefore invite Capitalists to join them in formation of a Company, to raise the required capital; the Proprietors will insist on retaining a considerable share of the concern in their own hands, as a permanent investment,[11] and such is their opinion of the intrinsic value of the concern, that if it should be continued under the management of the same parties, as now conduct it, they would give satisfactory guarantee that the average profit of the first five years shall not be less than seven and a half per cent, per annum upon the paid up capital, anticipating, as they confidently do, a much larger annual gain.

A memorandum of the terms on which the firm was ready to sell set forth that although the proprietors were so satisfied of the value of the concern, that they were desirous of disposing only of one half of it, yet, in view of all the circumstances, they offered the whole, including "an excellent family mansion, about 40 cottages, &c, &c, and the goodwill" at ten years' purchase of the nett profits for the year 1835, from which terms they would not recede. They were willing to guarantee the 7½ per cent, dividend, as said, provided that the same management were continued and the capital employed in working the present establishment; on these conditions "the managing partners would be willing to bind themselves to keep three-fourths of their shares for ten years, and the dormant partner would engage to keep half his shares for the same period." Additional points of interest in this paper are statements that the partners had themselves in view the establishment of plate glass works at a proper time;[2] that the extent of their property was now about thirteen

---

1   The manuscript has "instalment" evidently a slip.

2   Lucas Chance had expressed to his brother Henry on August 19, 1834, the view that " nothing

acres, freehold;[1] that in addition to the glass houses, alkali works were in full operation, producing about twenty tons of soda per week; that the firm turned out a sixth part of all the window glass made in the United Kingdom, and yet "have seldom a week's stock in hand and that" such is the quality of their glass, that they sell the largest portion of their manufacture at a great distance from this locality, and even in places where glass houses exist, notwithstanding the heavy percentage incurred in freight.[2] In another document the works are more exactly described, and their capacity to produce rather than their actual production given.

The Premises consist of a large mansion and pleasure grounds, a House for an under Manager, about 40 cottages for the Workmen. One single and two double crown Houses[3] capable of working from ten to twelve thousand tables per week. One House for working German sheet glass and shades. Lead chambers, and alkali works, in which about 25 tons of soda can be made per week. Carpenter's shop, a smith's shop on an extensive scale. Engines for grinding materials, &c. Gas apparatus, Warehouses, cutting rooms, pot rooms, &c, upon a very extensive scale. The Premises occupy 12 or 14 acres of ground, which is freehold and is bounded and intersected by the Birmingham Canal, both on the upper and lower level.

Then, after a statement of what was manufactured: "Independently of their country trade, they have an establishment at New York; established agencies in Canada, New South Wales, Van Diemen's Land, East Indies, &c, &c.; and the quality of their manufacture is known throughout the world." And further as to terms: the proprietors desired to hold half the shares of the new joint-stock company, did not feel at liberty to state what their profits were, but would base the purchase price on an amount of them to be agreed apart from interest and rent, proposing to sell everything for one sum, ten times the agreed profits and payable by instalments over a term of years. This principle accepted, they undertook to submit all necessary particulars.

---

but necessity would induce us to take up plate glass at present, but that should it be found expedient at any time that we should make plate glass, we were better prepared for it than any other Crown Glass Manufacturer."

1   As a fact, it approached 17 acres. Lucas Chance, as noticed below, made the extent 18 acres.

2   The rough draft of this document is in William Chance's hand.

3   The doubling of No. 1 house, however, seems not yet to have been carried out.

The Works, to realise the present profits, would require the same energetic and vigilant management as before, and the services of the present managing partners would be absolutely necessary; they would require salaries, which, in fact, they have hitherto had, being charged as an expense upon the trade before the profits have been reckoned. . . . The proprietors are of opinion that their knowledge of the glass trade generally would be of great use to the New Company in the management of their plans and in the erection of the works for manufacturing plate glass; the attention of the proprietors having for some time been called to this Branch of Business.[1]

Some further particulars of interest are contained in rough notes of "arguments" by Lucas Chance. He took the extent of the property to be eighteen acres. He insisted on the terms of purchase originally demanded, "say not less than £150,000 nor more than £210,000"; noted the guarantees given by "our strong desire to keep half the concern" and by the undertaking as to dividend; did not want a salary, but a percentage of the profits; observed the want of a coal mine to produce 20,000 tons; would "add 4,000 shares and bring in the Oldbury concern"; and would recommend his friends and relations to invest, and his partners to buy up the shares. He wanted the "bullion cup" patent conveyed to John Hartley, whose health was indifferent, "at least in regard to working in a Glass House" and would engage to make over to him shares to "satisfy him in case of his services being required." Other shares to be allotted to Clay (50 or 100), Withers (50), A. Dixon and Stamp (20).

A last paper contains "Observations on the proposed arrangements" and runs as follows:

R.L.C., W.C., J.H. and J.H. bind themselves that they will not directly or indirectly be engaged in or give their assistance to any concern manufacturing the articles we propose to or shall manufacture, say Glass and Alkalis generally, for ten years from the 31st December last, under a penalty of ten thousand pounds each to be levied by the Company on the shares the parties may hold respectively.

That Mr. R. L. Chance and Mr. James Hartley shall continue their status as heretofore.

That Mr. Jno. Hartley shall be appointed sub-manager, especially undertaking to attend during the working in one of the Houses, and to take the responsibility of the workmanship of the one, and two sets of men working in that house, for which he is to be paid a salary of £300 per annum.

That Mr. Chance shall be at liberty to bring up two of his sons to the business, with a

---

1   The latter part of this document also is a rough draft by William Chance.

view to qualify them to succeed him as superintendent.

That Mr. W. Chance shall be also at liberty to bring up one or two sons with a view of qualifying him or them to take upon himself the management of any department, which, in the rapidly encreasing prospects of the business, is sure to present itself.

That in case either Mr. R. L. Chance, Mr. James Hartley or Mr. Jno. Hartley should not fulfil their respective departments to the satisfaction of the Board of management for the time being, they shall each of them be subject to the privation of a part or the whole of the salary or percentage which they may have for management, but this shall require an absolute majority of two thirds of the whole number of members of the board: nevertheless, neither of the managing partners shall be removed from their situations, without two special meetings called for that purpose, and ten members being in favor of it at each of the two special meetings, at which the whole body of Directors shall be present.

The partners registered agreement among themselves to such terms as the above at a meeting on April 21. In view, however, of objections made on the part of the new company, it was proposed to limit the capital to £300,000; were that considered insufficient to include the Oldbury works, to make them a separate concern.[1] The reply was that the negotiation had already been broken off "by the withholding of the most important preliminary, namely, the amount and nature of the profit to be paid for, and by the prompt manner in which the Directors were required to decide"; that the allotment letters had already been issued; but that power to create additional shares could be reserved in the trust deed and negotiation be resumed "on an equitable basis, to avoid, if possible, the evils of a mutually injurious competition." William Chance, who conducted the correspondence, was thereon authorised by his partners to conclude on the basis of a payment of £150,000, to be raised to not more than £200,000, should the profits of the next three years show justification. Accordingly, he met the directors for discussion on April 25. The result was their unanimous resolution "That the new extended propositions of Messrs. Chance & Co. are inadmissible, and are consequently declined. The Directors assure Mr. Chance that every private matter mentioned during the conversation, shall not go beyond their Boardroom."

---

1   Letters of the same date.

CHAPTER II
1837 TO 1852

THE Hartleys gone, the firm adopted the style of Chance Brothers & Co. Lucas Chance added to his commercial direction that of the manufacturing, with William Withers under him as principal manager; William Chance, fully occupied with his own merchant's business and with municipal and social work in Birmingham, confined himself of necessity to attendance at the board meetings and to prudent counsel. It soon appeared that the addition to Lucas Chance's responsibilities exceeded even his powers of work; that in so large a concern both sale and manufacture could not be conducted by one man. Conversant as he was with every detail of the latter—witness his particular orders to Withers about the construction and working of furnaces after a visit to Rive-de-Gier in 1837, and many others of his memoranda—his special gifts fitted him peculiarly for the commercial side.[1] It became evident that someone must come in to help him in the manufacturing. Just completing his time at Cambridge—he graduated as seventh wrangler in 1838—was James Timmins Chance, William's eldest son, and the eyes of his uncle and father turned on him. He found himself obliged to abandon his intention of entering the legal profession, for which he had been admitted a student of Lincoln's Inn on November 17, 1836, and took up work at Spon Lane as soon as he was free of his tripos.

James Chance's intellectual and physical powers must have brought him to the front in any profession. Several were open to him; besides his high mathematical attainments he was versed in the classics and in theology, having read as a youth for Holy Orders, his knowledge of law was singularly wide and sound, and he was a very competent engineer. With his uncle's energy and initiative he combined his father's caution. He en-

---

1 "I am disposed to believe from the large profits made by us that some considerable portion of them must arise from the detailed attention given by your uncle to the sales, so as to gain profit wherever opportunity occurs; the mercantile department is certainly managed very profitably" William Chance to James Chance, August 5, 1848).

tered on his duties at Spon Lane with his accustomed ardour, always to be interested most in the scientific aspects of his work and sparing time, on his father's example, for the promotion of social reform. He was admitted to partnership as from January 1, 1839. His first achievement was to solve the problem of grinding and polishing the sheet glass, so as to give it the transparency of plate.

The cylinder-blown glass known in England as "blown plate" had been ground and polished for centuries. It was of much older date than cast plate, and in 1838 was still an article of usual manufacture, occupying under the excise laws a class by itself. But the processes employed for these plates, one-quarter to one-third of an inch thick[1] could not be applied to sheets of the thinness prescribed by those laws; from their want of perfect level they would have broken under the strain or have been worn in places almost into holes. Ways, indeed, of polishing, if not of grinding them were known; Lucas Chance was even now in treaty for machinery for the purpose devised by Robert Griffiths[2] although he had stated to the Excise Commissioners in 1835 that all attempts in this direction had failed. It was this, perhaps, that drew his nephew's attention to the matter. He hit upon a means of making the sheets lie perfectly flat, while under treatment, by causing them to adhere under atmospheric pressure closely and in every part to slates covered with leather or other suitable material soaked in water. Thus it was possible, in the words of his patent, "to remove the irregularities of surface without grinding away irregularities arising from any bendings or curves, which may exist in the general substance of the glass" and at the same time to obviate the risk of breakage, save by accident. The patent did not include the machinery that he designed, although that had novel and ingenious features. It claimed "not the grinding, smoothing,

---

1   See the *Guide du Verrier*, pp. 481-4. As seen by Reynell in Bavaria the plates were first fixed with plaster to a marble slab and ground with sand under a heavy wooden "upper work" faced either with iron or with another plate, then ground again with sand, glass upon glass, by hand, and finally polished with "a substance similar to what we put upon razor strops" by means of "a woollen substance similar to and about twice the thickness of the welt of a hat." He termed the process simple and would readily engage himself to direct it "without the aid of any foreign workmen and at a bagatelle of expense for machinery" (letter of February 22, 1833). By mechanical aids the cylinders were blown of great size. Reynell saw plates made from them from four to six feet long and proportionally wide, and examples at the great Exhibition of 1851 measured as much as 7½ feet by 5½.

2   English patent No. 7177, of September 1, 1836, granted to Robert Griffiths and John Gold, of Birmingham. On the occasion of presentation of plate to him in 1860 Lucas Chance said : "In 1838 ... I had just taken out a patent for Griffiths's mode of grinding and polishing sheet glass" when "Mr. James Chance invented his well-known beautiful machinery, which immediately superseded Griffiths's method."

and polishing thin glass generally, but only a mode of doing so, herein described."[1]

William Chance, in fear of anticipation from what he heard in London, wrote to his son on April 26, 1838: "It is necessary that we come into the market with the least delay possible, as several houses are preparing to enter the German sheet glass business. It therefore becomes exceedingly important to stimulate the completion of the building where our polishing is to go on to the utmost of your power." His injunction was obeyed; building was pushed on with the utmost vigour. The machinery was ordered of Messrs. Wren and Bennett, of Manchester, and James Chance was in almost daily correspondence with them about its details. Inquiry was made in all quarters for supplies of stones, slates, timber, emery and felt, the last found preferable to leather and before long replaced by thick "fustian or moleskin." In July 1830, James Chance wrote to his brother William at New York:

> My new process, as far as its extent at present, answers perfectly, but we have determined not to make any regular sales until we can ensure a large supply; and for this reason we are proceeding vigorously with new erections and machinery on the side of the canal opposite to our present works. The glass looks splendidly when silvered, in consequence of its great lustre.

Regular work was begun in May 1840 with eight polishing and twenty smoothing machines, worked by a 60 h.p. beam engine by Boulton, Watt & Co. Very soon it was found better to grind and smooth the glass on separate tables, one for grinding to eight for smoothing. In September James Chance wrote to Wren and Bennett:

> The great object you will please to keep in view now is to send us the remainder of the 80 smoothers, which have been ordered, as quickly as possible; for our polishing machines (now that the emery is good in consequence of a new man being here) are frequently obliged to stand for want of glass, and we ought to have a stock of smoothed glass beforehand. The slate bottom tables answer most perfectly; the iron ones we are now grinding down to a true face with glass, which of course is rather expensive, as we cannot avoid breakage.

---

1   Particulars of the process, *Guide du Verrier* pp. 485-6, and Charles Holtzappfel's *Turning and Mechanical Manipulation* pp. 1223-7. The patent was No. 7618 of April 21, 1838. It included a simple device for getting rid of the thick rims of crown tables, hitherto one of their faults.

And then early in October:

> By a quicker traversing motion accompanied with double weight we can polish in just
> half the time, in three hours on each side including stoppage. . . . You may imagine
> what an immense number of smoothers we shall want in order to absorb the power of
> two 60-horse engines. . . . Four glasses came off one of the machines (which has extra
> weights) to-day, finished in three hours on each side, the best polish I ever saw.

At least 500 smoothers, he went on, would be wanted; the last ten had "worked most beautifully." And but a few days later: "We have finished some glass well in two hours on each side." In January 1841 Lucas Chance could report from London: "our sheet plate is carrying all before it." In May 4,000 feet of glass per week were being turned out regularly, and it had become necessary to extend the buildings. The second engine was set to work in the summer of 1842. Forty-eight more grinding machines were on order in April of that year, and in September twelve more polishers.

In fact, the new "patent plate," at the same time brilliantly transparent and of light weight, supplied a want. It was the right thing for coach-windows, for choice paintings and engravings, for ornamental mirrors, and later for photographic plates.

Nevertheless, for some purposes, the glass was wanted of thicker substance. This had been apparent from the first. "We met yesterday at the Excise Office," says William Chance in the letter cited, "and had a long discussion; we were defeated in our attempt to maintain the intended alteration of the law, which would have permitted us to make our sheet glass of any thickness." Now, in 1841, it was resolved to make the glass in No. 5 house as blown plate, "it being distinctly understood," the minute runs (September 15, 1841),

> that no glass is to be made in the Blown Plate House which shall not be suitable for
> being ground and polished. The quantity of Blown Plate made is not to exceed the
> quantity of glass actually consumed in producing plate glass, the intention of this
> resolution having reference only to our Plate Glass Manufacture and its object being
> that we may be unrestricted as to the thickness of the article, as we And a most serious
> inconvenience in being confined to 1/9th of an inch so long as we make the glass as
> German Sheet.

Further it was resolved not to exceed 24 oz. substance without an order from the Board, to enter into negotiation with Bontemps about establishing the manufacture near Paris in partnership with him, to proceed with extension at Spon Lane gradually, and for a limited period to make the glass of any thickness that Lucas and James Chance should see fit in both the sheet houses at work.[1] A minute of December 7 ordained

> that glass be made under the blown plate duty for eight journeys commencing early next week to consist of the following qualities, viz., 20 sheets of thick plate in each journey, as much good glass for polishing of No. 1 thickness as possible and as much cylinder as can be made. Foundings to be in nine pots only. And that subsequently to the conclusion of the eight journeys German sheet be manufactured as usual.

Such shifts had to be resorted to until in 1845 repeal of the duties removed the difficulty.

Pending this change extension of the plant had been delayed, and immediately upon it set in a sudden and increased demand. One order, for which tenders were sent out, was for 28,000 feet of patent plate for the Houses of Parliament. In July 1845 James Chance was brought to book by his uncle for not attending properly to the manufacture, adjured to return from his honeymoon at once, and even told that of £60,000 or £70,000 invested £10,000 had been lost through his negligence. This, however, was in the excitement of the boom. William Chance asserted the plate works, under his son's direction, to be the best managed department of any. In October the latter could assure Lucas Chance that production would soon amount to 10,000 feet a week. In March 1846 eighteen polishing machines were at work, and he was instructing Wren and Bennett to push on as fast as possible with four more.

So much for the establishment of the patent plate. In other directions special attention was given to improvements in flattening and annealing the sheet glass, the weak point of the manufacture, as has been said. Very soon after the departure of the Hartleys Lucas Chance, in March 1837, undertook the journey to Rive-de-Gier specially to see Messrs. Hutter's latest modifications. He reported them to be "everything that was described and carried out most efficiently." The factory he pronounced "the best arranged I ever saw," yielding only in some respects to that of Bontemps, who remained supreme in pots and in economical arrangements as to coal. He

---

1   The second sheet glass house, No. 5, was built in 1838.

was certain "that the English mode of teazing is the cause of the glass being always of a bad colour. Our colour is like kelp colour, as compared with Hutter & Co's., who have not for a long time used a particle of carbonate of soda." The shade-making astonished him:

> They make probably 2,000 shades per week, and great numbers of much larger than André ever makes. They have 12 carts employed in taking shades to Paris all the year round. Their best shade maker gets 17 francs per day 22 times per month, so that André is manifestly well paid. The two first shade makers at Choisy are before André, and a third man makes 80 oval shades per day from 30 to 60 inches. We do everything on a small scale.[1]

As the result, it was resolved to pay Hutter & Co. 10,000 francs for setting their kiln to work at Spon Lane, and Lucas Chance included it in his patent for optical glass.[2] Then Bontemps offered for the like sum to establish there his own combination of the Hutter lear with the long annealing kiln of Houtard. The offer was not accepted; it was preferred to try something of the kind, so far as the excise regulations permitted, without his aid, and the result was failure.[3]

In August 1841 Lucas and James Chance visited Belgium to inspect factories there. Amongst much else they learnt of new modes of flattening that were being tried, but refrained from coming to any conclusion in the matter for the present. James Chance, on his return, apologised to A. Deby for delay in answering his polite invitation to inspect his new kiln, saying

> Before leaving Belgium we came to this conclusion in reference to the various plans which have been devised for flattening and annealing continuously, namely, that we had better wait a few months before we contemplated any change in our own methods. So that, as both my uncle and myself are not acquainted with the practical details of flattening, and could not therefore have given any opinion as to the merits of your system, we might have done you harm by visiting your works and by appearing to condemn your invention,

---

1   Instructions to Withers from Paris, April 1, 1837.
2   No. 7596, of March 19, 1838.
3   Bontemps, Lucas Chance reminded his brother in 1840 in the course of their discussion on the subject of optical glass (Chapter VIII), "proposed to us to agree that for the sum of 10,000 francs (less than £400) he would undertake to establish the union of the Houtard and Hutter modes of flattening, but previously to our laying a brick or going to any expense we should have the opportunity of seeing the plan in operation at Choisy on the large scale, and we should be at liberty not to adopt it if we did not approve of it. The total outlay attending the adoption was limited, and I think to £150.

because of our not adopting it. M. Drion, we were informed, has a patent in England for a new plan of flattening now being tried at Aniche.

In another letter James Chance informed M. Dorledot, of Couillet, near Charleroi, that Drion had acquainted them at Brussels with his new process, and just lately that it worked perfectly, asking them not to engage in establishing a continuous system of flattening until satisfied of the relative merits of his process and of Dorledot's.

In April 1842, on receipt of a letter from Bontemps notifying the success of his improved Houtard kiln, it was agreed to express in reply the pleasure that his information, long awaited, gave; to advise taking out an English patent at once "for the exclusive right of the continuous principle"; and to intimate desire to enter into negotiation for the establishment of the invention at Spon Lane or, were it not successful, of one of the others in operation.

The patent, No. 9407, was taken out by James Chance in July. It embraced two different systems: the one a rotatory arrangement of eight flattening stones, with a rotatory annealing kiln in which the sheets were

"We rejected this, Withers expressing an opinion that he could do the same thing, and James saying he would render him every assistance in his power. The most confident hopes were entertained that we should fully succeed. I believe James never gave any attention to the subject, and Withers's attempt was a miserable abortion, as bad as Hartley's attempt at Huter's, only attended with an expense direct and indirect (owing to the injury of our glass for months) of not £550, to which Bontemps was limited, but double that sum."

supported on wires; the other, clashing, it was said, much less with the excise restrictions, a long annealing arch connected with a lear of the ordinary kind. By this latter system the flattened sheets were piled in open carriages, and each carriage, when loaded, was moved away successively along the arch to the annealing kiln. By a supplementary patent of the same year, No. 9939, taken out in the name of John Withers, the sheets were piled in closed chambers of sheet-iron. In either case, as they passed away from the lear, they were subject to temperatures gradually decreasing. In November James Chance was authorised to erect, without Bontemps's aid, a lear on his second principle, "uniting a long annealing arch, containing railway carriages, with the present kind of flattening kiln," and at the same time Bontemps was requested to submit drawings of a "Monthermer" kiln and of his own. Finally, he says in his book, was adopted a mode of flattening practicable only in a country where the apparatus could be constructed very cheaply, a mode which was completely successful and remained in use for many years. There was no revolving arrangement, but a double system of *va et vient* on wheels, the sheets being finally piled for removal to the annealing arch in the sheet-iron chambers.[1]

On other observations made in Belgium Lucas Chance reported:

We found that the Belgian blowers work per month at least 50 per cent, more hours than ours and get about 50 to 60 per cent, of our wages, so that we pay much more than double wages. Originally I fix'd the wages at about 50 per cwt. more than the Belgian per piece, so that the Withers' have allowed this abuse to creep in. One of the most important things to do is to correct this. I propose to begin by altering the 6 hours working to 7, and as the first shade maker in France has offer'd to come to us, who will work as many hours as we please and on whose offer we must decide immediately, we must at once begin with Gaspard Andre. In fact, the Belgians and the French are such formidable rivals from their economy and activity that unless we manufacture on the best principles we can never sell our extra quantities abroad to a profit, but by diminishing the cost we may get a profit by all we export. In fact, the Belgians realise a capital profit from what they export. But if we work longer we must have larger pots, and hence the necessity of a speedy decision.

William Chance commented on this:

Should we require the erection of another house for working sheet glass, it should be built away from Smethwick, and with a view of commencing a cheap mode of working after the plan of the Belgians, but we shall never succeed in such a course at Smethwick without a risk of losing that pre-eminence in quality which we have attained, and must use every means to preserve.[2]

Before this, a revolutionary change in the sheet glass manufacture had been set on foot. Bontemps explains[3] how the Belgians introduced to help

---

1   Particulars of all these kilns, *Guide du Verrier*, pp. 285 *et seq.*
2   Letters of August 8 and 9, 1841.
3   *Guide du Verrier*, pp.117, 273-5

the blower a *gamin souffleur*, a youth who gathered the metal from the pots and began the blowing, and how, in consequence, the pots could be larger and economy result. But two men at one working-hole interfered with each other's operations, and larger pots meant keeping up the heat of the furnace for a longer time. If the working-holes had to be always open, and of the size necessary to complete the blowing, this was difficult. When the Belgian innovation came to be tried at Spon Lane, the remedy was obvious from the example of the crown houses—separate blowing holes. We have seen that the expedient had previously been adopted for shades, and Reynell had seen the like in use at Gernheim, near Minden, in 1834.[1]

First mention of these "blowing furnaces" occurs in a Board minute of April 1840, when they were ordered for No. 5 house on a plan agreed by James Chance and Withers. There followed in August a like order for the old sheet house, No. 2, and in January 1842 for two more in No. 5 and for six in place of the crown flashing furnace in No. 1, where half of the pots were now to be used for sheet glass. The working-holes of the founding furnaces, left to the gatherers, could now be reduced in size, shielded from currents of air and closed when not actually in use. Moreover, larger cylinders could be made and of heavier substance, while no more men were wanted to found, and the consumption of fuel in proportion to production was decreased.[2]

The number of cylinders, however, made remained for the present much the same, some 2,500 per week per furnace. For these were years of depression, and production was limited by demand and by agreement with the other manufacturers, as stated in Chapter XIV. In addition were produced a greatly varying quantity of shades, 1,000 perhaps in one week and but 100 in another. It may be noted that covered pots were ordered to be made for them in April 1840.

Crown glass, by like agreement, was confined at Spon Lane at this time to an average weekly production of some 5,400 tables. The particular

---

1   Letter of January 21, 1834: "I imagine you would have a finishing oven somewhat more like those of your Crown houses."

2   In July 1843 Lucas Chance wrote to James Chance, seeking sheet glass men in Belgium : " I now merely write to impress on your mind that the Belgian mode of working with gatherers is the improved mode and manifestly a much cheaper way of making glass than the French. Hence the Belgians make a much larger quantity in the same time and pay their blowers much less per foot than the French. There seems to be no reason why a gatherer should not gather as well as a blower, if trained to it. Even our French blowers make much more than Bontemps' in the same time. We shall therefore require 24 gatherers at least." And again: "I trust the *gamins* are considered by you as of scarcely less importance than the blowers."

number for each week varied with the number of pots and journeys worked; sometimes it was under 4,000, sometimes over 7,000. The "usual" weight of the tables still was 9 lb., and their minimum diameter 40, inches. In 1841 the pots were increased in size to 44 inches in height, and from 54 to 56 in diameter at the top, in the wet state. Inside measurements averaged 48 and 32 inches for the top and bottom diameters, and 40 for the depth. At the same time, as rendered possible by the new system, the sheet pots also were enlarged, from 34 inches in outside top diameter and the same in height, when wet, to 40 and 36 respectively.

Quality, in Lucas Chance's view, was far from being properly maintained. In a letter of June 26, 1840, he complained forcibly of the deterioration, and insisted on the necessity of the managers watching both the crown and the sheet glass men, and especially the flatteners, when at work.[1] In July 1842 all "Irish" crown glass was ordered to be broken down. It is noticeable that in the previous October it had been resolved to dispose of the firm's Dublin warehouse to Messrs. Dawson and Mitchell.

Of the furnaces, in 1838 No. 1 was at last made double and No. 5, to take ten pots for sheet glass, was built, apparently between Nos. 2 and 3.[2] In the following year No. 3 was ordered to be reconstructed in like fashion, so that the manufacture of sheet was already outstripping that of crown. In fact, its production amounted in 1839 to 1,150,000 feet, and in 1840 to no less than 2,200,000. In 1841 No. 4 also was reconstructed for ten pots, but remained a crown house. The figures for crown glass production in 1839 and 1840 had been 263,000 and 308,000 tables respectively.

Of materials, carbonate of soda was discarded from the crown houses in 1841 in favour of sulphate, and for the sheet glass used in the proportion of 6 per cent. only. Two years later, however, it was resolved to raise the amount to about a fourth. Sand was no longer being brought from the Isle of Wight, as formerly. Successful trial of samples from Bedfordshire in 1835 had resulted in a five-year contract for a supply from Dunstable. Then suitable sand was found at Heath and Reach, near Leighton Buzzard;

---

1   The Crown Glass has improved, but still the main thing is wanting. Withers is but little with the men during the time of working, hence no dependence can be placed on the workmanship. A set of crown workmen at work for 6 or 9 hours without any one in authority over them is manifestly not calculated to produce regular results. It was otherwise when the good work was produced after my blow up in November, and when Withers became anxious to prevent either Blackwell or Redford from spying the nakedness of the land."

2   Cf. pp. 18 (note), 83, 84, 87.

land was bought, and a first mine opened there in the spring of 1842.[1] These deposits remained the principal source of supply until worked out some fifty years later, when other had to be sought, as will be noticed.

Additions to the "acid works" in 1838 and 1839 were a second vitriol chamber and three saltcake furnaces. There is note of a laboratory to be built for one Upton. In March 1839 charge of the works was entrusted to Mr. Carpenter; afterwards a Mr. Cowper, whose duties extended to the Stained and Ornamental departments, was engaged as chemical adviser. In 1845 the whole of the manufacture was removed to Oldbury. It is only necessary to say further on this subject here that in 1842 advertisement for a practical manager brought from South Shields Edward Forster, on terms modified by the condition that "until he goes to Oldbury he is to receive £100 a year." He did not go there, but remained as an assistant manager at Spon Lane until, in later years, experience of his trustworthiness and capacity marked him out to succeed Withers in the highest post.

Another assistant manager was John Withers, a son of William, promoted in April 1841, from the charge of one or two glass-houses, to have the entire superintendence of and responsibility for the sheet glass manufacture in all its branches. He was bound for seven years at a salary of £200, with a "steven" for quality calculated to bring in £100 more. Another of the name, Samuel, replaced William Cutler in the same year in the charge of certain flattening and cutting departments, and in October 1842 was bound for seven years as a cutter. Other prominent men at this time were Edward Evans, sub-manager in the crown houses, and Thomas Gittins, charged for the present with the supply of coal. As founders, we find mention of William Parish, John Gittins, and James Bates, and of Taylor beginning as a "scorer down" at 6s. a week. Pot-makers were the brothers Robert and Zachariah Squires, the former in this employment since 1836 in place of two Candlishes.

A fourth member of the family working at Spon Lane, from 1841, was George Chance, brother of Lucas and William, returned from New York after termination of his partnership with the latter in their American business. He did not become a partner, but undertook various responsibilities until in 1847 he departed to take charge of fireclay works purchased at Oak Farm, Himley. These works were taken over very soon by Lucas Chance,

---

1   With the Board minutes of March and April 1842, concerning the purchase, are Withers full report on the property and a plan

and after passing in succession to his descendants of three generations are now a limited company, continuing to supply Spon Lane with clay, chiefly in the form of fire-bricks, of the finest quality.

On James Hartley's departure half of the House, the old "Hall" which he had occupied, was handed over to William Withers for a residence, the other half being reserved for the use of the partners. On the arrival of James Chance the library, dining-room and bedrooms therein were furnished, and a servant provided. He used the accommodation frequently, when passing the night at the works, watching the furnaces. In 1843 came new counting-houses and a board-room, a portion, presumably, of the present offices. In the same year a fire-brigade was instituted under the direction of James Chance. In 1840 a proper saw-mill had been erected by Messrs. Wren and Bennett. In 1844 more land was bought: from William Parkes and others 2¾ acres with two cottages for £850; from Samuel Pugh and Mallin 849 square yards with three cottages for £460; and from William Jones 561 square yards with nine cottages for £850. The piece first named, south of the lower canal, extended the property along Spon Lane halfway to the Oldbury Road. It provided at its southern end, in 1845, the site for the Schools. The other two filled the gaps left in the north-eastern corner of the Works after the purchases of 1835; the one that between Neale's property and Spon Lane, the other that between Kenrick's land and the original No. 1 glass house.

It is likely that these purchases had in view extension of plant to meet the increased demand foreseen on the impending repeal of the glass duties. The second actually gave room for the building of a new glass house, No. 6. Already there was strong recovery from the depression of late years, and new undertakings were springing up. Lucas Chance dated such from 1843, in anticipation of the repeal.[1] But no one had realised the immensity of the demand that would ensue; a veritable famine in glass set in. One result, long anticipated,[2] was the opening of the British market to Belgian and other foreign competition, but for the present that was swamped. New factories appeared in all directions[3] and the established ones set themselves

---

1   Letter to James Chance, July 31, 1849.

2   In 1833 Lucas Chance had recorded his opinion "that one of the effects of the repeal of the glass duties would be that the cheap green glass of Belgium would come into Great Britain in large quantities, unless a considerable duty by weight was laid upon it, and as the best quality of sheet glass would also come in, it will behove us in case of such repeal to do all we can with government to lay on not only a duty by weight but by the value also" (Board minute of October 22, 1833).

3   Henry Chance, in his paper of 1836, states that the number of crown, sheet and plate glass works in the United Kingdom rose from fourteen in 1844 to twenty-four in 1847.

to produce the utmost possible. At Spon Lane all the existing furnaces were put in full work, a sixth was built, as said, and there had been a seventh on the "Island" but for the veto of William and James Chance on spending more money, especially on so unsuitable a site. Prices rose high, and the profits made were great.

The main difficulty was to find workmen. The services of the small reserve of competent men were eagerly competed for. Several were suborned from Spon Lane, though not without reprisal. "This losing of men annoys me beyond anything, because it might have been prevented" Lucas Chance wrote. For the sheet glass it was necessary to find foreigners, for those employed had persisted in their refusal to teach Englishmen, and so but few of the latter had learnt the art. They had only been able to practise by themselves on what was left in the pots after the foreigners had finished. It is related that at least on one occasion a battle royal took place, to the discomfiture of the latter. In the depression of 1841 and 1842 some men had been discharged, among them Joseph André, who had found work with the Hartleys and was now employed in engaging compatriots of his for them. And whereas at the beginning the single sheet furnace had required but eleven blowers, now four had to be supplied with blowers and gatherers besides. In July 1845 they numbered at Spon Lane but twenty-seven.

Lucas Chance, grievously disturbed, set every means in motion to procure more. James Chance had to spend a portion of his honeymoon at Namur in treaty with Dorledot of Charleroi, while his cousin Robert Lucas Chance the younger[1] just back from Constantinople, was hurried off to Picardy and the Lyonnais. Says his father: "He takes up the journey *con amore*, and having read the volumes I have written on that subject, has his mind imbued with the importance of his mission." He succeeded in engaging some really good men at Chauny and Premontré, but at the latter place the managing director of the works had wind of his errand and stopped their departure. Dorledot secured at once no fewer than sixteen blowers, and gatherers besides, so many that it became a great difficulty to house them. But when the first of them arrived, they were found to be useless for good and heavy work. Dorledot was branded as a charlatan, and James Chance was bombarded with letters of exhortation and expostulation by every post.[2]

---

1   Hereinafter referred to as Robert Chance.
2   For instance: "In adding 21 blowers to our present numbers, and supposing that Everard, Drion

However, the Belgians turned out to be not such bad workmen, when they had mastered the difficulties of furnaces strange to them. And some first-class Frenchmen were procured, partly by the help of Gaspard André. Of gatherers, Dorledot was asked to send six more, if he could find men careful and proficient. More blowers, he was told, would not be wanted before the end of September.

The terms of engagement were necessarily high. Lowest wages for the blowers amounted to 400 francs a month for 30 hours' work per week, or 500 for 40 hours.[1] They were required to make in the former time 640 sheets of 40 inches by 30, and in the latter 800; namely, of 18 oz. glass 300 and 375 sheets respectively, of 27 oz. 200 and 250, of 36 oz. 140 and 175; if they averaged the proper quantity, it was agreed to pay their property tax (a question raised by them) as bonus. The gatherers mentioned were offered 120 or 150 francs per month for 30 or 40 hours per week, or 150 francs for a number of hours to be agreed upon.

By September houses Nos. 2, 3, 5 and 6 were at work for sheet glass. They accommodated over fifty blowers. But even so more of the glass was wanted; it was decided to make room for a seventh house by removing the "acid works" to Oldbury.

The great boom was followed in the autumn of 1846 by the inevitable reaction. The sudden demand satisfied, normal requirements proved unequal to the vastly increased supply. Of patent plate, for instance, but 20,000 feet were sent out in a month instead of 40,000. Other firms were found to be selling at much lower prices,[2] and reduction of costs became

and Castion leave us, only consider what a predicament we shall be in "We shall have, it is true, 45 blowers, but whom shall we have to make the plate glass ? The great deficiency of men to make 21 oz. glass is not supplied, and you propose to leave to Dorledot to secure the good men at Chauny and Premontré. Your arrangement is excellent, provided the men were good, but supposing they are otherwise, where is our good glass to come from? I can assure you I am utterly dismayed and cast down" And again: "Nothing since 1831 has made me so uneasy as the finding the 7 men sent by Dorledot to be such poor creatures. Rely upon it they are of quite a different grade to the men we previously had. Were we to have a furnace mounted with such men, how could we have good glass, good flattening and of such sizes as we require for our customers? And yet such men to be sent about their business might establish a thorn in our side. We can probably do with 7 of such men and turn them to some account, but with 17 or 20 of them our sheet glass would be ruin'd." To his son Lucas Chance wrote: "We must not mind a few hundred pounds extra expense to get a complete set of workmen, the only fear is lest they should not be all of the right stamp."

1   Bontemps wrote on April 14, 1846, that his best shade-maker was intending to go to works in the north of England on an engagement for five years at 10,000 francs a year. He thought the bargain a poor one, since the man was good only for shades and was often incapacitated by gout.

2   "Hartley and the South Tyne Co. are sending thousands of crates to Manchester at 22s. to 24s. per crate del'd, of size and weight much exceeding ours, and there is no doubt that they consider their

essential. In letter after letter Lucas Chance showed where saving could be made and leakage stopped. He was all for longer hours, lower wages, quality and maximum production. "Our profits depend entirely on quality" he wrote, but at the same time "to make large quantities is essential" and especially of good thick glass.[1]

Investigation made it plain that in the excitement of the boom management had been lax and economy thrown to the winds. On capital account Lucas Chance admitted, later, that £40,000 too much had been laid out.[2] Wages appeared to have been paid for half as much glass again as had been sold. Robert Chance, a partner since the beginning of the year, was set to examine the matter; there was a "vast deal" of saving to be effected in the wages department, his father declared.[3] Cases of neglect and of actual defalcation came to light; stocks smaller than accounted for, materials debited but not received. For instance, a calculation of the charcoal used in the glass mixtures during three years showed a difference from what had been paid for amounting to 13,000 bushels. Three workmen, convicted of the fraud, were discharged. To prevent the like for the future, clerks were told off to check and take record of everything entering or leaving the works by boat.[4] Lucas Chance denounced as "perfectly frightful" the waste and robberies in 1846. "Our profits are no longer to be calculated on; we must have about us a set of scoundrels." If the stock of timber, he asked, were found deficient, why should there not be thefts of everything else by the same parties? Why was not full value for labour and materials had in every department? Who was there to say that half of the 1,600,000 bricks ordered for the plate-works had been received? And how could the frauds happen without intimation from any quarter, with such men as Jennings and Forster and Withers in charge, and such large sums paid for

---

prices pay." (Lucas Chance, January 18, 1847).

1   " As we can obtain *far higher* prices in the United States for *thick* glass than in prop'n we can get for *thin* in England, we must send our surplus there. To make large quantities of thick of good quality and workmanship is of importance, *the very first importance.*

. . Better workmanship would save a vast expense in flattening and occasion greater demand. . . . Hitherto we have fail'd in making better glass by reducing the quantity, and if you were to fail in this respect the loss would be enormous. . . . Whatever you do make good extra thick, good in quality and good in workmanship. If any set make bad work of thick let them make no more. Make large glass when the metal is very good" (Memoranda for James Chance and for Withers, November 28 and December 2, 1846).

2   To James Chance, December 13, 1847.

3   To the same, November 13, 1846.

4   Board minutes of March and April 1847.

management? He did not want to blame, he told his nephew, but the system must be all wrong. "My habits lead me to attend too much to detail, but I attend so many hours that I also attend to general principles, and to the general management." One sixth of his time, he complained, was taken up by Board meetings, held, indeed, at this time two or even three times a week. If James Chance would undertake the supervision of the whole of the works, so many managers would not be wanted. Especially necessary were anticipation and prevention of fraud and abuses, particularly in regard to purchases. "Recollect Mr. Shutt's advice, that a partner should do nothing, but direct others.[1]

Principal opportunities for economy Lucas Chance saw in the use of sulphate only in the mixtures, in reduction of the consumption of coal, especially of large coal, in better workmanship, in "the various items constituting manufacturers' charges, which in 1845 were largely encreased," and in wages and salaries. Minor items in his list were straw, crates, boxes, and diamonds. The use of carbonate of soda he condemned as worse than a waste of money, as sulphate is the more regular article, and the sooner it is dispensed with the better.

> If you will recollect, in 1838 or soon after we used nothing but sulphate and did capitally well, afterwards we could not get on well and went back to a portion of W. Ash and at a cost of many thousand pounds. The different results must have arisen from worse furnaces or deficient charcoal or bad coal. The best manufacturers use sulphate only and I question whether we ever get paid for the use of W. Ash 1/10th what it has cost. . . . We formerly had a register of the mixtures of every journey, the appearance in the house, whether good metal or not, with the manager's remarks.

---

1   To James Chance, June 19, 1847. Thomas Shutt was one of the founders of the British Crown Glass Company.

Then on the work in the crown houses:

> The workmanship in No. 4 house averages good, but some journeys are very bad. Do.
> in No. 1 averages shamefully bad; we have two managers to pay and the quantity of
> bad workmanship is shameful, and the good work is far too crooked, not carried off
> well.

And lastly about the coal:

> From the reports relative to Coal we ought at present to get abundance of good. ... A
> very important saving may probably be made by the proportion of large coal being
> much reduced. Will not a return shewing the percentage of coal used for every depart-
> ment be a valuable document? . . . A full supply of capital lump coal is essential to
> good metal, and one of Withers' defects I take to be the not providing such a supply
> *at all times*. It requires constant and most vigilant attention and I rather expect Saml.
> Smith is the best man to insure it, or perhaps Thomas Gittins. I have ever found the
> latter full of energy, and he is certainly capable of any thing he undertakes.[1]

To proceed with this matter of the coal, the principal supply came from the firm's own colliery at Titford. But in this year, 1846, a fault encountered obliged the sinking of new shafts. To avoid the expense, search was made all around for suitable mines to hire or purchase. Particularly inviting, for reasons set forth in the Board minutes at length, were 80 acres of "thick yard coal" near Portway, the property of Lord Ward. Nothing, however, seems to have come of this or other like negotiations, nor of proposals to acquire the mines underlying the works at Oldbury, whether to work them or to prevent their dangerous extension.[2] Peter Ward, indeed, now the manager there, was authorised to purchase two 30 h.p. boilers to raise steam for the former purpose, subject to approval of them by John

---

1   Memorandum for James Chance, November 28, 1846, cited.

2   It was resolved on December 16, 1846: "That while it would not repay us to sink new shafts, at our own cost, it is the opinion of the Board and of Mr. Ward that we ought to negotiate with Mr. Houghton for working the above mines by deepening Nock's pits and that we should obtain the following advantages:

(1) Even if we should find no coal in Mr. Houghton's mines, still if we displeased Mr. Houghton by not renting his mines he might injure us by indicting us for our nuisances, and in various other ways.

(2) We shall be enabled to prove the nature of the mines under Willett's land; if there be coal under the Chemical Works we should have the advantage of working the mines in such a manner as to cause the least damage possible to our own buildings.

(3) We should have the coal delivered at once upon our Works."

Sanderson, but soon afterwards, on receipt of a letter from Mr. Dugdale Houghton desiring an immediate decision; for the reason that the state of the pits would not permit indefinite delay in the inspection of them, it was resolved to close the negotiation.[1] The shafts at Titford were sunk, and the colliery was in full working order again by the end of 1847. Its production in 1840, amounted to 600 tons per week, of which 50 of large coal, 30 of lumps and 200 of best slack were sent to Spon Lane, the whole of the slack, 150 tons, to Oldbury, and the remainder sold.[2]

In 1847 the use of lumps for founding was forbidden, and that of large coal ordered to be reduced. Thomas Gittins was promised a premium of £50 for carrying out a number of changes, provided that he rejected all commissions. In January 1848 it was resolved

> That no more slack be bought above 3s. per ton in future; that instead of black coal we are to try Round's bottoming coal; that of large coal we are to procure 3 boats per week from the Titford colliery and from Griffiths together, and of lumps about 3 boats per week from Bagnall and Jesson; and that no Brazils are to be bought except to save large coal and for Badger's staining kilns.

In November, Harecastle coal was reported replaced in No. 7 by lumps picked from the slack and its use in other houses reduced to one-fourth.[2]

The principal blame for negligence of management fell on Withers, with whose conduct at this time Lucas Chance was on other grounds gravely dissatisfied. Thus on November 28, 1846: "One of the great evils of our manufactory is the overwhelming influence of Mr. Withers. It is only a partner who can obtain any information, and that only when he enquires." Open dissension occurred about the engagement of crown hands turned off from glass works in Baggot Street, Aston. Lucas Chance was anxious to secure them, and particularly one Williamson, strongly recommended, to take charge of No. 1 house. "A master's man" he termed him, "one of the old set of managers, one who can gather, blow, flash and carry off, and has been a walking manager there ten years." The others, he said, "can have places in abundance, but they are anxious to have permanent places" Yet Withers had rejected nearly all; "in fact, our place is unpopular from Withers' system of favouritism." Williamson declared the reason to

---

1   Board minutes of January 8 and February 13, 1847.
2   Board minute of February 13, 1849.

be private enmity to himself, and Lucas Chance held that he, like others, could uphold his authority against Withers, if properly supported by the partners.[1]

However, three days later he confessed that he had been mistaken about Williamson. "A failure would be a considerable expense and a great annoyance, and encrease Withers' influence." He hoped that the agreements drawn up might not have been signed the previous night.[2]

The root of the trouble was a scheme of Lucas Chance for working the two crown furnaces with five sets of men, partly in order to have substitutes at hand for cases of sickness or accident. Experience, he said, showed that "the only way to get good work is to have spare men in every department," and he was confident that to work with five sets would be a very great advantage. Withers objected a strike, whereon Lucas Chance declared that in that case he would stand no nonsense, and would send the ringleaders to Stafford. He considered the danger to be "the invention of the timid mind of Mr. Withers, whose want of resolution is the cause of the present discussion" and who would himself be the author of the strike, if it took place. He had understood, he said, James Chance and Withers to have advised making up a fifth set of men from the best of the others, and a late note from the former had not altered his opinion.[3]

It was resolved accordingly, on January 18, to form of the surplus crown hands a fifth set to work in rotation with the existing four, "the flashing furnace men and kiln men and boys and other men of each present set remaining and working with the fifth partial set." James Chance, however, while consenting to carry out his uncle's wishes, protested against a step that he deemed opposed to the interests both of the firm and of the men and "most impolitic." He declined to be responsible for deterioration in quality and workmanship that might result, and while he would never shrink from a contest with the men on a proper issue, in this case, he declared, "we injure both the men and ourselves" They were "not mere machines, but have sensitive feelings like ourselves."[4]

Apart from this question, reduction of wages was clearly imperative. Lucas Chance held it to be "not the real interest of workmen that wages should be higher with us than elsewhere, when losing prices prevail." He

---

1   Board minutes of May 20, April 8 and 13, 1847, January 18 and November 28, 1848.
2   To James Chance, January 2 and 5, 1847.
3   To the same, January 15.
4   To Lucas Chance, January 21.

asked: "Can we not require the Frenchmen to make for £16 all that they make in thirty hours, I mean all those not paid by the piece?" In crown glass he noted that the South Tyne Company's men, although engaged at 35*s*. and 27*s*. 6*d* for five years, had readily come down to 27*s*. 6*d*. and 22*s*. 6*d*, and afterwards to a lower rate by the piece. And he showed that the Hartleys, whose books he had seen, were paying much lower wages than themselves.

> If the men do not meet us to-day let us at all events reduce the wages of the boys to a proper standard; instead of £18 I hope we can bring it down to £10 for the 4 sets. . . We are to meet on 7th Sept. at Derby and I recommend you to attend the meeting with reference to making further enquiries as to wages. Not a word to be said about the meeting even to Withers.[1]

A beginning had been made in January by a resolution to engage no more crown hands at wages higher than 25*s*. for blowers and 20*s*. for gatherers for their standard make of 1,200 tables, which, it may be noted, were now of larger dimensions, 51-52 inches in diameter, and 11-12 lb. in weight. But most of the men already in employment had agreements covering a term of years. In March 1848 these bound men were called together and addressed by Lucas Chance on the necessity of a reduction of the terms. He pointed out that sales could not otherwise be extended, that at all other glass works reduction had been voluntarily accepted, and that the increased amount of work would much more than compensate. And further, that as a large number of the agreements were about to expire, it would be much better for all to accede at once to such terms "as might induce the firm to place the whole body on the same footing"—25*s*. a week for blowers, piece-openers, etc., and 21*s*. for gatherers, with overwork at the same rates, and the £11 per annum bonus as hitherto. Those cancelling their agreements to receive compensation at the rate of £2 per annum for the period cancelled.[2]

As there is no further reference to the subject for the present, it may be presumed that the proposals were accepted. The state of the crown glass manufacture is shown by resolutions of 1847 to work but three journeys a week in No. 4, and to give up the whole of No. 7 and half of No. 1 to

---

1   To James Chance, July 23 and August 16.
2   Board minute of March 30, 1848.

sheet glass.

The production of two crown houses in 1846 and 1847 amounted to about 4,000,000 lb., in 1848 to 3,600,000, and in 1849 to 4,250,000. The first figure, at 11-12 lb. per table, gives some 1,675 of them per week for each of the four sets of men. In a particular week in April 1846, the number amounted to from 1,850 to 2,000 per set. To make up a set were required four blowers, four gatherers, two (or three) pontystickers, a piece-opener, a carrier off, a piler, a kiln assistant, a skimmer, and several boys.

In the sheet houses also economy was envisaged in reduction of wages as well as in increase of production, saving of coal, and diminution of breakage. Especially desirable was the substitution of English for the more highly paid foreign blowers. In February 1847, for instance, it was resolved that "in order to promote the increase of the produce in No. 7, where they work only four times a week, this house shall be manned with English only." John Withers was desired to do all he could to instruct certain English gatherers, named, in the art of blowing, so that they might be ready to replace foreigners leaving. In April orders for them to work out the pots, when the blowers had finished, were repeated.

In spite of discharges, a list of July 1847 shows twenty-four foreign blowers still employed out of a total of forty-six, though of fifty-three gatherers all but three were English. Under a new arrangement fifteen of each were appointed to No. 2 house, ten to No. 3, fourteen to No. 6, and to the half of No. 7 that now was being worked for sheet glass seven blowers with but six gatherers, one of each a "relay" man. The remaining eight gatherers were made a "travelling set," so that in each journey two might be put off in No. 3, and three each in Nos. 1 and 6. Apart from Gaspard André and the other shade-makers, the blowers sent to No. 2 were the least efficient. Those in No. 3 were to make 26 and 30 oz. glass, and those in No. 7, the large sizes.[1]

In September four foreign blowers could be paid off at 1,000 francs each and replaced by Englishmen. In May 1848 the whole number had risen to fifty-two, and fifty-five were wanted—for the half of No. 1 nine, for No. 3 twelve, for No. 5 eighteen, for No. 7 sixteen. In August we find the number of "cylinder blowers" stated as forty, so that the rest must have been employed for shades and other purposes. More than half of them made only 13 and 16 oz. glass. In the next month Gaspard André

---

1   Board minutes of July 23 and 27, 1847.

and his son had notice to terminate their agreements, as also another of the best men, Eperthener. But the former were to be permitted to renew at £6 10s. per week for the one and £3 10s. for the other, with allowance for property tax, rent, and firing.[1]

In spite of the intention to be rid of the foreigners, revival of trade in 1850 obliged more of them to be engaged, and a number remained in the firm's employ for many years yet. The following passage, descriptive of them, is taken from an elaborate account of the glass works published in 1862.[2]

Amongst the "glass-house crew," bearded, blue-bloused, and with dark eyes and olive cheeks under their heavy, flapped, leather hats, are a little colony of French workmen, who, with their families, occupy a row of houses adjoining the works and support a native "cabaret" of their own, where *ordinaire* is to be bought by the "chopine." Several of these men accumulate enough from their savings in the course of a few years to return to their native departments as small farmers or landed proprietors.

One way by which the manufacturers proposed to retrieve their position, was by obtaining repeal of the window tax and so renewing demand. Lucas Chance, formerly a strong advocate of this, now disapproved; he looked to reduction of wages and general output first. He set forth his views in the following memorandum of November 24, 1847.

> The stirring of this question at present, from whatever quarter it may come, would be most prejudicial to us. The main, immediate, matter with us is reduction of cost, and the diminution of make of other manufacturers. Is it to be supposed that the new manufacturers will give up making, or those who have suspended their works will not be roused into fresh activity, or that the Glass-makers in April next will come to terms with us, in the prospect of the Window-Tax being removed? The expectations of all these classes will be raised far beyond all reasonable calculations, and the reaction after such an excitement would be far worse than the present state of the trade. . . The difficulty which we must surmount is this: that so long as we give a week's wages of 24 hours' work so long will Foreigners, whether we have a Window-Tax, or not, supply us in a great measure at home, and supplant us abroad. Under no circumstances can the present manufacturers of Glass be supported by Home Consumption, even if the window-tax be removed.

1   Board minutes of May to September 1848.
2   *Illustrated Times*, June 21, 1862.

To agitate the Window-tax removal before that we are in a position to compete with Foreigners is to put the cart before the horse: it is worse,—the agitation of the question would so excite the minds of Masters and workpeople as to prevent any settlement of the question of wages. Our policy is to remain quiet for the present.

P.S.—I am certain that Ministers would not now take off the Window-Tax: they are overdone with difficulties.

In contrast to other departments the patent plate continued to be profitable. In 1847 the buildings were further extended, machinery in store was brought into use, new smoothing and examining rooms were provided, and a number of the cottages in "Scotch Row" were taken over for the Silvering department. John Withers was instructed never to fail, when he had good metal, "to make for the Plate Works long glass from 46 to 52 inches long of 16, 21, 26 and 32 oz. in preference to any other sizes, unless there be some special sizes required.[1] Lucas Chance declared that he could sell 10,000 feet a week, were the quality good: "the patent plate at present is our sheet anchor." He expected, if slack were used generally instead of lump coal, to produce 16 oz. glass for the department at a cost of 2*d*. per foot.[2]

One branch, however, of the manufacture, that of looking-glasses, was abandoned in 1848. A first resolution was to limit their size to 16 inches by 12, a second to stop the make, a third that Baillie, the manager, should submit a scheme for working on the most economical basis until the manufacture could be disposed of. Kenway, the firm's traveller, was ordered no longer to solicit orders for looking-glasses.[3]

As has been said, the repeal of the excise duties largely increased the number of glass-works in Great Britain. Many of them succumbed to the subsequent depression, and among them the crown glass factory of T. and W. Stock at Eccleston, near St. Helens, and that in Baggot Street, Aston, mentioned. Lucas Chance, in November 1847, proposed to purchase the former works, for otherwise, he wrote, "they will be the means of keeping prices low for many a long year." He had the idea of getting rid of With-

---

1   Board minutes of February to July 1847.
2   A memorandum of June 26, 1848, showed the respective costs of the patent plate in 1846 and 1847 to have been: for smoothing, £1,535 and £1,235, for grinding £2,316 and £1,920, for polishing £1,920 and £1,766, for cutting £610 and £614. But Lucas Chance still wanted to get back to the standard of 1844.
3   Board minutes of 1848-9.

ers by sending him to manage them, but discarded it in view of the risk of his becoming a rival manufacturer. It would seem that the works were bought, for on February 1, 1848, Lucas, William, and James Chance signed an agreement to close them and to transfer their quota of crown glass to Spon Lane; for one reason that the Lancashire manufacturers might not be initiated into Spon Lane methods.

The Baggot Street works were taken over from the then leaseholders, John Silvester, William Henry Powell, and William Marshall, on a sub-lease of £250 a year as from March 31, 1850. They remained in the hands of Chance Brothers & Co. till 1876, from 1869 at a halved rental.

The reforming work of 1847 included much experiment in novel construction of the furnaces. In January, No. 3 was ordered to be rebuilt "with a deep grate-room after the French fashion." In April it was resolved to have in all of them flues between every two pots as well as at the corners. James Chance took out a patent on this principle:[21] "the fire is placed below the pots, and the heat and flame rise up on either side of each." The new No. 3 having been tried, orders were issued to rebuild Nos. 6 and 4 on similar lines—No. 6 to have six flues about fifteen feet high on each side, No. 4, designed to take the largest pots possible, grate-rooms narrower, deeper and sloping, five high flues to the bottoms of the pots on each side, and a chimney to create a strong draught from the working-holes. Another experiment was to divide an eight-pot furnace (No. 5) in the middle, and the houses generally were furnished with iron doors to keep out the external air.[2]

The patent mentioned included an arrangement for annealing sheets of glass on shelves, and mechanical mixing of the materials. It is not un-natural to suppose that both this latter innovation and the reconstruction of the furnaces were consequences of the intercourse with Henry Bessemer, of which in the next chapter. Heating the pots from below was an impor-tant principle of his patent of 1841, bought by the firm and mechanical mixing was the first claim in one of 1847, of somewhat later date than that of James Chance. The improvement, says Bessemer, suggested itself immediately that he saw the imperfect methods employed.

Changes of 1848 included a six-pot furnace for No. 2 on the plan patented, and the reconstruction of No. 7 for eight pots after the same

---

1   English patent No. 11749, of June 15, 1847.
2   3 Board minutes of 1847.

mode; which done, No. 2 was directed to be rebuilt for eight pots also. But it was now recorded to be "the decided opinion of the Board that shallow grate-rooms are preferable to deep grate-rooms for furnaces such as ours which are enclosed with doors; the former causing much less wear and tear both of the sieges and the pots." Next year, in April, No. 2 was ordered to be rebuilt anew, this time for ten pots, and yet after four months No. 5 to be repaired and lighted as soon as possible to replace it. Then again, very soon, it was resolved to make the shades and coloured in a new furnace, with twelve 38-inch pots, in No. 3 house. This fell in from faulty construction as soon as it was built, whereon it was decided to make it an eight-pot furnace, to retain No. 2 (now under the charge of Bontemps) for the shades and coloured, and when worn out to reconstruct it for twelve pots of a size "to hold the greatest quantity of metal that can be founded six times a week."

A minute of June 1849 ordained "that four pots in No. 1 glass-house be appropriated to sheet glass . . . without increasing the number of sheet glass blowers, as it is the decided opinion of this Board that we ought to work expressly for *quality* and not for *quantity*." Of the thirty-two ordinary blowers at present employed, ten were to work in No. 2 house, besides the shade-makers, fourteen in No. 7, and eight in No. 1. When No. 5 was ordered to replace No. 2, the object stated was "to employ more sheet glass blowers in the shade furnace," No. 1 being now again appropriated entirely to crown glass.

Consumption of pots in 1846 was found to have amounted to 235 in the crown glass houses, and 21 large and 260 usual in the sheet. The department was now reorganised to produce 312 of each in the year, twelve per week. Three makers were employed, the two Squires and Piers Barnes, each with his temperer and roll-maker; a fourth, Edwin Brettell, was reduced in the interest of economy to be a temperer. The sheet pots were allotted to Robert Squires, the crown to the two others; to Zachariah Squires any required for special purposes in substitution.

Among sundry articles produced at this period were such things as milk-pans and propagating glasses. A minute of 1847, concerning these, directed them never to be made before blowing, but

in one of the sheet glass blowing holes during the time of working, by having the hole enlarged, when required for this purpose; and in no case must any such articles be made except from metal which is too bad for cylinders of a quality above the coarse or horticultural.

Other minutes, the last of October 1850, limited their make to the end of each journey from the bottoms of the pots, and ordered the erection of a blowing hole and kiln for them at No. 6 house.[1]

Another manufacture of the firm was the exceedingly thin glass known as "microscopic," up to about 1880 made by them alone, and still by them alone in England. On examples shown at the Great Exhibition of 1851 the Jury reported:

> They also exhibit some extremely thin glass, 200 or 300 to the inch, for purposes connected with the use of the microscope and for experiments relating to the polarization of light, the want of which had formerly been found to be a great disadvantage in researches of this nature. This thin glass was introduced by Messrs. Chance as far back as the year 1840; and the Jury were informed by Mr. Ross that by its use microscopes were made of very far higher power, than could otherwise have been produced. [2]

Possibly the manufacture was taken up originally on the initiative of Sir David Brewster, who was in frequent correspondence with the firm, and whose researches on the polarisation of light are classical; or, and more probably, it was first made for its chief use, glazing microscopic slides. Its value for polarization lies in the fact that in a pile of glass plates for reflecting light the use of very thin glass practically eliminates absorption. Employed to cover microscopic slides it enables lenses of high power and short focus to be used. In modern times, for a period, there was a very large demand for it for phonographs. The standard thicknesses now in use are 70, 130, and 200 to the inch, but it has been supplied to a limit of 350.

Further acquisitions of land in 1846 had been 519 yards with four cottages purchased for £260 from Benjamin Johnson and others, and 1,557 yards with sixteen cottages for £1,800 from Daniel Grigg. These properties are bracketed in the schedule drawn up by James Chance in 1859 as

---

1   In 1868 a number of these miscellaneous articles were being made. A list left by John Chance includes preserve jars, propagating glasses, ship lights, milkpans, pastry pans, fern stands, bee glasses, fish globes, aquaria, rolling pins, cucumber frames, and gauge tubes. They were worked in various places as occasion served by men named Tandy and Cooksey, usually known as Tandy and Co.

2   *Reports of the Juries* 1851, p. 533. As stated by Ross in a paper read before the Society of Arts in 1837 the thinnest glass then obtainable measured 100 thicknesses to the inch. Whether this was produced at Spon Lane, or elsewhere, he does not say.

in "Spon Lane and Union Street, West Bromwich, Belgium Row" The former cottages were to be converted for extension of plant, the latter, no doubt, obtained to house the crowd of foreign workmen. Further, 7,040 yards on the far side of Spon Lane were bought from George Unett for £704, of which land 2,920 yards, between George Street and the canal, were subsequently sold to the Stour Valley Railway Company. For other property—a right of way at the back of "Scotch Row," two cottages adjoining the sign of the "Glassmakers Rest," and other land in that vicinity, on the west of the works—Withers was instructed to negotiate with Mr. Simpson.[1] But this purchase does not appear to have been effected at the time.

In 1845 the gas-works were enlarged and the road past "Scotch Row" ordered to be closed and a cottage or lodge placed at the gate on the Oldbury Road.[2] Two years later, on the site of warehouses appertaining to No. 1 house removed elsewhere, was erected the seven-storey building, at an estimated cost of £2,000. The ground floor was to be used for a packing-room and storehouse for filled crates, the second for splitting and cutting, the upper ones for stowing cylinders and crates, or for potmaking, as from time to time might seem desirable.[3] Over the offices, now rearranged, it was proposed to have a reading-room and library, but this was not carried out at present. In 1848 the part of the House appropriated to the partners was let to Henry Badger, manager of the Ornamental department, and James Chance was authorised to make such alterations in roads and buildings as the coming of the new railway through the works might necessitate. Also, since the lease of the London office in Coleman Street Buildings was expiring, Macartney, in charge there, was desired to look out for new premises near Oxford Street, "chiefly with a view to an eligible position for exhibiting Ornamental glass."[4]

The new railway was the Birmingham, Wolverhampton and Stour Valley, talked of for years before the issue of its prospectus on September 1, 1845. This followed on a meeting of landowners and manufacturers at West Bromwich, in the long list of whom the name of William Chance is distinguished by the title of Esquire. James Chance was appointed a Director in March 1846, and the Act received the Royal Assent on August 3.

1   Board minute of January 17, 1846.
2   Board minutes of November and December 1845.
3   Board minute of February 9, 1847.
4   Board minutes of 1848.

The line was to follow through the works the southern bank of the lower canal; there was to be a goods station east of the Spon Lane bridge; and a siding was to be made thence as far as Hartley Bridge, at the company's expense, but to remain the property of Chance Brothers & Co. for ever. An amending Act of July 2, 1847, altered the course of the railway between Birmingham and Spon Lane, but did not affect it beyond that point. Seven years were allowed for completion of the line. The first locomotive is recorded to have run at Spon Lane on February 5, 1849.[1]

By the arrangement made, goods arriving by rail were delivered at or across Hartley bridge, those wanted on the new side of the works being taken across the line on the level. To obviate this inconvenience, foreseen, it was proposed to build for their transfer a high bridge, provided with a hoist at its north end and sloped so that the loaded wagons might run down it. By a provisional agreement of June 13, 1846, the railway company undertook, amongst other things, to pay £500 towards the cost of this. However, the matter was left to slumber, until in 1853 the great extension of manufacture on the new side had rendered the blocking of traffic by the use of the level crossing intolerable. After correspondence on the subject the bridge scheme, unfortunately, was abandoned; the difficulty was met by construction of a branch line for direct delivery into the new side of the works.

Another member of the family now assisting at the works was Edward, fifth son of William Chance, towards the end of 1846 given principal charge at Oldbury. Four years later came John Homer Chance, third and youngest son of Lucas, and Henry Chance, sixth son of William. The last-named, like his brother James, was a Scholar of Trinity College, Cambridge, and of brilliant ability; he should have taken a higher place than tenth in the Classical Tripos, had he had his brother's love of work. The accessions, however, though valuable, could not compare at the time with that of George Bontemps, whom circumstances connected with the French Revolution obliged to quit France for a while. No one alive knew more than he about every branch of glass manufacture, whether in theory or in practice. Coming to Birmingham in April 1848 he agreed to superintend the Coloured and Ornamental departments, to carry out the manufacture of optical glass in accordance with Lucas Chance's patent of 1838, and "generally to advise and assist in the glass business" of

---

1   Robert Chance's diary.

the firm ;this for a salary of £500 a year, five-twelfths of the nett profits of the Optical department, and one-tenth of those of the Ornamental. It was stipulated that, while he was "to devote his exclusive services" to the firm, nothing in the agreement should constitute a partnership. Its term was from July 1848 to the end of 1854.[1]

Returned to Choisy-le-Roi, Bontemps wrote that he expected to be at the firm's disposition early in June. He would not have it known that he was coming, until about a month beforehand. His work on the optical and ornamental glass will be noticed in its proper place.

Other important changes followed on the death of Withers, in May 1849. His eldest son John was not fit to succeed him; his services were soon dispensed with, and he died in July 1851. The two younger, Joseph and Samuel, received charge—the one of two crown, the other of two sheet houses. Joseph we find often in receipt of bounties for good work, but there is note of indifferent health hindering the full performance of his duties. Other trusted managers were James Grant, in the Coloured department, Joseph Neale, Isaac Mallin and Edward Forster, the last in superior charge of everything that pertained to furnace and other building, pot-making, supply of clay, sand, coals, and the rest, and soon to begin his long and efficient service as principal manager.

Among prominent workmen were John Ault, long in charge of the hauling, William Maltby, chief millwright, Peter Rigby, at present entrusted with the grinding of the patent plate, Thomas Smith, in the same department. Maltby had been one of the men sent down by Messrs. Wren and Bennett to put up the patent plate machinery. He gave such satisfaction, that in 1847 Chance Brothers & Co. obtained his transfer to their own service. Founders were John Tandy, John Gittins, S. Hadley, S. Grigg, James Bates, W. Parish. Others noticed in the minutes as men to be depended upon, or who were rewarded by a rise in wages, were Samuel Smith, Thomas Gittins, Averill, John Sanderson, two Wynnes (in the Ornamental department), Jacob Green (in charge of the mixing), Lashley (furnace-keeper), and Leighton (cylinder cutter). Lucas Chance's opinion of Thomas Gittins has been noticed. In the offices Edward Jackson, ledger clerk, incapacitated after 21 years' service, was allowed a pension of £2 a week; principal men there were Gillon, an autocrat, and O'Brien, the cashier.

---

1   Agreement of April 13, 1848.

As a symptom of the revival of trade, in 1850 came a trouble happily rare in the history of the firm—a strike. The crown glass makers, led by one Slater, demanded an advance of wages. The crown furnaces were promptly shut down, but the men stood firm; a conference led to no agreement. However, they held out only for a month, returning at the end of July to work on the old terms. In November, prices having continued to rise, on renewed demand of the men the Manufacturers' Association agreed to an advance.

The revival also obliged more workmen, as said, to be sought for the sheet glass. Bontemps in May of this year undertook a business tour on the Continent on behalf of the firm, and part of his commission was to engage blowers. He succeeded in binding twelve, among them the two first-class men, Louis Eperthener and Jules André, guaranteeing them £18 a month, with lodging and firing. The engagement of these men was fortunate, for very soon came the question of glass for the Great Exhibition building of 1851, Joseph Paxton's "Crystal Palace" The firm tendered, and Robert Chance was hurried off to Rive-de-Gier with one of the French blowers, Stengre, to secure yet more workmen. Reaching Lyons on July 25 he was able to engage thirteen blowers, and then heard from his father that the tender was accepted and that thirty men were wanted. Eight of those engaged arrived at Spon Lane in Stengre's charge on August 8.

There was also, in the spring of 1850, a great demand for patent plate. Lucas Chance was insistent on employing more hands and working night and day. Some of his arguments in answer to his partners' objection to night work perhaps were specious, for example:

> The extension of the p. plate is the certain way to reduce the cost and thereby make it continuous, and the number of persons thereby employed will be of a class to whom it is peculiarly valuable; and moreover, it being an addition to the demand for labour in the World, makes it peculiarly our duty to carry this manufacture out to the greatest practicable extent.

In July he had to complain of produce being hampered by the small quantity of good glass available. Previously, in March, another polishing machine had been ordered to be erected, and the room to be enlarged to receive it. There appears to have been no further extension of the plant until 1856.

Glazing the "Crystal Palace" was one of the great achievements of the firm. The contract demanded production in a few months of some 200 tons of sheet glass over and above the ordinary output. All the furnaces had to be put in at extra pressure. Lucas Chance looked to get 30,000 lb. per week from No. 2, and 50,000 from each of the four others. Would not, he asked, seven hours five times a week give the guaranteed Frenchmen more than their guarantee, say 800 feet of 16 oz. glass each per journey? "Good metal, and pots thoroughly work'd out, are the two grand things to aim at. ... I am confident that 50,000 lb. should come weekly out of our 8-pot furnaces."[1]

There was much trouble both with the new Frenchmen and with the Belgians already in employment, called upon for extra work. Also a good deal of the glass delivered at the Palace was smashed by the carelessness of the workmen; one of them thought fit to use a pile of packed crates for standing on. The question of re-starting the Baggot Street works was discussed, and production was pushed up until, in January 1851, 63,000 panes of 16 oz. glass were turned out in a fortnight. The whole work was completed by the end of the month, 100,000 feet beyond the contract. The panes were of dimensions 49 inches by 10, cut from sheets three times that width. The quantity put in aggregated in area nearly a million square feet.[2] When the Palace was removed to Sydenham three-quarters of a million feet more were supplied.

At the Exhibition, besides what are elsewhere noticed—Bontemps' magnificent optical discs, Tabouret's lighthouse apparatus, the microscopic, coloured, and ornamental glass—the firm showed tables of crown glass, of ordinary diameter 52 inches, but one of them of 66; cylinders and sheets of 13 to 32 oz. substance; patent plate; shades, round, oval and square; and several forms of Baillie's ventilators. The shades, while other English glass was found to be much greener than the French, were lauded as remarkably pure, white, and of irreproachable regularity, as shown by the closeness with which they could be packed in "nests."[3] Two, made by an English blower, Pearsall, were gigantic—83 and 78 inches in height[4] It is

---

1   Memorandum on the glass for the Great Exhibition, July 30, 1850.

2   The exact figure returned was 956,194 feet, 11 inches.

3   Official Catalogue, ii, 700-1. *Reports of the Juries*, p. 533; *Travaux de la Commission française*, vi 12; Bontemps, *Examen historique et critique*, p. 36. He terms the patent plate, save that it was not white enough, the perfection of sheet glass, possessing all the qualities of plate glass, excepting its strength.

4   Robert Chance's diary. Mechanical aids were employed, such, no doubt, as are described by Bontemps (*Guide du Verrier* pp. 325-7). Some considered the shades, "certainly the largest ever made"

related that a boy, placed inside one of them to show its size, was all but asphyxiated.

Benjamin Baillie's well-known window ventilators, formed of glass louvres that could be opened and shut, had been taken up by the firm in 1844, when James Chance expressed a very high opinion of their practical simplicity and expected a large demand for them. A patent, No. 10402 of that year, was taken out in Baillie's name, as an improvement on an earlier patent of his, No. 7307 of 1837.

What the Jury thought of the firm's work they expressed as follows:

> The name of CHANCE occurs so frequently in the preceding observations, and is so honourably connected with every branch of the manufacture, that we cannot but regret that, according to the regulations laid down by the Commissioners, their firm is precluded from entering into competition for the Medals by the fact of one of the partners having consented to act as a Member of our Jury. But though Mr. R. L. Chance is thus disqualified by his own act, he has entitled himself still more to the consideration of the Jury by the valuable assistance which his practical experience and intimate knowledge of the details of the subjects committed to our investigation have enabled him to afford.
>
> When we witness the magnitude and variety of the operations undertaken by this firm, the merit of their works, the liberality, intelligence and spirit of enterprise which they have manifested at great cost and risk in experiments tried for the purpose of introducing into this country branches of manufacture almost exclusively practised by continental enterprise,—when we consider the advantage of inducing men so eminent in their occupation as M. Bontemps and M. Tabouret to settle in this country and superintend our works,—we feel that we should not act with justice by Messrs. Chance, or do our duty by the Commissioners and the public, if we did not call their attention, in a special manner, to the merits of the firm.[1]

Bontemps' commissions for his continental journey of 1850 were to bind sheet glass blowers, as said, to open markets for the firm at Hamburg and elsewhere, to investigate the use of gas in furnaces and lears, to examine German methods of making coloured glasses, particularly ruby that would "stand the fire,"[2] and to introduce the firm's optical glass to German opti-

---

to be the firm's "most remarkable production" (The Crystal Palace and the Great Exhibition, p.126).

1   *Reports of the Juries*, p. 533.

2   That is, Bontemps explains in his official report on the glass at the Paris Exhibition of 1855 (*Reports* ii. 394), the second firing to fix the enamel on stained windows.

cians. Here it is necessary to write only of the third commission, noting that in 1846 Chance Brothers & Co. had acquired, for the sum of £100, all the rights of James Alexander Forrest, of Liverpool, to an invention of his for heating kilns by gas for flattening, bending, annealing, or staining glass. In pursuit of his commission Bontemps visited Müllensiefert's works at Witten, near Cregeldanz, and those of Fickentscher, the inventor of the process, near Zwickau. A third works, at Ritberg, where the gas was obtained from turf fuel, he learnt were closed. He did not find the Witten process satisfactory either in production of the gas or in its application, and noticed that the sheets were coated with a bluish smoke. This, he learnt subsequently, was due to the action of sulphurous acid, produced when the air for combustion was in excess. At Zwickau he found gas used not for the sake of economy but because the pot clay was inferior, and the Bohemian workmen experienced only in burning wood. Using coal, Fickentscher had had excessive breakage of pots and melting of the furnace crown. He worked at a low heat, and though mixing 50 parts of alkali with 100 of sand took 45 hours to found. He confessed that for gas his consumption of coal was greater and that he could not use it small. All, therefore, that Bontemps could report in favour of gas heating was that it might obviate the noxious effect of coal smoke on white and coloured glasses.

In general conclusion, he was unable to observe much progress made in glass manufacture either in France or in Belgium. Separate blowing holes had nowhere been adopted, nor in the south of France, nor at Chauny, were there gatherers. A blower at Chauny, gathering his own metal, made no more than 500 feet of 16 oz. glass per journey; the pots were very small and cooled quickly; the top metal was generally seedy and the bottom stringy and blistery. At Valenciennes and in Belgium there were gatherers, but the blowers worked at the same holes, with the consequent inconvenience. Alkali being dear, and coal cheap, mixtures were usually very hard, so that the metal required to be worked very hot and was never properly settled. For the installation of separate blowing holes and larger pots the houses were too small. But fortunately for the manufacturers, the chief demand in France and Belgium was for inferior glass. Wages were very high in proportion to production, and so long as the present mode of manufacturing was maintained there seemed little to be feared in England from French and Belgian competition, especially if prices were kept moderate.

In Germany Bontemps found everything transacted on a "micro-scopical" scale; furnaces small, tools small, pots holding but two or three hundredweight. Otherwise the Bohemians, he thought, with their cheap labour, materials, and wood-fuel, might be dangerous competitors. But manufacturers were not ambitious. One Bippart, for instance, whose orders for his thin plate glass were so extensive that they could not be executed for six months, did not care to extend his works, but was content with a moderate profit and a quiet life.

On another journey, in 1852, Bontemps observed that Müllensiefert, flattening his glass by Hutter's improved method at a low temperature and very slowly, used no lagres, but surfaces of beautifully polished brick, which lasted about two months. In regard to glass required for the pictures in the Dresden Art Gallery he found that the tariff, 9s. per piece, had been prohibitory against Chance Brothers & Co.'s patent plate, and that samples of their ordinary sheet glass sent had been so bad as to be quite out of competition.

APPENDIX I

REMARKS AS TO THE EFFECT OF TAKING OFF THE EXCISE UPON GLASS
By LUCAS CHANCE, January 24, 1860

I HAVE known more or less of the manufacture of Crown Glass since 1801.
1.   At this time and until the abolition of charging by the Gauge the frauds on the revenue were enormous. The drawbacks on the Exportation were usually so regulated that the exported obtained a great deal more than the Duty actually paid.
In many cases the bonus went, not into the pockets of manufacturers, but of the chief dealers.
I expect these remarks apply equally to Plate, Flint and Broad Glass. One Plate Glass manufacturer told me that they had realized Fifty to Seventy thousand pounds by exporting German sheet at 2s. 9d p. foot drawback, the duty being not more than 9d., without any infringement of the Act of Parliament regulating the duties and drawbacks on Plate Glass. This led to the limitation of thickness to 1/8th of an inch and upwards.

2. When Mr. Wood became Chairman of the Board of Excise a great
change took place in the intercourse between the Board and the manufacturers— previously the manufacturers rarely saw the Board or the
Chairman but the Solicitors to the Excise, and so long as Mr. Vivian was
the solicitor the manufacturers were treated as so many persons whose
business it was to defraud the Revenue.

Mr. Wood took immediate steps to make the acquaintance of the manufacturers with a view to ascertain who paid the Excise Duty in full and who
did not—from the former he sought to obtain information and before the
tax was taken off Glass in 1845 the frauds on the Revenue had been greatly
reduced. Probably the net revenue of the Excise duty on Glass, as long as it
was charged by gauge, was not more than one-half that would have been
netted had the drawbacks been properly regulated and the full duty been
paid by each manufacturer. After the duty was paid from the mouth of
the kiln probably three 4ths of the net duty was paid into the Exchequer,
but to the last I have reason to think that by more than one house a large
portion of their income was derived from frauds on the revenue.

The effect of these frauds was to interfere with the regular course of trade,
and deeply to injure those who paid the Crown all that was due.[1]

3. Another pernicious effect of the Excise Duty was that of causing the
manufacturers to make their glass as thin as possible, and since the duty
has been removed the ordinary Crown Glass has been encreased in substance from 40 to 50 per cent, and Sheet Glass is on the average probably
encreased as much.

4. It is very difficult to say what the encreased quantity in measurement
has been made in Great Britain since the duty came off. Before then none,
or next to none was imported. We had a monopoly of the Home market
and of some of the Colonies, but the quantity consumed in Great Britain
is probably encreased fivefold, of which probably one third or one fourth
is imported at about 20 per cent, below what we can sell it for here.

5. The manufacturers of Great Britain, however, make a good deal of
heavier Glass than they could make before owing to the smaller cost, and
probably they make five times the Weight that they made before 1845.

---

1   The Spon Lane firm was of the latter class. A Board minute of March 26, 1834, runs: " Mr.
Chance having reported that Mr. Dehany told him he had no intention of altering the drawback on
sheet glass, it was resolved that it is expedient to keep back at all times such a quantity of sheet glass
as to prevent the drawbacks exceeding the duty paid, until such a home trade can be established as
will remove the difficulty."

6.   No Sheet Glass was manufactured in Great Britain until it was made at our Works by the importation of a body of Workmen in 1832 from France and Belgium, and it is well that we did then commence and teach a considerable body of workmen, for even now we have not been able to teach a sufficient body of workmen to enable us to dispense with foreign workmen. Hence we have to pay higher wages than those paid in France and Belgium, which with other advantages they have in materials and costs enables the manufacturers of those two countries to export Sheet Glass to London and all the ports of this country and all over the world, of better colour and at lower prices than the manufacturers of Great Britain. Whilst our glass is prohibited in France and charged with a high duty in Belgium foreign Glass comes into this country duty free—but I hope this will shortly be remedied, for although the foreign ordinary glass is of better colour than ours, yet the surface of our glass being better we beat the foreigners in the real quality and import the French sand to make the higher qualities.

8. The best proof that great frauds were made by manufacturers before 1845 is that the number of manufacturers is diminished one half.

R. L. C.

APPENDIX II

THE following instructions from Lucas Chance to his son John Homer, on the latter going to the Works in 1850, contain many points of interest.

MEM^A FOR J. H. C. RELATIVE TO THE MANAGEMENT
OF THE SHEET GLASS MANUFACTORY
July 2, 1850

1.   The first thing is to ascertain the substances and sizes required for plate.

2.   The second do. do. for orders.

3.   Then to watch every journey that the managers of each house understand exactly what size and substance each man is to make.

4.   To enable you to make every kind at the lowest rate of wages, you must make yourself acquainted with the different tariffs at which each man works, and know therefore what kind to appropriate to each man.

5.   At the end of every journey ascertain the quantities made by each, and if any man makes less than his proper quantity enquire why and wherefore,

and if it proceeds from idleness or neglect have him before you, and if necessary before me or J. T. C. Pass over no case of neglect, and visit the defendant at his house rather than not know the cause of absence. Give no heed to pleas of sickness without the strictest enquiry.

6.   It often happens that the shade blowers are obliged to make cylinders because their gatherers are absent, or from the want of proper previous arrangement. The same with the coloured and flashed glass. I have to-day given written instructions to Mr. Bontemps that we hold him responsible for proper arrangements for shades and flash'd and colour'd glass. But don't depend on this; see that no case of neglect occurs, so that the managers will all know that you are aware of all the losses arising from bad arrangements.

7.   Look at the journals of each manager daily, and any defect you see, that does not appear in their journal, you will add, or else keep a journal of your own to register every thing that goes wrong.

8.   We lose immensely by metal being made and not work'd out, owing to the non-attention of workmen or [their] being absent, or idle, or blowers lacking gatherers, and gatherers lacking blowers, from a want of rigid discipline; and this is an especial evil now we are short of men.

9.   There is the greatest difficulty in getting 13 oz. made, and even 16 oz., because they get more wages in making 21 and 26 oz. You must insist on the coarse glass being made into 13 oz., and no inferior workmen should on any account make any substance beyond 16 oz., because all the breakage of 21 oz. and upwards is a most serious burden and a very great loss, indeed we don't know what to do with it.

10. To encrease the quantity of 16 oz. of good, and 13 oz. of the coarse, is of primary importance, which will have the double effect of increasing the quantity of saleable, and of diminishing the unsaleable glass made.

11. You must have before you at all times the quantities required of each kind for orders generally, and for especial orders. At the present moment we want an encreased quantity for plate, that we may work night and day, of which a large portion should be 16 oz. I think a certain quantity of 32 oz. is also required, but Cashmore can tell you best.

12. The large sizes for New York, Boston and Philadelphia must have your especial attention. Aim at one order at a time. Finish off the absolute orders, Schrack's and Lamon's, &c, and for consignments prepare Henry Holland's in preference to any other.

13. Watch the quality of the metal, and have reports and samples of the colour in each house daily, and should there be any bad founding, or stones in many of the pots, or broken pots, or inferior colour, take care I know it, or J. T. C. in my absence, and especially have the founder before you, that he may not repeat the same thing over and over again.

14. Broken pots are the worst of all evils, and a report of every case should be instantly laid before me or J. T. C, and don't rely on that, but tell me also.

15. Many of my remarks are equally applicable to each manager, and you will therefore make such abstracts for them as you think necessary.

16. There are three or four men engaged at £3 per week as blowers, and 30s. per week as gatherers. Whenever we can dispense with them as blowers they must be employed as gatherers, *viz.* Thos. Thompson, Thos. Oakes, and Parish. They are bound as gatherers at 30s. per week without limitation of hours.

17. Zellar, the shade maker, must always be employed at shades, as he is bound at high tariff as blower.

18. To have good shade metal every day is of primary importance, and therefore a first rate founder should always be in the shade furnace.

19. No. 3 is so much farther from the shade arch than No. 2 that I doubt the expediency of making shades there. Ascertain the loss of quantity made thereby, and any other evil that may appear to arise from this extra distance, and make a special report on this point.

20. It is said that the metal in the small pots in No. 3 is not habitually so good as in No. 2, and the metal in No. 2 not so good as in the 8-pot furnaces. These are points to be carefully ascertained.

21. At present, the colour is not so good in No. 2 and No. 3 as in Nos. 5 and 6. You must endeavour to learn why and wherefore.

22. It will be an important part of your duty to see that the workmanship of each man is good, and therefore you will make yourself thoroughly acquainted with what constitutes good work, for which purpose you must visit the warehouse daily and examine every man's work for yourself, and report to each manager the defects of each workman.

23. The defect which is of primary importance is breakage, because it converts a saleable into an unsaleable article. Take every possible pains to diminish this in blowing, in cutting in the cylinder room, in carrying, in the cutting, and in the packing.

24. The next defect is scarcely of less importance, that of having the sheets of equal substance throughout the sheet, without which it will not do for plate.

25. The 3rd defect is blisters, which also diminishes vastly the quantity which will do for plate.

26. Don't forget to note when the metal is seedy, as that prevents the metal being at all fit for plate glass.

27. The flattening is now under Mr. Sam'l Withers's management, and all defects in that department you will report to him as well as to me. You will have to learn to distinguish between the bad flattening proceeding from bad work in the blowing, and that from over-firing, not rubbing down, &c. &c, by the flattener. A great deal of good glass is spoiled by the flatteners.

28. As we have plenty of flatteners, no bad flattener should be employ'd, and if any of the engaged men are inferior flatteners we must get rid of their engagements by purchase or otherwise.

29. At St. Helen's and Sunderland they don't pay the flatteners for bad work, and we must adopt the same plan as soon as we can.

30. Treat the managers with great respect, go hand in hand with them, don't send for the men without consulting them and do everything in your power to support their authority.

31. As Robert took this department for some time, I recommend you to copy my remarks and send them to him, that you may have his corrections and additions.

32. You must pay special attention to the having all the men to whom we guarantee a certain payment per calendar month fully employ'd and you must notice especially whenever they don't make the full quantities each day from their own default, from absence, from being unwell, or any other cause. You must take especial care that gatherers are provided for them whenever their own gatherer is absent.

33. Be especially watchful to provide additional gatherers for Mondays, as the men are very apt to be absent Monday mornings. It is especially the business of each manager to provide for these contingencies, but it will be your place to enquire whether they are fully prepared.

34. Above all, let no man absent himself, or neglect his work, from the highest to the lowest, without reporting it in your journal, and without endeavouring to prevent a repetition of it.

35. I have omitted to say anything as to the patent plate works, but it will be quite necessary that you are in constant communication with the heads of the respective departments with reference to a full supply of glass.

(1) Lawton for the proper quantities going to be flattened.

(2) Wm. Cutler and Holloway do. do.

(3) Jordan do. do.

(4) Sam'l. Withers, to ascertain that he gets an abundant supply to be reflatten'd.

(5) Rigby, do., to be ground, of each kind necessary.

(6) Wm. Shirley, do., smoothed.

(7) Jno. Shirley, do., to be polish'd.

(8) Jos. Gittins, do., for his orders.

36. Coloured department. Ascertain the quantities made by each man and the quality, and have the whole measured by Solomon Cutler jun'r and reported each week, and go into the coloured room constantly.

37. Stain'd. Do. do. Hawkins measures this up.

38. Leguay. See him constantly and report how he gets on, he and his son.

39. Ornamental dept. Go there constantly with the object of ascertaining that each party employ'd is fully occupied.

CHAPTER III
PLATE GLASS—BESSEMER—ROLLED PLATE

CHANCE BROTHERS & Co. had always in contemplation, as has been said, the manufacture of cast plate glass at a suitable time, and when, in February 1845, word came to Spon Lane that the Birmingham Plate Glass Works were to be had on a lease of seven years at £1,500 a year, with option of purchase for £20,000, that time seemed to have arrived.

Conduct of these works appears to have passed into the hands of two brothers, Brueton and Joseph Gibbins, who had added, or were about to add, the manufacture of sheet and crown glass to that of cast plate. After preliminary correspondence, definite negotiations were opened with them and Mr. J. F. Ledsam, acting as a Committee of the Directors. On March 7 James Chance wrote to the last-named that, in spite of the uncertainty as to what might result from the abolition of the excise duties, yet, in view of the advantages enjoyed by his firm for disposal of cast plate in connexion with their other manufactures, they were disposed to consider a lease with power to purchase, "or to join with the present proprietors in forming a new public company" reserving to themselves half the shares, and entire management of the concern. He anticipated that with his firm's "intimate knowledge of the best modes of manufacturing and polishing glass . . . the new Company would soon rank in point of quality with the best houses in the Kingdom." He laid it down that the works must be sold to the new company at their present value, "which, in our opinion, is, in common with all similar property, greatly reduced by the recent measures."[1]

A counter-offer was to sell for £25,000, payable in five annual instalments, and to hand over the lease of the London premises on its present terms. But now the matter had been reconsidered. James advised his partners that to purchase the works outright and set them on a proper footing would cost certainly £31,000, and probably £50,000; that a great increase of output generally, and continuance of low prices were in prospect; and

---

1   In the Birmingham Post Office Directory for 1845 the works are styled "Gibbins & Co., Plate and Crown Glass Manufacturers, Smethwick," but commonly they were known at this time as the "Birmingham Plate and Crown Glass Works." Employed there in 1846 were three blowers and five gatherers for crown glass, and at least three blowers, working without gatherers, for sheet.

that there was quite enough to do at Spon Lane for some years to come without taking up new responsibilities, unless with the certainty of good profit. It was replied, therefore, on March 18, that it would not answer the firm's purpose to purchase the works on the terms proposed.

Nevertheless the attraction remained and, as the great boom developed, it increased. One firm, Isaac Cookson & Co., took the opportunity to sell their works,[1] but generally there was a rush to manufacture. Chance Brothers & Co. determined to erect cast plate works of their own at Spon Lane. Plans were prepared; one, dated September 1845, shows a site selected near "Scotch Row." But their prosecution was interrupted by renewed negotiation with the Birmingham Plate Glass Company—negotiation this time all but carried through. By an agreement signed on February 6, 1846, Chance Brothers & Co. and the Gibbins brothers agreed to purchase the works for the sum of £57,500, each party paying £24,750 for joint ownership of the plate works, and Chance Brothers & Co: £8,000 in addition for exclusive possession of what were styled the "Smethwick Crown and Sheet Works." They proposed among themselves that William Chance and his family should have the entire management, Lucas Chance not wishing to have "any further plague in the counting-house," and further so to arrange matters that the two Gibbins could be bought out as soon as sufficient profits had accumulated.[2]

What the attraction was may be seen from a return of this year, evidently prepared for the firm's consideration. It showed the vast change of conditions that had taken place in twenty years: coals in London 13s. per ton instead of 30s.; pearl ashes £23 instead of £43; wages higher, but in proportion to production, much less; in place of four covered pots in a furnace, of about 12 cwt. capacity, six open ones customary now, each large enough to hold a ton; £500 a year saved on each casting-table by heating from below with cinders instead of on the top with charcoal; improved

---

1   To R. W. Swinburne & Co., a company whose members included George Hudson, "the Napoleon of Railways," and the famous engineer Robert Stephenson (Lucas Chance to James Chance, July 31, 1845).

2   James Chance to Lucas Chance, and the latter to William Chance, February 5, 1846. James Chance wrote: "I entirely accord with you as to the management of the Plate Works, that none shall devolve on your family: most sincerely do I wish that you could induce my father to substitute the Smethwick Works for Railways: he must be employed, as a matter, merely, of health, to say nothing of his activity of mind, and the Smethwick Works will become of such magnitude as quite to merit his attention." And Lucas Chance: "I look entirely to you and your family for managing the Smethwick concern. I cannot undertake a particle of the responsibility of management further than attending the meetings of the Board."

grinding and polishing machinery, driven by engines of doubled power, turning out at least three times as much in a given time; kilns capable of annealing 400 feet of glass in five days, as against half the quantity in two or three weeks; large plates made "with perfect facility" at an average cost of less than 3s. per foot instead of "with great difficulty" at about 10s. In 1827 large stocks on hand unsold and short time worked; in 1846 manufacturers, working night and day, able to supply their customers only "from hand to mouth, and that very inadequately" In 1829 a plate glass works selling at 12s. per foot closed for want of reasonable profit; in 1845 one with a paid-up capital of £125,000 realising at 5s. to 6s. per foot from £20,000 to £30,000 nett. Weekly productions of a furnace are given as follows: in 1827 5,000 feet at an average selling price of 12s. per foot; in 1836 7,000 at 8s. to 9s.; in 1844 23,000 at 6s. to 7s.; in 1846 45,000 at 5s. to 6s. The return concludes: "Looking at the extraordinary increase that has taken place notwithstanding the severity of Excise restrictions, and seeing that the demand now progresses more rapidly than ever even at 5s. to 6s. per foot, if the price were reduced to 4s. or 3s. 6d. per foot (which, free as the Trade now is from Excise interference would afford ample profit), what must then be the demand!"

The agreement with the Plate Glass Company was looked upon as settled. Lucas Chance indited a long memorandum on matters that required special attention at the works; agreements with the managers and workmen there were put in hand; James Chance planned to double the plant for a weekly production of 9,000 to 10,000 feet by the addition of a 60-h.p. engine and nine or ten new grinding and polishing machines. Then, suddenly, difficulties arose. There were objectors to the transaction among the proprietors, and it appeared that the directors had not, as they supposed, authority to sell the works. It was clear that there must be long delay before the formalities required could be completed, and this was not what had been bargained for. William Chance wrote to his firm's solicitor, Mr. Whateley, on February 25:

We all wish the proposed engagement cancelled, but without any pledge as to its renewal, for the following among many reasons, namely, that the delay is pregnant with evil, inasmuch as at present remunerative prices are obtained for plate glass, the result of which is that every house is increasing its manufacture; our object therefore has been to commence instantly the proper and energetic working of the concern to insure profits whilst they are to be had. To wait for all the arrangements rendered necessary by the articles of the Company preparatory to a sale would render the value of the concern quite a different affair, and therefore we must not be led into any pledge as to the future.

On these considerations the agreement was cancelled by mutual consent, and thereon Chance Brothers & Co. reverted to their project of manufacturing plate glass at Spon Lane. Preparations were pushed forward energetically. In April James Chance opened correspondence with Messrs. Wren and Bennett respecting engines and machinery, and Mr. Wren undertook to bring to Spon Lane himself a general plan and particular drawings of what would be required. Agreements were signed for supplies of bricks at the rate of 166,000 weekly, an estimate for cutting a canal basin was prepared, and particulars were obtained of plant at the Union Plate Glass Company's and other works. It was proposed to begin with one eight-pot furnace for founding and another for refining.

About a man to direct the casting Bontemps was consulted. He proposed a foreman of his, by name Dupont, to be aided by his two sons, one of whom was a mechanic. He considered that arrangements for the grinding and polishing could not be in better hands than those of James Chance; advised on the kind of glass to be made, not too white, but hard; and supplied valuable particulars as to plant, production and costs, agreeing to come to Spon Lane himself and bring Dupont with him. He strongly approved of the undertaking: "Je trouve que vous avez grandement raison de preferer former une enterprise nouvelle pour une fabrique de glaces coulees, plutot que de continuer l'exploitation d'une fabrique ou vous auriez a heriter d'un personnel et d'un materiel impropre a un bon usage. "[1] When he came over in May plans were concerted for a production of 5,000 feet per week, in plates twelve feet by six. In June James Chance presented estimates for the machinery for the "New British Plate Works," which were approved.

But it was already apparent that supply was overtaking demand and that the boom was coming to an end. Besides which, knowledge was had of novel processes of manufacture proposed by Henry Bessemer, and firstly of a patent in his name and that of Charles Louis Schonberg, taken out

---

1   Bontemps to Lucas Chance, March 16 to May 9, 1846.

in 1841.[1]

Bessemer was a man of extraordinary ingenuity and quickness of perception, compelled by nature always to be inventing. The patents that he took out probably outdo in number and in variety of subject those of any other individual. Some were successful, the one that made his name enormously so; others, as his optical furnace (to be mentioned) or his bath of fusible metal on which to anneal plate glass, were weird. An experimenter from boyhood, he had made himself master of mechanical appliances, and he had no qualms about venturing into provinces of industry of which he knew nothing. He expected to learn as he investigated them, and held it, indeed, an advantage to be clear of the ruts of experience. As with all habitual inventors, one experiment immediately suggested another, so that he was accustomed to have several on hand at the same time. In his "Autobiography" he rates it a misfortune that inventions continually sprang up in his mind without being sought and gave him no peace till he was at them.

The principal feature of the patent of 1841 was an arrangement for heating each pot from below and separately. The pots were ranged round a solid centre, each with its independent grate beneath. It was claimed that by this arrangement any one of them could be removed from the furnace for casting without lowering the heat for the others, whereas under the existing system it was necessary to complete the casting from all in the shortest possible time. Other advantages were recited, and in accordance with Bessemer's habit a number of other devices were included in the patent. For better conduction of the heat each pot was to be mounted on a platinum disc. Refining was to be assisted by a vacuum apparatus. The annealing oven was to have a floor of hollow fire-bricks bolted together and ground to perfect level. For making optical glass a whole furnace was to be hung so that it could be made to oscillate regularly and give a slow rolling motion to a closed crucible, containing the glass, placed in its top.[2]

The present owner of the patent was J. Alfred Novello. It was resolved to purchase it from him, if his title were good, and to experiment at once with a furnace for a single pot.[3] Not that James Chance thought much of the principle; he doubted its being either likely to succeed or new; the

---

1   English patent No. 9100.
2   Figured and described also in Bessemer's Autobiography, pp. 105-6.
3   Board minute of June 25, 1846.

valuable parts of the patent he held to be the brick annealing floor (chiefly), the vacuum idea (possibly), and the optical glass device (if practicable), and he would have liked to have all else away.[1] However, the furnace was put up, but at the first trial with unfortunate result. In James Chance's words: "The heat underneath the pot broke the bottom of it in a manner different from anything hitherto experienced. . . . The indications of our first founding were not favourable, but at present I form no opinion whatever." [2]

Before agreement with Novello was concluded attention was called to a new invention of Bessemer, which promised to bring about a revolution in the manufacture of glass. After correspondence with him in July 1846, it was resolved to subscribe £250 towards the cost of putting up experimental plant at his residence in London, but at the same time that no step should be taken without authority signed by the two senior partners and that on no account should Bessemer be admitted to the firm's premises. It was decided, further, to postpone erection of the cast plate works, until the results of this and other experiments should be seen. It was desired, James Chance informed Messrs. Wren and Bennett (July 20), "to have more time for the development of certain improvements, which we now have under experiment, and which, if successful, would modify materially our plans and erections." Other reasons advanced for the decision were the propriety of waiting to know exactly where the new railway through the works would come, avoidance of building during the winter, and the necessity of suspending all extension for the present, with a view to reduction of wages and general economy.

The principal proposal of Bessemer's new patent (No. 11317, of July 30, 1846) was to produce sheets of glass by passing the molten metal between rolls. For melting his glass he proposed a rectangular basin, heated from above and having at the bottom of one side a slit. Through this, when opened, the glass was to stream out upon the rolls, the upper one being provided with a knife or knives for cutting the continuous sheet into lengths. When formed, the sheets were to slide down an inclined plane straight into the annealing kiln. The great results that he anticipated are set forth in his "Autobiography " as follows.

---

1   James Chance to William Carpmael, informing him of agreement to purchase the patent, if it were quite valid, and desiring his expert opinion of it, August 7, 1846.
2   To Lucas Chance, July 2, 1846.

Up to this period the fusion of glass in large crucibles was universal, and the reverberatory furnace which I had erected at Baxter House for this purpose was the first in which glass was made on an open hearth, and the parent of all the bottle furnaces in which the fusion of glass is carried on in open tanks.

Claiming to have reduced both time of founding and consumption of fuel by about two-thirds, and to have got rid of the "slow, laborious, and expensive" operations of the cylinder process, he goes on:

I aimed at converting the whole contents of the furnace into one continuous sheet of glass in ten or fifteen minutes, wholly without skilled manipulation of any kind or the employment of the other furnaces, which are necessary for opening and spreading the blown cylinders. It will be obvious that the continuous sheet as it passed from the rolls might be cut into any desired lengths, and thus very much larger sheet glass could be made than it was possible to obtain by blowing it into cylinders.

Then after a description of the furnace and machinery:

From this general description of the process, and of the simple mechanism employed, it will be seen that a large quantity of glass could be produced with a very small plant.

Thus, suppose that the glass materials are melted in five hours and that the time of casting is, say, fifteen minutes, a cast would easily be made every six hours, or four times per day. A bath only 4 ft. by 3 ft. in area, and 12 in. deep, when making strong horticultural glass $1/10^{th}$ in. thick, would yield theoretically 5,760 ft. per day (say 5,000), equal to, at least, 400 blown cylinders 4 ft. long by 1 ft. in diameter.

The patent, as usual, included much else in its eighteen claims. One was for a singular device for flattening crown glass tables—a good example of untutored ingenuity. A revolving disc of iron, faced with brass, was perforated with small holes through which air could be exhausted. The hot table, applied to this, was to adhere to it firmly by the action of the vacuum created. The bullion having been cut out, the table would be removed to the annealing kiln perfectly flat and suitable for close piling. If required, the disc might be shaped like a watch glass, or be ribbed or fluted.

There is nothing more on the subject of the machine in the "Autobiography," saving an account of what purports to have been Lucas Chance's first introduction to it and to Bessemer. That cannot be right; erection of the plant was subsequent to the correspondence of July 1846, and we know from a letter of Bontemps that they had met early in that month. About the payment of the £6,000, also, from what is said below, Bessemer's memory must have betrayed him; he wrote the "Autobiography" in his old age. But the story is amusing, and may be taken for what it is worth. What is true in it probably relates to the experiment of June 14, 1847, after which the payment was agreed. It runs :

I was quietly pursuing my experiments with the apparatus described, when I was unexpectedly called upon by an eminent glass manufacturer. He said that he had heard that I was doing something novel in the production of sheet glass, and if my patent was secured he should much like to know what was the nature of the invention. I told him my patent was secure, and that I should be happy to give him a general idea of the scheme. . . . He was very desirous to see the experimental apparatus, and knowing that my guest, Mr. Lucas Chance, was at the head of the largest glass-works in the Kingdom, and worthy of all confidence, I acquiesced in his strongly-expressed desire, and said that if he would call again the next day at noon, I would have a charge of glass ready to roll into sheet in his presence. . . .

There were myself, Mr. W. D. Allen, my eldest son Henry, a carpenter, and my engine-driver present in the small room in which the furnace and machine had been erected. As soon as the bar retaining the charge was removed, and the tenacious semi-fluid glass touched the lower roll, the thick round edge of the slowly-moving mass became engaged in the narrow space, where the second roll took hold of it, and the bright continuous sheet descended the inclined surface, darkening as it cooled slowly. I had intentionally omitted the cutter in the roll so as to make a continuous sheet, so this had to be pulled away, for my little room was not half long enough to accommodate it.

The heat suddenly thrown off from so large a white-hot surface threatened our garments if we stood too near, and unfortunately some oily cotton waste took fire, causing a momentary panic. Mr. Chance called out, "Cease the operation, cease the operation." We were all in a perspiration, and the long adhesive sheet of glass, 70 ft. long by 2½ ft. wide, was gathered up before the door. The heat was very great, and throwing the rolls out of gear, we all beat a hasty retreat. However, as far as the rapid formation of thin sheet-glass was concerned there could be no doubt whatever, and I and my visitor sat down quietly to cool ourselves and think over what had taken place.

To continue the story, Lucas Chance admitted the mechanical success of the invention, but pointed out that its commercial development was quite a different affair; this he proposed to undertake with the advantages to hand at Spon Lane. The business, Bessemer goes on, was settled at dinner the same evening; Lucas Chance agreed to purchase the patent for £6,000; next day the agreement was drawn up and signed, and the money paid, a cheque for £1,000 and a short-dated bill for £5,000. "We parted very good friends, mutually pleased with our bargain."

A first thing done was to consult Bontemps, who was horrified. "Je m'effraie," he replied to Robert Chance's communication, "de vous voir tenter des essais dans cette direction. . . . Je regrette vivement de vous voir engage dans des semblables essais."

Evidently, he said, Bessemer had no notion of the inherent properties of glass. Should he succeed, it must be by some new principle that he had discovered; in which case he would have shown himself superior in genius to Arkwright or Watt, and his own admiration would be in proportion to his present want of faith. James Chance also was a doubter. He wrote to Bontemps: "I perceive that you consider that the experiments connected with rolling glass will fail: I think so too: but a trial will be most satisfactory to all parties, and we have limited to a moderate amount the sum of money which we are willing to expend upon the experiments." He asked for particulars of the various plans for rolling glass in this way which had been tried in France and elsewhere, so far as Bontemps knew of them, and for a statement of the reasons of failure in each case.[1]

On the night of October 7, 1846, Lucas Chance was able to see the process at work. Greatly impressed, he wrote offhand:

> As he considers he has perfectly succeeded in getting rid of all smoke, will greatly diminish the consumption of coal, and get a much greater heat than we do, and as I think his furnace far superior to one-pot furnaces, or any other that we hear of, and that it may be tried on the large scale at a very moderate expense,

---

1   Bontemps, August 2 and 19; James Chance to him, August 9, 1846.

I withdraw my consent to the experimental furnace we talk'd of. In three weeks he expects to have a furnace ready by which he will be able to found about half a ton of metal, and connected with which he will have a rolling machine to cast plates 24 in. wide by any length required. However disagreeable it is to wait, we have no choice. At his request I paid him the balance of £50, he says he has already paid more than the sum allow'd him.

I was surprised to find that he is bound by us not to introduce any plan of flattening into his patent, and as he says in his investigations into glass he has discovered a plan of annealing superior to any other, which he wishes to secure to us by including it in his patent of 31 [sic] July, I think we ought to abandon that, but if we do not permit him to include it in his patent then he considers he will be at liberty to sell his invention to whomsoever he pleases. . . . We ought to secure him, for he certainly will be a very dangerous opponent. . . .

He is perfectly satisfied that he can make plates of any size on his plan, and consequently that a casting table will be quite unnecessary.

There is no doubt that he is directing all the energies of his mind to the improvement of the manufacture of Glass and he has satisfied me that there is a necessary connexion between his new plan of furnace and the successfully carrying out his plan of rolling.

Besides which, Lucas Chance had to tell of Bessemer's idea of having furnace-crowns and grate-bars thin, in order that they should be less destructible.[1]

James Chance would have nothing to say to the last proposition. He thought it also "almost impossible that there can be any necessary connection between the new principle of Bessemer's furnace and the success of the plan of rolling glass. ... I am still entirely sceptical as to the new plan of rolling being better than casting." Nor did he know what could prevent Bessemer from introducing into his patent anything that did not infringe existing ones.

I am astonished that he should talk of being at liberty to negotiate with other parties after the agreement signed in the presence of Mr. Hind, and after his having received £250 from us. ... I do not like this. You will not of course exceed the £250 already advanced. I should not be surprised if Bessemer is disappointed at not receiving 2 or £300 for Novello's patent; and I expect that he may ask for further advances from us,

---

1   To James Chance, October 8, 1846, 12.30 a.m.

but of course nothing can be done until after my father's return. Last July I wrote to Bessemer for a copy of the agreement referred to above, but my application was not answered. Please not to be very communicative to Bessemer, however much with him, for I do not anticipate any ultimate alliance with him.' [1]

Meanwhile, experiment with the one-pot furnace at Spon Lane had gone on. There was further misfortune at an early stage, for the crown fell in; but nevertheless it promised well. James Chance proposed to found in it every day and to have six cuvettes to refine and cast, each in its own furnace, or all in one together with one or two single-pot furnaces to supplement. Thus he expected to produce at convenience plates of any size required up to an aggregate of 600 feet per day, a capital scale, he thought, for a beginning. [2] The former plate glass project was now practically given up. [3]

Several weeks elapsed before Bessemer's new plant was ready. Lucas Chance tried to persuade his nephew to see it at work, with him or alone, if he thought the operations likely to succeed. "Every day," he wrote, "is of importance." In December he was arranging to confer with Bessemer about the Irish, Scotch, and foreign patents. James Chance informed him that he would be glad to see "any fresh improvements of Mr. Bessemer" and to pronounce thereon, but he refused consent to taking out patents or advancing money for the purpose "in the present state of his plans, for I view them as altogether unsuccessful." On this Lucas Chance adverted to a suggestion that he should take up the process independently, whereon his nephew: "I never contemplated your taking a separate and individual interest in Bessemer's operations in any other manner than in the form of a concern entirely distinct in every respect from that of Chance Brothers & Co., and whose works, as in Hartley's case, would not be allowed to be within a given distance of our glass works." He was prepared to concur on these conditions, chiefly in view of his uncle's alarm lest Bessemer should set up works without him. [4]

William Chance, spending the winter at Paris, was of like mind. He trusted that in agreeing to take out patents for Bessemer's plans care had

1   James Chance in reply, October 9.
2   Memorandum of October 3.
3   Entailing difficulties with Dupont, staying on at Spon Lane on full wages but doing nothing, and wanting his wife and sons to be allowed to come over, in accordance with his agreement.
4   Correspondence of November and December 1846.

VIEW OF THE GLASSWORKS 1857
(To face page 80)

been taken not to involve the firm by implication in anything further. He understood that nothing more had been done, and yet that his brother was "as determined as ever to support him and is bent on paying the £6,000"; an odd thing, when the account at the bank was so much overdrawn "and our pecuniary position is much more unfavourable than he himself calculated on." And then, after the £6,000, a further outlay of £20,000.[1]

Bessemer had lately written: "The last experiment which I made since you were here renders any more at present unnecessary." His next letters were concerned with the patents and with a new method of silvering mirrors that he intended to include in his specification. Then he had "rather a formidable alteration to make in the machine and furnace," but was certain of the result. Again: all was in hand; he would have six new pots delivered by March 13 (no tank for founding, it will be observed), and had little doubt of success, being firmly convinced that his system would entirely supersede the present mode of manufacture. He was "quite content to stand or fall by the result," claimed "the most perfect control over the operation of casting" and ability to make small plates without any cutting, and undertook in nine or ten days to get the machine in order for James Chance to see, expecting it, though it could not be perfect, to suffice to show the soundness of the principle. But he complained of the unnecessary obligation laid upon him to prove his process further at additional expense, when he was already nearly £100 out of pocket; he consented, because he would not have it said that he had not acted with perfect fairness. "The opinions of men of experience agree that anything beyond that which I have in hand cannot be required to prove the principle." [2]

Lucas Chance now had his doubts,[3] and was desirous of bringing Bontemps over to see the process at work. "Considering the variety of claims in Bessemer's patent," he wrote, "and especially, that if we don't employ him, a very formidable opposition will almost certainly rise up, I think, as Bontemps has offer'd to come in answer to my enquiry, that we should act unwisely in not having the benefit of his opinion. If you concur in this, Robert shall write to him, in case Bessemer agrees to shew him his experiment." Then, on receipt of a letter from Bontemps, he went so far as to declare that the patent must be invalid, because it did

1   William Chance to James Chance, Paris, January 18, 1847.
2   Bessemer to Lucas Chance, January 14, foll., 1847.
3   "From what you say, Bessemer is at a discount, not only in your mind but in your uncle's" (William to James Chance, March 9).

not provide for overcoming difficulties which "every glass manufacturer would encounter in carrying out the patent, and as you have arrived at the same conclusion, there can, I think, be no question of the fact. I quite approve of erecting one of Bessemer's furnaces, as you suggest. ... I cannot but think that Bessemer is deluded about his rolling glass successfully by some blockhead whom he consults seeing no difficulty, probably Ensell (he told me he had shewn it to him). . . The better plan will be to await Bontemps' opinion after he has seen the specification, for from the remarks in his letter it would appear that your remark to Robert as to your hope that he (B.) would succeed, has so changed his mind, that he thought so clever a mechanic as Bessemer might accomplish the object."[1]

James Chance had seen the machine at work on April 9, after word from Bessemer that it had been "working most satisfactorily for the last two days," and had formed a decided opinion about it. He wrote to his uncle: "So far as Bessemer's patent refers to plate glass I am content to abandon all thoughts of making plate glass, and I know that my father would be delighted at such a decision."[2] William Chance, indeed, wrote in thorough agreement.

> I am much pleased to find that you have sent the specification to Poole and Carpmael, to ascertain whether it is valid, and that you are using every endeavour to determine in your own mind the nature and extent of the improvements made by Bessemer: and as you say that your uncle is very anxious to pay that gentleman £6,000, I will endeavour to ascertain from your uncle what we are to have for it. While I wish to do what is right, I am just as unwilling to pay the £6,000, as your uncle is anxious to pay it, and I certainly shall not consent unless a good case is made out. In the first instance it cannot, of course, be necessary to call in Bontemps for his opinion, and I think, knowing his bias towards your uncle and the influence that the latter has over him, it would be a dangerous experiment at any time, in as much as a qualified opinion in favour of the plan would be considered by your uncle as entitling Bessemer to the premium. . . . He will under any circumstances require Bessemer to be paid, and I am equally determined to resist it unless you are satisfied.[3]

Lucas Chance held the distrust of Bontemps evinced to be "the more extraordinary, because I have never known a more fair and upright and

---

1   To James Chance, April 10 and 12.
2   To Lucas Chance, April 12.
3   To James Chance, April 14.

less selfish man than Bontemps in his transactions, for the long period of 15 years." And, indeed, his verdict need not have been feared. What he had written in the previous August he repeats in his book, declaring lamination of glass between rollers to be by nature impossible, and the idea an unhappy one, not original with Bessemer. Of his advice to Lucas Chance he writes:

> The proprietor of a very important glass factory consulted me on the merits of the invention and on the opportunity of purchasing the patent. . . . The whole system showed, unquestionably, immense mechanical ability, but in my opinion complete ignorance of the qualities inherent in the substance of glass. I tried, therefore, to dissuade the manufacturer from the enterprise, but he objected that Bessemer was also in negotiation with the proprietor of another glassworks, and that if the process were successful in his hands it would be a matter of fatal consequence to himself. To this I replied that for his rival to treat with Bessemer would be the happiest thing for him. However, Bessemer's influence was stronger than mine. My friend bought the patent at a very high price and spent more money in exploiting it firstly under Bessemer's direction, afterwards under that of an engineer, very clever, but who could not but fail in the execution of an enterprise contrary to all the principles of the properties of glass.[1]

> In his general assertion Bontemps was wrong; glass now is laminated between rollers with perfect success. But not after Bessemer's continuous method, nor with the complications that he introduced. He could not be satisfied with one problem at a time; must have also novel forms of furnace, machinery to displace hand-labour, annealing kilns of fanciful design. All could not be perfected at once; some accident stopped every trial; and the remedies that he instantly devised took weeks or months to put in operation.

After James Chance's visit, Bessemer wrote regretting ignorance of his verdict: "Mr. J. C. does not seem prone to give opinions generally." He was sure, however, that "so shrewd an observer with a perfect knowledge of the capabilities of his business cannot entertain the slightest doubt of the perfect success of my invention," while on his own part all diffidence was removed by the results and by the opinions of competent persons. The fact that the metal did not work quite properly between the rolls he attributed to imperfect founding. To show his process to Bontemps he

---

1   *Guide du Verrier*, pp.37, 436-7, translated

expressed perfect willingness. A pregnant sentence in his letter runs, in reference to a difficulty mastered by having his rollers unpolished: "Trifling as this circumstance may appear, it serves to shew me how requisite it is for me to get as much knowledge of the state of the glass manufacture as possible" He desired also to know what success had been obtained with a new experimental furnace erected at Spon Lane. This had three pots heated from above. It succeeded so far that a similar one for six sheet pots of the largest size was ordered. At the same time the sum of £50 was directed to be paid to Bessemer " towards the expense of connecting our annealing kiln with his present furnace and apparatus for rolling glass, in order to make his experiments complete."[1]

By June Bessemer had developed new ideas and expected to produce "in Sheet Glass an article not hitherto attempted. . . . The plans I am about to patent will give extraordinary facilities in this department." He was to have a second pair of rollers, an annealing kiln arranged in continuation of the inclined plane, and adaptation of the rollers for producing patterns on or fluting the sheets. He claimed besides in his new patent, No. 11794, of July 17, 1847, mechanical mixing of the materials for founding, and a singular device for making crown tables without bullions.

After again seeing the apparatus at work, on June 14, Lucas Chance made up his mind. He wrote next day: "The rolling succeeded last night and five plates are now in the kiln. I saw quite sufficient to satisfy me that the patent is valid, as far as respects the rolling and annealing.'

His brother and nephew still refusing to have part in the venture, it was agreed that he and his son Robert should take over all rights and li-abilities and work the process on their own account in London, separately from the firm. The document signed (June 24) stated that Lucas Chance took "the affair upon himself as his individual speculation." Deeds were to be executed vesting the whole in him, and if the "lear patent" or any others belonging to the firm were required for carrying out the process, he was to be entitled to their use on terms to be agreed upon or settled by arbitration. It was agreed, further, to purchase Bessemer's patents of 1846 and 1847 for the sum of £6,000, and for £550, and for protection, one by Benjamin Aingworth, of Birmingham, whereby, instead of the roller

---

1   Board minutes of April 13 and 29, 1847.

moving along a stationary casting table, the arrangement was reversed.[1]
The purchases were to be made in the name of the firm, but subject to
the terms of the agreement.

As for the Bessemer-Schonberg patent, it was sought now merely to obtain
a licence under it. But Novello refused this, threatening to raise his terms
if his present offer to sell were not accepted. There was a good deal of
trouble and dispute about the matter, but eventually the patent appears
to have been purchased for £1,250.

Lucas Chance now proposed to commute the £6,000 for an annuity
of £1,000 for fourteen years, and Bessemer (June 28) expressed his willingness to accept this arrangement. He hoped that "we shall not be so long
in bringing any other matters to an issue as in this instance, for there were
so many things in my Patent besides the rolling." He went on :

> I quite agree with you that before December 1848 you will have time to judge how far
> it will be prudent to make any further engagement, and certainly before that time you
> will be in receipt of the profits of the invention. I am most happy to learn that you will
> individually arrange all matters relating to Plate Glass, as I shall have the full advantage
> of your energetic co-operation. ... I think, if you are waiting, you had better put up
> one of your old furnaces, and we shall see what will be developed in carrying out my
> plans on the large scale. . . . I can try both modes of annealing here, if you wish it, but
> upon consideration I find strong reasons for adopting my own.

It had been understood, we learn from a letter of James Chance of July
23, that the new Bessemer works might be built on land rented from the
firm at Smethwick, and all ordinary purchases and sales conducted in the
firm's offices for the convenience and separate account of Lucas and Robert
Chance. However, it was decided to be preferable, on Bessemer's side for
his convenience, and on that of the firm in order that he should not have
constant opportunity to observe what was being done at Spon Lane, to erect
the works in London. After long search suitable premises were found on the
south bank of the canal in Camden Town, hard by the goods station of the
London and Birmingham Railway. Plant was ordered and buildings put in

---

1   English patent No. 9378, of June 4 1842. James Chance had written to Carpmael in May: "My
uncle is very desirous of having as quickly as possible your opinion as to the using of Bessemer's rolling depending on a licence of Aingworth's plan, Aingworth being supposed to be the first to patent a
fixed roller." And Lucas Chance now, on June 15: "Even should there be any doubt about Aingworth's
patent, I consider it to be of the first importance to secure it."

hand, but there was endless delay. Certain brass tubes, for instance, by which Bessemer proposed to conduct the waste heat of the furnace to his boilers, took months to make. New ideas, embodied in two more complicated patents,[1] entailed constant alterations. Help was sent from Spon Lane; pots were supplied, and Robert Squires and other experienced workmen went to assist or to advise. But the year 1848 passed, and yet no start was made.

To Bessemer's admission to the Spon Lane works on any occasion William and James Chance made special objection. They knew how keen was his thirst for information, and how sharp his eye. Says William Chance in one of his letters:

> My main objection to your uncle's connexion with Bessemer is the fact of his being able to obtain thro' him every information regarding our manufactory and a knowledge of every improvement in furnace, pots, etc., etc., as they are adopted by us, and in our own defence we must adopt every necessary precaution to prevent this knowledge being used to our prejudice.

In accordance was James Chance's refusal to consent to trial at Spon Lane of Bessemer's notion about crown glass tables. A year later, in reference to the arrangement that Lucas Chance should take up the rolling process separately, he wrote:

> I felt that I could never act with Bessemer. ... I was certain that it would be less injurious to our united interests for you to undertake a separate concern than for the harmony of the general concern to be broken by the introduction into its management of a party with whom I could never have cordially co-operated.

And then, in reference to the proposed annuity:

> Under any circumstances I should consider it most unwise on your part to enter into any such contract. But as you may think otherwise, it is the wish of my father and myself that you should now understand, beforehand, most unequivocally, that the fact of your entering into any such agreement will never be admitted by either of us as an argument why Bessemer should be allowed to try any experiments in connection with the general concern, or in any way to interfere therein.[2]

---

1   Nos. 12101, of March 22, 1848, and 12450, of January 31, 1849.

2   Memorandum of November 27, 1848.

At last, towards the end of January 1849, Robert Chance, who had undertaken supervision of the Camden Town works, was able to report a satisfactory trial and everything almost ready for a proper beginning. In March, that he might be constantly at hand, he transferred his residence to London. But failure followed failure, with yet always the promise of success. At one time air-vessels leaked, and new ones had to be provided; at another a furnace built after a patent by John Juckes collapsed within a week; at a third there was a "regular mess" from break-down of the machinery. Something or other was always giving way. In August Lucas Chance came up, and it was resolved to close the works until matters had been well considered. A letter of William Chance states his brother to have told Bessemer that he would do nothing more unless a public company were formed.[1]

Discussions had no result; difficulties, says Robert Chance, appeared to increase. Bessemer himself was more interested now in a machine for sugar refining. Trials were resumed during 1850, but always with the same result; a sheet or two might go well, then some collapse. In April 1851 Bessemer's interests were liquidated and final separation from him effected.

Previously, management of the works had been entrusted to a Belgian named Biver, a man of whom it was said that anything to which he set his hand succeeded. A new furnace and new machinery were installed, and all was ready by the beginning of August. But Biver could no more make the process work than Bessemer. In November he presented a report, with the following conclusions :

1.   The reverberatory furnace in use was uneconomical, costing too much in fuel. He could not guarantee, either, more than one pot of good glass for each casting operation, the others being always too forward or too backward.

2.   Very great difficulties in practical execution were involved by the complication of the rolling machinery, in which, in addition, new and irremediable defects continually developed. Glass thinner than 3/16ths of an inch could not be produced at all, whereas by casting that was easy.[2]

3.   No glass thicker than 1/8th of an inch could be received on tables formed of bars placed near together for direct delivery into the annealing

---

1   To James Chance, August 21, 1849.
2   It may be noted that Biver, shortly before, had furnished a full description of his plan and plant for casting small thin plates.

kiln. The resulting surfaces were fluted and full of cracks and crevices. The surface of the receiving table must be plane.

It was painful for him, Biver concluded, to write thus, but it was the truth. He had done with experiment, and was perfectly convinced on all points. And so the venture came to an unhappy end. The works were closed, and the premises let for other purposes. Early in 1854 a final clearance was effected, part of the plant being sold, part sent to Birmingham.[1]

In singular contrast with the failure of Bessemer's complicated devices was the contemporaneous success of one in the same province of the simplest character imaginable. James Hartley under his patent of October 7, 1847, made sheets of thin cast plate by ladling metal from the founding pot directly on to the casting table, instead of by way of the refining cuvette.[2] The glass thus cast could be limited to any size desired, and sheets turned out down to one-eighth of an inch in thickness. For their principal uses they did not require to be polished. They were just the thing for skylights and glass roofing; they could be fluted, or impressed with patterns; coloured, they were admirably suited for church windows. They could be piled in the kiln like crown or sheet glass, and the costly usual process of annealing be thus dispensed with.

The cheapness of this glass enabled Messrs. Hartley to tender for the Great Exhibition building of 1851 at a price per foot but slightly in excess of that of Chance Brothers & Co.'s 16 oz. sheet glass accepted.[3] The size of the sheets, 62 inches by 21 instead of 49 by 10, would have greatly reduced the extent of framing required, but the weight would have been half as much again. It was considered inadvisable, in the case of a building stored with contents of such inestimable value, to adopt a mode of roofing not yet submitted to ordeal.[4]

The official catalogue described examples of the glass shown as "patent rough plate glass of improved surface, for roofing factories, conservatories, &c., 1/8|th of an inch thick and obtainable in sizes larger than in any glass of less than ¼ of an inch." Its obvious value at once determined other firms to embark upon the manufacture. A Mr. Hadland, already at work on it at

---

1   Most of the above particulars are from Robert Chance's diary. He returned to reside at Edgbaston at the end of November 1850.

2   English patent No. 11891. See, for the process, the *Guide du Verrier*, pp. 489-492.

3   The figures, Guide du Verrier, pp. 491-2.

4   Bontemps doubted whether the sheets could be so well annealed as to be reliable, and considered that the Commissioners would have incurred great responsibility in accepting the glass (*Examen historique et critique*, p. 73).

Eccleston, came to Spon Lane in December and took Biver back with him to inspect his plant.[1] After successful trials made on Biver's return, it was resolved to put up on the new side near the timber-yard an eight-pot furnace of the ordinary kind (No. 9) with twelve kilns, three pot arches and a warehouse.[2] To avoid infringing Hartley's patent, the glass was to be transferred from the ladle into an intermediary vessel, answering to the old cuvette.[3]

Very soon, however, it was found better to arrange with the Hartleys, who had brought an action against Hadland for infringement of their patent. In the course of the, year licences at £500 per annum were granted to Chance Brothers & Co. and the Messrs. Pilkington. Hadland losing his case, the manufacture was left in the hands of the three firms.[4] At Spon Lane were ordered more casting tables, a building 150 feet by 40 for cutting and packing, and annealing and reflattening kilns for the thicker substances.

Towards the end of 1854 the Hartleys proposed to grant licences to others besides the Chances and the Pilkingtons, and in particular to Hadland, who had lately taken a lease of the Nailsea works. Letters of December from James Hartley to William Pilkington the younger, of which copies were sent to Spon Lane, explain the situation. He stated that originally he had made two propositions; the one that he should be precluded from granting further licences, on the two firms undertaking to bear their share of costs in protecting the patent, the other the opposite; which latter, being preferred by them, had been embodied in the agreements. He had thought then, he said, and still thought, that they had made a mistake, and was still willing to carry out the first proposition.

I have no desire to license the patent to any one, it is no advantage to me to receive £500 a year from a House, I am out of pocket by what you and Messrs. Chance pay

1   Robert Chance's diary.
2   Board minute of February 26, 1852.
3   Robert Chance's diary, April 12, 1852. A Board minute of April 15 runs: "Resolved that the adoption of the intermediate vessel, as not infringing Mr. Hartley's patent, having been considered; and the question of the expediency of at once informing Mr. Hartley of our manufacturing and selling rough plate glass, being the produce of the use of the intermediate vessel, having been considered; it was Resolved that no such communication ought to be made, as it might assume the appearance of a challenge and originate a contest, which perhaps might never otherwise take place."
4   Hadland's defeat involved his bankruptcy. Lucas Chance wanted to acquire his works by compounding for him with his creditors for 5s. in the pound. James Chance, on the other hand, was all for settling their fate in concert Pilkingtons. While the matter was yet under discussion, word came that they had bought the works for £13,500 (letters of April and May 1852).

me, as compared to what I should be if you did not make rolled glass, but it is not good policy for a patentee to grasp too much, but to be satisfied with a fair amount of reasonable compensation, hence my satisfaction with our present agreement.

This, he went on, had led him to the conclusion that he ought to grant a licence to every one on equal terms, however injurious to his own interest; he could not be expected to take on himself the onus of refusing a licence to other parties and run the risk of a lawsuit. He was quite willing that the two firms should have a veto on granting licences, if they would take their share of the burden proper to the privilege.

The position with Hadland he stated to be as follows. When, after the verdict had been obtained against him, he was endeavouring to arrange with his creditors, he had come to Sunderland and agreed to abandon all further opposition to the patent and to render what service he could in its support, if he were released from his liability for costs. Making a merit of necessity, since there was little prospect of getting anything out of him, the agreement had been made of which a copy was enclosed. When James Kayll, Hartley's partner, was from home he had met Hadland and another, and was told of their intention to make rolled plate at Nailsea, but correspondence which ensued had led to no result. If Hadland attempted to break his agreement Hartley could put in an execution on his property or "take his person."

The negotiation resulted in the Chances and Pilkingtons assenting to Hartley's first proposition. Robert Chance wrote to William Pilkington on May 2, 1855:

We have now no objection to agree to the principle laid down by Mr. Hartley in his letter of 11th Decr, 1854, as the basis of the new agreement. On the contrary, we are really very glad to be able to meet his views, perfectly coinciding as we do with your father in the opinion he expresses as to the very handsome and liberal spirit exhibited by Mr. Hartley in this negotiation. We understand that Mr. H. will grant no licence without the joint concurrence of the three Houses, and that the expenses of defending the patent will be shared equally by them (your House, ours and Mr. Hartley).

In the following year, when Isaac White threatened to take up the manufacture at Nailsea, James Chance wrote to James Hartley (November 15, 1856):

We quite understand that, in conformity with your correspondence in December 1854 with Mr. Pilkington junr., the defence of the rolled glass patent is to be at the joint expense of the three houses.

The alliance thus formed was strong enough to keep the process, during the term of the patent, in the hands of three firms. In 1858 production of rolled plate at Spon Lane, as given in an estimate of cost of ¼ inch fluted for the railway station at Genoa, amounted to 2,912,498 lb. as drawn from the kilns, equivalent at 3½ lb. per foot to 832,142 square feet.

CHAPTER IV

1852 TO 1869

A new partnership of 1853 brought in two of the three junior members of the family assisting at the works, Edward Chance and John Homer Chance. The third, Henry Chance, was admitted two years later, as from January 1, 1855. The capital of the firm was fixed at £360,000, equally divided between the two sides of the family.

Trade was now prosperous. Increased demand was seconded by reduced supply, for several of the works started on the removal of the excise duties had failed to survive the period of depression. At Spon Lane in 1852 and again in 1854 wages were raised and more foreign blowers were sought, while no fewer than five new furnaces were built. Of these, No. 8 was specially for lighthouse glass and No. 9 for the new rolled plate, though arrangements were made to use it for sheet glass, if required. Before long, however, when the new manufacture had proved a success, the blowing holes erected were ordered to be replaced by kilns. Nos. 10 and 11 were for sheet glass, No. 12 for crown. All but No. 8 were full-sized eight-pot furnaces on the new side of the works.

No. 8 was a four-pot furnace on the old side, near No. 6. Its institution was due to the inferior colour of the lighthouse glass as compared with the French, as seen at the Great Exhibition. If there was no particular optical prejudice, there was obvious disadvantage in appearance. In the new furnace the pots were covered, and especially pure materials were employed. The minute directing its establishment ran:

> Resolved, that a furnace, blowing-holes and kilns be provided for making the glass for Lighthouses, at the spare ground adjoining No. 1 basin, and in such a manner that the furnace will be available for Coloured Glass in open pots, when the covered pots are not required, and that a furnace for flashed glass be connected therewith, so as to be equally distant from the said furnace and No. 6 furnace.[1]

---

1   Board minute of February 26, 1852. From a later one, of October 26, it appears that this "monkey" flashing furnace was intended to replace the old one near No. 2.

The genesis of No. 12 was as follows. It was resolved in the first place "That an extra half crown side should be provided in connection with one of the furnaces, from which the glass can be worked into sheet glass, when not required for crown, since by this arrangement four sets of crown hands can always be employed during the renewal of the crown furnaces." A week later it was proposed to erect a double crown house on the Island, "to avoid the heightening of No. 1 cone and with the view of dispensing with the old side of No. 1 house, to afford extra room for the mixing department and other purposes." When, however, it was found that no safe foundation could be secured on the Island, even at great cost; that an outlay of at least £1,500 would be required to convert No 1 into a good double crown house; "and as an extra crown side cannot be provided without a sacrifice of £1,000 in removing No. 3 lear"; it was decided to build a new house by the Oldbury Road, the cost thereof not being greatly in excess of what would otherwise be required, as above. The site appointed was in the south-west corner of the firm's land adjoining the said road; the cave to run north and south and the kilns and blowing-holes to be in the same building as the cone.[1] Thus came into being one of the best furnaces on the works and the one in which crown glass, in such small quantities as are now occasionally required, continues to be made.

No. 1 furnace it was decided, in October 1853, to work for the future for sheet glass only, unless during temporary suspension in one or other of the crown houses.[2] Blowing was to be done in the same holes as No. 6, with considerable saving, it was expected, in coal, and the space gained to be used to extend the mixing accommodation and for storing 80 to 90 pots. Next year, however, fourteen blowing-holes were ordered to replace the old crown side of No. 1.[3]

Bontemps about this time opened a memorandum: "Messrs. C. B. and Co's furnaces seem to work now on this principle, namely, to consume the greatest quantity of coal to produce a certain quantity of glass." He blamed the height of the crowns, the size of the working holes, left entirely open, and the increase in thickness of the walls of the pots from three to five inches. "The necessity of filling with flame the immense capacity of the furnace, in spite of the width of the working holes, and in spite of

---

1 Board minutes of February and March 1853.
2 "The five sets of crown-glass makers being ordinarily employed in Nos. 4, 7 and 12."
3 Board minutes of October 17, 1853, and October 18, 1854.

the draught produced by the iron doors, creates naturally a maximum of consumption of coal." There would be no necessity, he held, to build the crowns so high, to avoid droppings, but for the great heat obliged by the other arrangements. The glass had been just as free from seed before the iron doors were used, and so when the heat was less. The doors had their advantages, but with them the furnace openings ought to be reduced in size. The result of their use had been "to increase the heat, the consumption of coal, the thickness of pots, the height of crowns, all points which are connected together." He recommended that the openings should be more than two-thirds closed during the time of working, that six to ten flues should be constructed at the corners and fronts of the furnaces, and that proper teazing, putting on the coal in small quantities at a time, should be insisted on.[1]

It was decided, after full discussion of the above, that the furnace crowns were not too high, that the coal used would not heat the furnace properly, unless the working holes were open, that the advantages and disadvantages of flues were not yet settled, that under existing conditions the pots could not safely be made thinner, and that "the minimum of coal is used with us by using fine slack, which prevents heavy firing."

One change, however, was made, and with success, filling the pots by five instalments. It was found that three hours were thus saved in founding, and resolved to have the fillings as frequent as possible without injuring the quality of the glass. To prove the matter, No. 7 was ordered to be worked five times a week. On the other hand, attempt to make sheet glass with all the working-holes of the furnace open at the same time, failed. Blisters were diminished, but production was decreased.

Notable in this decade was the increase in the size of the pots. For the crown glass they had been made of 54 to 56 inches outside top diameter, when wet, as far back as 1841, and by 1850 some of 50 inches were in use for the sheet glass. Two years later a diameter of 56 inches was fixed upon for both, for ordinary glass, and then, in November 1853, 58 and 60 inches. Next year four furnaces, Nos. 1, 9, 10 and 12, had 60-inch pots, and all the others were to have them, it is stated, excepting Nos. 2 and 8. In some cases, the number also in a furnace was raised to ten, with a view to maximum production. When this had been done in No. 12, in 1855, for the second journey the number of tables drawn from the

---

1   Memorandum of September 29, 1853.

kilns amounted to 1,574. In 1858 the dimension was yet further raised to 64 inches; an extra cylinder room had to be provided for No. 10 to take the produce. James and Henry Chance, indeed, considered these very large pots, with their thicker walls, to militate against quality; the former wanted to return to outside top diameters of 54 to 56 inches, and eight pots only in a furnace. The latter change was made, at least in No. 12, and for size it was agreed, as a compromise, to taper the pots to 52 inches at the bottom instead of 56, and then, in February 1860, to have them of respective diameters 62 and 54 inches. But in November, when no improvement had been observed, and it was seen that the reduction did not materially affect the length of the found, there was reversion to a measurement of 64 to 65 inches. During 1861 all the sheet and crown glass furnaces had pots of this size. It was calculated that the 64-inch held 2¼ tons of glass, three times as much as a 42-inch pot. Their inside diameters were 54½ inches at the top, and 45½ at the bottom.

Among the pots made in 1861 were some for which a pound of asbestos was mixed with every hundredweight of clay and potsherds. And others by a casting process which Forster had seen in use at Messrs. Swinburne's works, and reported very simple. The asbestos pots were a failure, the two tried broke, the one in the first, the other in the second journey. Says Henry Chance: "The asbestos melted out where it could and where it could not ran into glass."

The average production from eight pots in 1853 was 10,000 lb. per journey. In 1856, by working the metal lower, this was increased to 13,000 lb. For the larger pots, figures for November 1863 show yields of 15,000 lb. for the sheet furnaces and 17,500 for the crown, 13 inches of metal being left in each case. It was decided again to work the pots lower, for yet greater produce; a journey in No. 7, in January 1864, gave over 17,000 lb. of sheet glass, and one in February, in No. 12, 1,730 crown tables, weighing 23,061 lb. But this did not answer, at all events for the sheet glass. It was determined not to attempt a yield of more than 14,000 lb. of it per journey in the new gas furnaces, since, with more, four journeys could not be worked in the week, unless at the expense of quality. For gross production a half is to be added to the above amounts, a third being waste.

Presumably, it was the increased production from the large pots that enabled houses 2 and 6 to be dispensed with for ordinary sheet glass, for a time. As the cost of founding was but little increased, there must have been great economy in thus obtaining the required yield from fewer furnaces. No. 6 was turned over in 1860 for lighthouse and extra white glass, first with eight covered 42-inch pots, then with six 49-inch, its cone being heightened by 25 feet. No. 2, the old original sheet furnace, was tried in 1856 with four 60-inch pots for coloured glass, flues being provided with a view to making " crystal sheet," when coloured was not wanted. In this case a fan-blast was employed to increase the draught. But in 1859 the furnace was condemned as useless for good glass; additional blowing-holes for No. 5, small ones for shades, and a new lear were ordered to occupy its site. The order, however, apparently was not carried out, although repeated in 1860. We then find first a new four-pot furnace resolved upon, secondly one with two covered pots for green glass, and thirdly, in 1862, one with eight pots for "crystal" metal, heated on the new Siemens gas-heating system.

Of the other furnaces, No. 8 was converted in 1855 for eight covered pots, it being understood, the Board minute runs, that green glass might be made there occasionally instead of lighthouse and extra white, but all other kinds of coloured in No. 2. In 1858 its number of pots was reduced to two, to revert to eight in 1860, before selection of the furnace for first trial of the Siemens system. No. 3, after being tried with four pots in 1857 for the "new ruby" (noticed in Chapter IX), had in the next years successively ten and six 42-inch covered pots, and then six 64-inch. Besides all this, there was continual experiment and change in all the furnaces in such matters as flues and depth of grate-rooms.

At the same time that the pots began to be so large, there was a general increase in the sizes of the glass made. Thus in October 1854 the minimum diameter of the crown tables, now weighing 11 lb., was made 52 inches, and in March 1855, the tariff for the sheet glass gatherers was based on 100 cylinders of 16 oz. glass 43 inches by 32, instead of 41 by 31, as heretofore.

A minute of January 1855 is interesting as showing that houses 2, 3, 5 and 7 stood in near proximity, and also where the first optical furnace was. It ordered the "monkey" furnace—there would seem at this date to have been but one of these—to be removed to the site of the old optical furnace, near No. 2 lear, so as to be available for all those four houses named. Another object stated was to give opportunity for making flashed ruby in the daytime, that is to say, when its colour could be properly observed.

Besides the twelve main furnaces there were two small ones, Nos. 13 and 14, the latter built to try casting thin plates from small covered pots. Some of these small pots were but 20 inches wide and 18 high, others 29 and 24 respectively. No. 13, a two-pot furnace, was used for "antique" and for signal green. Special furnaces also were built from time to time, for instance, in 1857 one with four open 42-inch pots for the "new ruby" above-mentioned and a similar one for extra white, green, purple and flashed opal, and at the end of 1859 one with six 33-inch covered pots for casting lighthouse glass. A patent was taken out in the names of Henry Chance and Thomas Howell for fitting copper nozzles to the mouthpieces of the pots, to facilitate the pouring.[1]

An experiment to improve the quality of the sheet glass was to leave more metal in the pots and reduce the number of blowers to a furnace. But it failed of effect and was soon abandoned.[2] Another was the use of rings 40 inches in diameter divided down the middle. Their object was to reduce the gatherers' blisters and to get the metal worked oat, when required, more quickly. But they were found to produce much string, after the top metal had been worked, and were retained only for the shade pots, where their use had been found very advantageous. In 1861 again was tried in No. 7 the experiment of having all the working-holes open at the same time, to reduce the blisters, a damper being placed in the cone to reduce or stop the draught. The damper proved so useful that in the following March the like was ordered for the other sheet houses. It was found to obviate the practice of finishing the journey with the iron doors of the house open. Yet in September the old method was adopted again in No. 1.

At the Paris Exhibition of 1855 the glass shown by the firm was pronounced to be well founded and refined, but very green in colour. Bontemps, indeed, in his official report, held it to be inferior in this respect to the least white of the Continental glass, for which great care was taken to use materials free from iron. Should the latter come to equal the English in quality, he expected it to prevail. On the other hand, he said,

---

1   No. 54 of 1860, January 7. The nozzles remained in use at such times as the lighthouse glass was poured.

2   "A week's trial of leaving bottoms in the pots and producing only 10,000 lb. per journey in the sheet houses instead of 12,000 lb. has shown that the reduction in quantity does not perceptibly diminish the number of blisters, or improve the quality in any respect, while such reduction is attended with great expense. It is therefore advisable to make the maximum quantities." (Henry Chance's note book, August 1856.)

some extra white patent plate shown by Chance Brothers & Co. "really attained the ideal perfection of glass making."[1] Of the Spon Lane sheet glass shown at the London Exhibition of 1862 it was reported:

> This is considered by all who have seen it to be very much superior to any glass made on the Continent. The flattening was very much admired, both as regards the perfect flatness and the freedom from marks of firing, the smoothness and brilliancy of the surface being retained, whereas that of the foreign glass was nearly destroyed. The jurors, who examined this glass officially, said that our third quality was quite as good as foreign best. The colour and the price are the only objections to it. If the price could be brought down to something nearer the prices of foreign, even with the present colour it would obtain a ready sale in France and Germany, where it is almost impossible to get sheet glass of good quality. This is the opinion of Continental merchants. . . . The extra white was much liked also, but is much too dear for ordinary purposes.

And about the shades:

> Ours are much preferred to foreign for quality and shape, but the price is very much higher, and the colour not so good. Foreign shades are imported at a price which defies competition, where quality and form are not great considerations.[2]

From the year 1861 dates the revolution in glass founding brought about by the Siemens regenerative gas furnace. Of so well known a system it is unnecessary to state more than the principle. The furnace is heated by gas produced from coal, instead of by coal directly, and both the gas and the air for its combustion are brought to a high temperature before entering the furnace by passing through "regenerators" underneath it; chambers packed with a network of bricks and previously heated by the gases leaving the furnace. A great amount of heat is thus obtained almost gratuitously, and, since the gas producers require only slack, the use of expensive large coal is obviated. The master patents in England were Nos. 2861, of December 2, 1856, and 1320, of May 11, 1857, granted respectively to Frederic and to Charles William (in later life Sir William) Siemens. Frederick, the younger brother, had been known to the Chance

---

1   *Reports of the Paris Universal Exhibition*, 1855, ii. 391-3.
2   General opinion of foreigners on the glass exhibited at the Exhibition of 1862—a private report by A. F. Jack.

family for some ten years; Robert Chance records entertaining him in 1850 and 1851, and thought him a very nice young man.

Forster saw a regenerative furnace at work at Rotherham in November 1860. It was experimental, and by his report by no means a success. Nevertheless it was determined to make a trial at Spon Lane, and No. 8 house was chosen. To ensure proper working, a damper was fitted in the cone. Six 42-inch open pots were set on March 28, 1861. Difficulties natural to first attempts having been surmounted—difficulties such as the gas and air mixing and taking fire before entering the furnace[1] — the trials resulted in an agreement being concluded with the Siemens brothers on April 12, and on the 25th it was determined to apply the system to a full-sized furnace, blowing-holes and a lear. It was debated later whether to purchase the exclusive right to the patent for crown, sheet and rolled plate glass, but decided only to complete the licence founded on the said agreement and on sundry correspondence.[2] This under a new patent of January 22, 1861 (No. 167 of that year), with special application to glass founding, taken out in the joint names of the two brothers. The licence bore date February 13, 1862.

The gas blowing-holes and lear appear not to have given satisfaction, for before long coal was used for them again. But the large furnace, No. 6, ready by the end of 1861, was quite successful. Adoption of the Siemens principle followed in 1862 for Nos. 9, 10 and 2, and in 1863 for Nos. 4 and 7, all with eight 64-inch pots. Conversion of No. 11, proposed, appears not to have been carried out before 1868. No. 10 worked extremely well. A report for a week in November 1863 gave 3,011 cylinders from the "crystal" mixture fit for patent plate, and only 5 coarse out of about 5,000. The average for seven months to that date showed 36J per cent, of plate and less than one per cent, of coarse.[3] The founder, Joseph Wheeler, received a gratuity of £10 in consideration of the large quantity of good glass made, a gift repeated for the year 1864.[4]

The changes involved the disappearance of No. 5 house; its room was wanted, according to the Board minute, for producers for "No. 7, No. 4,

---

1   Siemens almost despaired, but the difficulties were overcome by Forster.

2   Robert Chance's diary, January 30, 1862.

3   Henry Chance's note book. "The plate works depend upon No. 10," John Chance wrote in 1865.

4   Previously, in January 1863, he had had £3 in view of his constant and successful attendance to his duties as founder in No. 6.

etc."[1] Its number was used again for a furnace built in 1886 for casting lighthouse glass, on the new side of the works. At about the same time No. 8 was ordered to be pulled down for complete rebuilding, cone and all, and No. 1 to be tried with a double crown, one of Welsh, the other of Stourbridge clay, 12 inches apart, for the reason that it had been found difficult to "keep" the furnace with a single crown of Welsh bricks.

Various improvements were tried with the gas furnaces as they were successively built, for instance, mixing the gas and air before their entry, to obtain more perfect combustion. And the question of the size of the pots came again to the fore. In 1865 were set in No. 4 eight of 60 inches outside top diameter, then Nos. 1, 2 and 12 were rebuilt to take ten of 56 inches. On the other hand, for the rolled plate in No. 9, 66-inch pots were tried. These changes were quite temporary; in the autumn of 1866 both the 66- and the 56-inch pots were given up, excepting for No. 12 in the latter case. Forster had noted in the previous November: "64-inch are again in the ascendant." In 1867 were tried cylindrical pots, of equal diameter at top and bottom. These were soon given up, and all were ordered to taper from 64 to 60 inches, a fashion retained for some ten years.

In 1865 is mentioned a new "monkey" furnace, to take pots of larger size. In 1868 was ordered a cuvette furnace in connection with No. 14, now in use for lighthouse glass and heated by gas. Previously had been contemplated erection of a lighthouse furnace and kilns on the land bought from Benjamin Darby, below mentioned. A resolution of 1869 for a new eight-pot gas furnace on the new side of the works does not appear to have been carried out. In the same year, 48-inch pots were ordered for trial in No. 8, 42-inch having been found too small to give a profitable result. And No. 3 house, which four years before had been altered to take six 42-inch covered pots for extra white and signal green, was ordered to be pulled down, and first a double lear, then two pot-arches, to be erected on its site. In 1874, however, it reappears as a coal furnace with similar pots for extra white.

The crown glass manufacture was in rapid decadence. Already in 1863-4 the average number of tables made—their regular weight now was 14 lb.—was not much more than 6,000 a week, enough only to employ three sets of men working No. 12 furnace and a half of No. 4. Then the

---

1   December 4, 1862. It may be conjectured that it was intended to place the producers for these two houses between them.

three sets had to be reduced to two, and at the end of 1870 to one. John Chance noted thereon: "Crown glass is evidently dying a natural death. We can easily re-form a second set, when wanted." While production of sheet glass rose to 15 million lb. in 1860, and to 16½ million in 1871, and that of rolled plate in the same years to 4 and 5½ million respectively, the corresponding figures for crown glass were 5,624,141 and 2,415,916 lb., about 400,000 and 172,500 tables respectively.

Lears and kilns underwent but little development, excepting in capacity. For instance, in April 1855 all the reflattening kilns were ordered to be doubled in size. And there was gradual substitution of double sheet lears, turning out some 2,700 sheets per week, for single ones with a capacity for 1,800. The former were somewhat more costly to work; a calculation of December 1863 showed 17s. 5d. per 100 cylinders for them against 16s. 8d. for the single lears; but they had compensating advantages, one being that fewer were required, when room was difficult to find and yet flattening power fell short of furnace production by a third. It was decided to build two new double lears on the old side of the works, on the site of the box and timber departments, and to convert the single No. 3. After which, says John Chance, "doubling No. 5 would balance the two rival powers." In 1867 two more double lears were required on the old side to meet the same deficiency still existing there, though on the new side the powers about balanced. To quote John Chance again: "Should we work No. 12 into all sheet, we should require another double lear. But the new Belgian one will just meet the case."

This Belgian lear was the invention of one Bievez. It was shown at the Paris Exhibition of 1867, and was highly commended by Bontemps in his report for its simplicity and other merits. He stated that the sheets, rapidly cooled by being raised successively on iron bars and so kept apart, could be turned out in 25 to 30 minutes as against seven or eight hours under the system of piling them.[1] However, the lear failed to give satisfaction, and after a short trial was pulled down. Nor do earlier attempts in the direction of novelties appear to have fared better. In October 1851 James Chance is recorded in his cousin's diary to have been in high spirits at the success of an invention by Jules Frison, for which he had applied for an English patent,[2] but we hear no more of it. In April 1855

---

1   *Rapports du Jury Internationale*, Paris, 1867, III. 81. A description of the lear, *Guide du Verrier*, pp. 294—6.
2   No. 13699, of July 28, 1851. It included an improved fork for lifting and transferring the sheets. Bontemps' description of the lear, with his own suggested improvements, *Guide du Verrier*, pp. 291-4.

there is mention of a double-arch arrangement by Leguay, a man to whom it had been resolved a year before to pay £100 for his obscuring machine, provided that a patent were obtained for it, and that it continued in use after September. The only improvement of value seems to have been one for which James Chance took out a patent in 1856 (No. 283, of February 1), substituting rollers for the wheels on which the stones were moved and thereby rendering the movement much more easy.

Experiments with fuel aimed at the substitution of slack for lumps. The economy was obvious, if its use could be extended; slack cost in 1861 2s. 7½d., in 1862 2s. 9d., and in 1865 3s. 9d. per ton, as against large Cannock and other coal at 8s. to 10s. It may be noted that the lease of the Titford Colliery, completely worked out and exhausted, had been surrendered in 1853. Washing the slack was tried in 1855, with such success that Lucas Chance proposed to take out a patent. But it was determined first to try the effect of screening, and this was found both to be cheaper and to give better results. When after four years the screening was ordered to be abandoned by degrees, decrease of quality was ascribed to the change. In 1861 the purchase of lumps was forbidden; what were required were to be picked from the slack. But repetition of the order later, more than once, shows that at least it was not well observed. Indeed, in 1868 the quantity of lump coal purchased had risen to 5,500 tons, in addition to 5,000 picked from the slack. Deterioration of the quality of the latter was alleged in excuse.

With the sand supply there was trouble, for the quarries in use at Heath and Reach were beginning to give out. Attention was directed a second time to the white sandstone at Sweeney mountain, near Oswestry. After a visit of inspection by Forster it was agreed to take 1,000 tons of it, for experiment on the large scale, at 9s. a ton delivered. Next year, however (1858), when sand of proper quality was found on other land at Heath and Reach, four acres there were purchased of Mr. Thomas Swinstead for £1,250. Work then went on without hindrance, until in 1865 it was stopped by an injunction in Chancery at the suit of Colonel Hanmer, a local landowner. The firm, fighting on behalf of the local copyholders, won their case on appeal[1] and work was resumed, but in the interval large quantities of the Sweeney sandstone were again employed.

---

1   They lost in the lower court, their leading counsel failing to put in an appearance. Their action gave them great credit with the other copyholders.

Of other materials, refined saltcake was tried in 1855 to improve the colour of the glass, but any effect it may have had was not sufficient to retain it in use, at least in the crown houses, when falling off in quality there was attributed to the innovation. "Extra-refined" saltcake fared no better in 1863. Two years later, however, Bontemps, consulted, advised that purified sulphate had displaced carbonate at St.-Gobain without detriment to colour and with a saving of £12,000 a year, whereon Lucas Chance urged every effort to produce at Oldbury saltcake free from iron, as "the last thing to make our glass equal to the best made in the world."[1] For many years refined saltcake was used habitually.

There was experiment also with various kinds of limestone and chalk, and further with addition of a small quantity of pipeclay to the mixture. This for several months in 1856-7, and again in 1865. No effect was observed, whether beneficial or harmful.

In July 1855 an alteration in the rate of deduction for bad work brought about a strike of the sheet glass men. It was not of long duration; in September they expressed their willingness to return to work unconditionally as soon as a furnace could be put in. No 11 was therefore ordered to be restarted at once, with fourteen blowers, and it was resolved to draw up a revised tariff for the unbound men. By this their wages were reduced and the rates of stoppage for bad work were fixed as follows:

All blowers usually making cylinders of a weight when delivered into the cylinder room of 18 lb. and upwards to be stopped for bad work only the excess in value of 1s. 6d. in the £1 on the gross amount of their wages. All those usually making cylinders of a weight under 18 lb. when delivered into the cylinder room to be stopped for an excess of 1s. in the £1 as above.

This settled, the gatherers struck against the new penalties for bad work imposed on them, but were pacified by the following arrangement:

Gatherers to be fined who have above 20 per cent, of blisters from the tops of the pots (say 45 pieces heavy and the corresponding number of light) and above 40 per cent, in all pieces gathered after the above. For less than 18 per cent. from the tops of the pots one shilling premium, for less than 15 per cent, two shillings premium. One per

---

1   Bontemps to Lucas Chance, February 3, the latter to Henry Chance, in charge at Oldbury, February 19, 1865.

cent, of pipe blisters to be allowed, the excess to be added to the gatherers' blisters. Gatherers' bad work, all making less than 18 lb. cut off usually to be allowed 6 (six) bad pieces per hundred, all making 18 lb. and upwards cut off usually to be allowed eight bad pieces per hundred.

In 1857 there was another strike, this time by the crown glass men against a reduction of wages made in accord with the Pilkingtons and Hartleys. It lasted during June and July. Its settlement is recorded in a minute of August 10:

> Resolved, that having received information from Mr. Pilkington of his intention to return this day to the rates of wages paid to his Crown Glass makers previous to the recent reduction, we offer to our men the rates paid by us when they left off work, as the reduction made in consequence of a previous reduction on the part of Mr. Pilkington is now rendered unnecessary by Mr. Pilkington's return to his old rates. Memorandum. Having had an interview with the Crown men, who expressed their willingness to return to work at the old rates, we ordered Nos. 4 and 12 to be lighted as soon as possible.

It may be noted that average wages per day in 1858, given in a memorandum by James Chance, were 9s. 4d. for blowers, 4s. 10d. for gatherers, 7s. 3d. for flatteners, 2s. 8d. to 3s. for labourers, 1s. 6d. for boys and women.

The addition of five new furnaces obliged provision of warehouses to accommodate their produce. In 1852 were built one of two storeys at the north end of Hartley bridge, another, 200 feet in length and with capacity for 2,100 crates of sheet glass when piled double, on the canal bank between the bridges. To provide a level foundation for it a retaining wall was constructed. At the same time the carpenters' premises were removed to "the garden adjoining the smiths' shop," that is to say, to a site by the entrance road to the works, not far from the schools' playground. Other warehouses built in 1853 and 1854 were one of two storeys, 150 feet by 30, at the back of No. 9 cutting room, for storage of pots above and for packing rough plate below; a similar one of twice the length on the canal bank adjoining No. 4 house for empty packages; one of three storeys, 100 feet by 30, adjoining the clay-mill, the two lower storeys for clay and furnace bricks, the upper one for pots; and one of two storeys as an addition to the Ornamental department, constructed over the road that passed there.

Other changes of 1853 affected Maltby's fitting and smiths' shops the branch railway into the new side of the works, mentioned in Chapter II, having to be carried through the latter. The Board minute runs:

"Resolved that Vanderkiste's cutting room be extended over Maltby's present shop and the blacksmiths' shop, through which the branch railway is to pass."[1] Other resolutions, also of May 10, were to appropriate the present stables (between which and the House the line passed) for Maltby's department, and to erect new stables, as also a dispensary, elsewhere. Effect, however, at least as regards the stables, was not given to them. In the same year there was extensive alteration of the offices, additions including the new Board room. Previously, in December 1852, a reading-room and library had been opened for the use of the clerks and others, as related in Chapter XII.[2]

Outlays for the year 1855 included two more rooms at the offices, and enlargement of the Obscuring department by roofing over, with basaltic slabs and rolled plate, about 2,000 square feet of ground enclosed by the polishing and grinding rooms. Next year the patent plate cutting room was ordered to be lengthened by 100 feet. In January 1857 was completed a two-storey building, 100 feet by 20, for the rolled plate, near No. 9 house. In April of the same year it was resolved to build a second seven-storey building, between the existing one and the offices, at a cost of £2,000; the various floors to be appropriated to packing and storing sheet glass, shades and coloured. In 1858 it was decided to lower the road under the railway so as to allow the passage of carts, hitherto using the level crossing at much delay and inconvenience. Other orders of the year were for extending the mixing department by a building of two storeys, 50 feet by 22, and for raising the roof of the room adjoining No. 3 house, occupied by Field the "crown" carpenter and Evans the saddler, and adding a storey, so as to provide a two-floor cylinder room for that house. A shop for Field was to be provided over the obscuring room, facing the crown cutting rooms. It was also proposed to remove the emery house to a more suitable site, so as to enlarge the fitting shop and prevent the injury done there by the emery dust. This, it is noted, was not done then, but the change was made later.

In 1859 the brass foundry was enlarged; next year it was resolved to extend the boiler-makers' shop so as to make it available for a good

---

1   T. Vanderkiste is mentioned in 1865 as foreman of the "plate cutting room" (cf. p. 100).
2   Robert Chance recorded in his diary that the first Board meeting in the new room was held on April 21, 1853, and his father's next Sunday meeting in the reading-room, the large room upstairs being still in confusion.

smiths' shop, and in 1861 to erect a saw-mill near the carpenters' shop for their use, the existing one, on the old side of the works, to be confined to sawing round timber. In December 1863 the square-timber yard was ordered to be removed to the new side, adjoining the carpenters' shop, to make room for two double lears, as said. Six months later it was resolved to build a new carpenters' shop, 100 feet long, 30 wide and 11 high, at an estimated cost of £300. The site of this was farther to the east, against the playground. The old shop was appropriated to box-making.

Also is to be noticed the decision of 1859 to build a wall 410 feet long and 16 feet high along the Oldbury road, at an estimated cost of £180.

Acquisitions of land in 1853 and 1854 were from Solomon Simpson, 7,088 square yards at 3s. a yard; from Patrick Allan Fraser and his wife, just an acre for £1,936; and from Richard Pearsall, for £614, nearly an acre of "Rick Leasow," with two cottages and the appertaining half of "Fordrove," the way that passed there to the Oldbury road. Simpson's land was part of "Grove Pleck," with "Spon Croft; it ran with the firm's property on the east northwards from the Oldbury road. "Rick Leasow," and Fraser's land extending northward from it to the railway, lay on the west side and gave room for extension of the patent plate and lighthouse works. In January 1855 it was resolved to construct a basin on Fraser's land, near the aqueduct.

There were also negotiations for a field of 3½ acres beyond the Oldbury road, in which was a canal feeder. The plan was to widen this for the passage of boats, and to bring into the works that way all the coal from the Tipton district. The land was held by trustees of Thomas Webb Hodgetts on behalf of his son, a minor. A price was agreed in 1855, £400 per acre, but there was no sale. Five years later, on renewal of the negotiation, it was found that the trustees had no power to sell; nor, as was proposed, would they grant a 99 years' lease of so much of the land as was required. Correspondence went on for several years with no result, one difficulty being that of obtaining permission to carry a tramway across the Oldbury road; carting the coal was held to be out of the question.

The remainder of "Rick Leasow," nearly an acre in extent and costing with its half of "Fordrove" £457 2s., was bought from Richard Pearsall in 1859. Erected on it two years later were a three-storey building for a store and pot-rooms—replacing a store-room in the lighthouse works and enabling the fitting shop there to be extended—and a sheet glass lear.

Other purchases of land had directly in view the very necessary extension of the lighthouse works. First was acquired in 1859 from William Bagley, for £120, a narrow strip of 100 square yards running east of "Scotch Row," with a public-house known as the "Scotch Tavern"; then from Benjamin Darby the "Barn Meadow," a plot of an acre and three quarters extending westwards from " Rick Leasow" to the brook and southwards to about 40 yards from the Oldbury road. The price agreed for this, on August 27, 1860, was £500 per acre, but legal difficulties delayed completion of the purchase until 1866. Between this property and the Oldbury road lay from east to west firstly a piece of 1,040 square yards, with four cottages, also belonging to Benjamin Darby, and then, beyond an occupation road known as "Darby Street," "Yew Tree Place" and "Yew Tree Cottage" the residences of Miss Mary Ann Gift and of the Misses Darby. The piece first mentioned was conveyed for £300 in September 1867, and at its south end was built the new lighthouse erecting room. "Yew Tree Place," 1,200 square yards in area, with eleven messuages, was bought in 1873 for £1,200, and "Yew Tree Cottage," mortgaged to the firm from 1867, in 1881.

Besides these extensions there were negotiations for the purchase of other works, for instance, those of Joshua Bower, at Hunslet, near Leeds, which had stopped work in 1860. By agreements of May 31 and December 27, 1861, the Hartleys, Chances and Pilkingtons took over in equal shares all his plant and stocks and various liabilities, undertaking to pay him £1,000 a year for fourteen years on condition of his abstaining from the manufacture of or trade in glass in the counties with which they were concerned. It was no great matter; the plant and stocks were valued at about £3,200. Towards the end of the term, in May 1873 and again in July 1874, Chance Brothers & Co. were approached from the Bower side about renewing the agreement. It was replied that before coming to a decision, the other parties must be consulted.

Another negotiation, concerning the Stourbridge Glass Works, is recorded as follows:

Mr. Gwilliam (Secretary to the Window Glass Manufacturers' Association) having explained to the Board the circumstances under which the Stourbridge Glass Works had passed into the hands of Messrs. J. Hartley & Co., Resolved that we agree to the proposal made by Messrs. Hartley through Mr. Gwilliam that we should bear two-fifths of the loss arising from the purchase of the Works (without having anything to do with the deeds) on condition that Messrs. Hartley and the St. Helen's Co. bear the remaining three-fifths (Messrs. Hartley being willing to bear one-fifth).

This was in 1867; by a further arrangement of February 1870 it was agreed to pay Messrs. Hartley on account of the works the sum of £300 a year for not less than five years, and to take over such half of the available utensils there as could be turned to use. This, the minute runs, was on the supposition of the Stourbridge Works lying dormant, but Messrs. J. Hartley & Co. and ourselves reserve the power of availing ourselves of any opportunity of turning the Works to such account as by common consent we may determine upon.

In the previous year it had been resolved to renew the lease of the Baggot Street Works at £125 per annum on the same conditions, but an offer of Platt's Glass Works, closed, was declined. The purchase of the Nailsea Works will open another chapter.

At Paris in 1867 Chance Brothers & Co. were the only English exhibitors of window glass. Their lighthouse apparatus was triumphant, and they had a fine show of optical glass, although outdone in this by Feil, of Paris. Among their other exhibits were examples of coloured and ornamental glass and of extra white rolled plate. Being represented on the Jury, they were "hors de concours" as regarded medals. Two years later came recognition of their deserts in the form of election to membership of the "Academie Nationale Agricole, Manufacturière et Commerciale," with award of its" Medaille d'Honneur" (June 1869).

Head of the firm now was James Chance. His father had died in February 1856, his uncle, at the age of 82, in March 1865. Neither for some years before his death had been able, from failing health, to take much part in the conduct of the works. In spite, or perhaps in consequence, of their marked difference of temperament the two had worked well together, and were always the best of friends. In contrast to his brother's activity and enterprising restlessness William Chance, conservative and shy of adventure, pursued his way in calm and quiet. The distinguishing feature of his character was prudence. He did not make business the engrossing end of his life, but took a leading part in the civic affairs of his native town and was ever forward in support of schemes of social progress. It was written of him truly on his death:

Birmingham has lost a man who was universally respected, whose kindness endeared him to many, and whose charity was felt in all directions. . . . His life was like that of the good men of all parties, and by the good men of all parties his loss will be deplored.[1]

Or again:

Alike from the weight of his character and his social position, Mr. Chance was one of the most prominent amongst the many citizens of whom Birmingham has reason to be proud, [a man of] high principle, exhaustless beneficence, large hearted kindness, and deep Christian piety, . . . equally estimable in every relation of public and private life.[2]

Of the energy of Lucas Chance, his enterprise and the largeness of his conceptions, it is unnecessary to write again. From boyhood he had devoted himself almost wholly to business, and for forty years was the acknowledged head of the glass trade. It was his principle to be always occupied, to waste no minute of the day. Such time as he could spare from business he spent in reading. He said of himself in 1860, on the occasion of a presentation of plate in recognition of his services to the glass trade:[3]

---

1   Birmingham Daily Press, February 11, 1856.

2   To quote further from the notice: "In all the efforts made during the last forty years in this great community, having for their object the glory of God and the well-being of his fellow-creatures, Mr. Chance took a prominent part, and his deep religious principle was alike exemplified in the fulness of his generosity, in the devotion of his time to public labours, in the uprightness of his commercial dealings, and in his deep and tender sympathy with the difficulties and sorrows which came under his notice (and he was quick to discover them) in private life. The spread of the Gospel both at home and abroad, the diffusion of the Holy Scriptures, the promotion of education among all classes, especially the most destitute, the emancipation of the slave, and the restoration of the outcast, in short every scheme of philanthropic piety found in him a most earnest advocate and a liberal supporter. . . . The General Hospital and other local charities, the British and Foreign Bible Society, the Church Missionary Society, the Church Pastoral-Aid Society, and the Town Mission, with many others, shared his warm interest, and the first and last of these Societies especially, from their very commencement, his strenuous exertions. The Ragged School, the first of its class in this town, which he established and maintained at his own cost, and the Prize which he founded at King Edward's School for the study of Holy Scripture, will perpetuate the name of one whose life demands that it be handed down among the benefactors of Birmingham." (Aris's Birmingham Gazette, same date).

3   "I suppose you have heard that the country dealers of England, and those of Scotland and Ireland, have given me a piece of plate for which they paid 200 guineas as a testimonial of their respect and esteem for the benefit I have been to the Window Glass trade in improving the articles and introducing new articles, and from the manner in which the trade has been carried on." He explained that the London dealers had not been asked to participate, since the others feared that their subscriptions would be eclipsed. (To his brother Henry, October 3, 1860.)

In 1832 I came down to Birmingham and myself undertook the chief management of the Glass Works. . . . I had great energy, great perseverance, and though not a scientific man myself I was deeply impressed with the great value of such knowledge, and therefore secured the assistance of scientific men, and I took care that everyone I had in my employ was well looked after. At the same time I kept my eyes open, and if I saw any species of glass made upon the Continent which I thought might advantageously be introduced into England, I at once introduced it. These are the principles which are still adopted by our firm, and as long as these principles are adhered to, so long will our firm continue to flourish.

On his death, the following notice appeared in the *Birmingham Daily Post* (March 9, 1865):

We shall not be wrong in saying that to his remarkable ability, far-sighted enterprise, sound judgment, and unwearied industry the progress of that House [Chance Brothers & Co.] is greatly due. Partly on account of deafness, and partly from his habit of mind, Mr. Chance avoided all public business. But there was one great public duty, which few men have more thoroughly or unostentatiously discharged; on a settled principle of benevolence, and as a point of conscience, he acted rather as the steward than as the owner of his well-earned wealth. His public munificence, great as it was, relatively bore a small proportion to his private beneficence. . . . Mr. Chance, by natural gifts and acquirements, was eminently fitted for a high place among public men. . . . He had read much, and thought deeply.

In 1861 a new partnership had been formed, as from January 1, 1860. It was agreed that the capital of the firm should be made up to £420,000, equally divided, as before, between the two sides of the family. On Lucas Chance's death his share, a quarter of the whole, passed to his three sons. The second of them, Frank, was not engaged in business; after graduating at Cambridge he had entered the medical profession. Apart from that his interest lay in the study of languages, with a dozen or more of which he was conversant. His knowledge of Hebrew gave him a place in 1875 among the Old Testament Revisers. He was included in a new partnership of 1867, but three years later his share was bought out by his brothers, as had been agreed.

Edward Chance remained a partner, though he had retired from participation in the management towards the end of 1858.

Junior members of the family now helping at Spon Lane were William Edward and Alexander Macomb, the younger sons of George Chance. The former began work there in 1860, the latter a few years later. William Edward for several years, until 1872, had charge of the optical department. In 1870 it was resolved to make an agreement with him to undertake the manufacture and sale of coloured, antique and optical glass, or should the last be given up by the firm, such other work as might be agreed upon.[1] Previously he had undertaken other charges, such as the flattening, where he successfully introduced the use of lagres made of common glass,[2] and investigation into the settling and gathering of the metal, with a view to diminution of blisters. In 1873 he left to conduct the manufacture of shades and "antique" given up to him by the firm, at works established by himself at Oldbury.

Alexander Chance in 1866 was in charge of the Ornamental department, and two years later entered on his brilliant management of the alkali works at Oldbury. His achievements there will be noticed in the chapter dealing with that subject.

With managers the works were overdone, for the old system of different men to manage particular glass-houses, men highly paid and dignified by the title of Mr., was maintained. Principal among them, and with general responsibilities besides,[3] was Edward Forster, given for his residence in

---

1   Board minute of June 24, 1870.

2   John Chance's note on this subject runs (November 20, 1866): "W. E. Chance has been engaged during the past two months in proving the practicability and pecuniary advantage of flattening our sheet glass upon lagres introduced into the kiln as sheets and not as cylinders. The sheets are obscured on both sides, thus enabling us to use 4ths instead of 2nds, as on the old plan. The sheets are also crocused before being delivered to the flattener. The experiment has been quite successful, and we are now only prevented from using the flat lagres (as they are called) all over the Works by the limited powers we have of obscuring by machinery. Thos. Brown, the flattener, has been W. E. C.'s active coadjutor in this matter and has had an opportunity of trying his favorite hobbies." In spite of increased cost, John Chance estimated the nett gain, owing to the better glass being kept for sale, at 2*d.* a foot, equivalent on 240 lagres per week, 52 inches by 38, to £1,409 a year. The difficulty of obscuring was very soon got over by a suggestion of James Chance that it was not necessary, which proved to be the case. Says his cousin further: "Certain spots having been scoured, the flat lagres now are thoroughly carried out and answer capitally." Forster in 1874 supplies the information that the sheets, after being flattened, were taken "to Brown's place, where the blisters and string are rubbed off by running a smooth and level piece of glass all over, where there are blisters and string," a round piece, namely, about 3 inches in diameter and half an inch thick. Also that one side of the sheet, that on which the "crocus" was put on, was roughened by fine emery, so that the crocus might stick.

3   John Chance records under date September 26, 1864, that in consequence of the unsatisfactory management of the sheet houses of late, the three managers, Forster, Howell and Brown, having no definite responsibility, Forster was to become general manager again, with No. 10, as the special house, under his own charge and power to go into any of the others, as formerly.

1854 the part of the House occupied successively before him by William Withers, Davies the surgeon, and Badger. From 1862 it became an annual practice to reward his efficient service by presents of £100 or £150. Other managers surviving from the forties were Joseph and Samuel Withers (the latter in 1859 transferred to superintend the coal supply), Isaac Mallin, and John Neale; new appointments those of James Anderson, George Neale, Joseph Grant, Thomas Howell, and William Brown. Joseph Withers appears as recipient of £20 in 1857 in acknowledgment of savings that he had effected in No. 12 house, and of £25 in 1862 for like service in No. 4, particularly in regard to coal. Mallin was the other crown house manager. From the beginning of 1854 he was to have an extra 10s. per week and £6 quarterly "for managing No. 12 house and (generally) succeeding to Edw. Evans' department." In 1858 he had a present of £10 for carrying out in No. 4 a system of founding uniformly successful during four months. Joseph Neale was given in November 1855 charge of all the sheet houses excepting No. 5. That house—it was the time that shades and coloured glass began to be made there—was entrusted to Anderson, previously in charge of them at No 2. But a year later it was found necessary to remove him; Forster was given the responsibility in his place. Grant was transferred for a time to be head of the cylinder department, but in 1857 was replaced in that office by John Neale. He appears to have been sent back to the glass-making; one, at least, of the name was in charge under Forster of No. 9 in 1868, at the end of which year he was discharged.[1] Howell until 1865 superintended the manufacture of rolled plate in No. 9 house. He was a valuable man, and it was desired to retain him, but his repeated irregularities during 1864 made it impossible. Brown succeeded George Neale in 1861 in charge of No. 6, under Forster's superintendence. In consideration of extra work during 1863 he had a present of £25, and his wages were raised to £3 a week. In 1868 he and William Bird, storekeeper, received £5 each for introducing a substitute for beeswax calculated to save £30 a year.

John Neale, above mentioned, was another man of value. On taking over the cylinder department he had his wages raised by £10 a quarter, to cover all claims for attending also to the obscuring. When re-engaged at the end of 1869 for a further term of five years he was given a salary of

---

1   Probably it was he who in 1869 was offered the post of librarian at the reading room in the place of his son, to be employed elsewhere. John Chance notes that he had been kept on in the glass houses from Forster's "kindly disposition, i.e. a mixture of philanthropy and business, which never pays."

£30 a month, with a bonus of £50 "in recognition of the very satisfactory manner in which he has discharged his duties."

Among other prominent men Job Lawton continued to look after the sheet flattening. In March 1865 his salary was raised to £226 a year, and in following years he received £20 for work in connexion with the push-fork, of which below, £25 on general considerations and especially for his exertions in connexion with the "flat lagres," and £20 for inventing and carrying out the movable push hole. A man who came rapidly to the front was Peter Rigby. In April 1864 he received £20 in consideration of improvement effected in the quality of the rouge for the patent plate. In January 1865 his salary was raised to £20 a month, and in April 1866 he signed an agreement to serve as foreman of the lighthouse grinding and polishing for £250 a year. Maltby, chief millwright and engineer, died in the spring of 1868 and was succeeded by his son Walter.

Others whose services were acknowledged by bounties or by a rise of wages were: William Dawley, presented with £25 for "attentive performance of his duties during the erection of additional buildings in 1852 and 1853"; Thomas Brown, sheet flattener, a man of inventive fertility generally known in the works as "Professor," who after giving up flattening worked as a moulder in the optical department; Enoch Hall; W. Pemberton, foreman of the coloured glass cutting room; Thomas Barratt, crown glass splitter (1861); John Neale, junior, whose wages were advanced in 1862 to 30s. a week and £11 a year in consideration of his efficiency in attending to bent glass; Joseph Field, foreman of the carpenters; Daniel Smith, foreman of the coal department on the new side; Mann and Cox, head ladlers in No. 9, accorded £2 10s. each quarterly in addition to their wages (January 1865); William Griffiths and George Stone, given £6 10s. quarterly money instead of £2 15s.; Thomas Field, of the box department; and Samuel Smith, given in February 1866 £2 for bringing to notice the use of slack for the pot-arches and other purposes at 2s. 6d. per ton instead of 7s., and promised £2 more when the change should be accomplished. Thomas Brown had bonuses of £10 on more than one occasion for improvements in connexion with the lears, in particular for his push-fork.[1]

---

1   "Thomas Brown, sheet flattener, was paid a premium of £10 for having invented and made practicable a fork for introducing the cylinders into the lear, thereby avoiding the old plan of pushing them in on rails, the jarring consequent on the latter system often causing the cylinders to lap and drop bits of glass on to the inside of the cylinder. The fork ... is suspended by a chain" (Henry Chance's notebook, October 1857).

At another time he was given £5 for his plan of shutting up the lears on Saturday nights, whereby a great deal of Sunday labour was avoided. Enoch Hall received £5 in 1858 for improvements in reflattening rolled plate, and £10 in January 1865 in consideration of special exertions during the past year. In October, when he replaced T. Vanderkiste as foreman of the plate cutting rooms, his wages were raised to 42s. weekly and £2 10s. quarterly. Joseph Field was given his cottage rent-free in April 1863, and in September 1865 an advance of wages to 42s. a week, with a present of £5 in lieu of previous claims for an advance not granted. Daniel Smith's reward was of £10 in January 1865 for his exertions during the late colliery strike.

Besides these may be noticed Joseph Adams, storekeeper, in June 1863 placed in charge of the sawing; Clarke, foreman of the Ornamental department; Martin Muscott; John Silk, appointed in July 1867 to superintend the blacksmiths, boiler-makers, gas fitters and jobbing fitters; another Joseph Grant, and John Sheldon. The two latter were taken into employment in 1860 on the terms of their agreements with Mr. Bower, of the Hunslet Works. Grant was appointed to superintend the sheet blowing in Nos. 1 and 7, but only stayed a year, entirely failing in his charge. Sheldon's agreement as foreman of the bricklayers, as renewed for three years in September 1865, was on the same terms as before, £3 a week. Next year he received a bonus of £25 for his good work.

Among the founders were Edward Cashmore, Richardson and Gibbs at different times in No. 2, the efficient Joseph Wheeler in No. 10, Joseph White in No. 6, W. Deans and then John Lloyd in No. 7, W. Butcher in No. 1 and then in No. 11 (1867). Wheeler in 1869 was promoted to manage No. 7 house, with a rise of wages to 50s. a week to compensate him for quitting his former post. In 1868 Price was founding in No. 7 and Whitehead in No. 11; but the former having broken seven pots they exchanged places. Next year Whitehead was sent back to No. 11 and replaced in No. 7 by John Jakeways, management of that house, as said, being entrusted to Joseph Wheeler.

Of the potmakers, Robert Squires ceased to do much work from 1860 and left early in 1870, but Zachariah kept on, and two other members of the family, Thomas and Joseph, came in. Edwin Brettell was still employed in 1856. Of Johnsons there were three, James, John, and George. James in 1867-8 received premiums amounting to £20 for improvements in arching pots and for his share in a plan of turning their bottoms by machinery.

A census of August 1871 showed exactly 1,800 persons employed at Spon Lane and 440 at Oldbury. Of the former 1,309 were men and youths over eighteen, 99 women and girls, 392 youths under eighteen and boys. For Oldbury these figures were 385, 7, and 48. The sheet glass manufacture employed by far the most, 600, 20, and 156 respectively—a total of 776. Next came the lighthouse works, with 129 men and 93 boys. The patent plate employed 104 persons, the crown glass 93, the coloured and ornamental 79, the rolled plate 55. In the first of these departments there were 37 and in the third 38 women and girls; boys in the four together numbered 74. The optical glass required but 2 men and 2 youths. For the coal were employed 34 men and 11 boys, and for the hauling 25 and 1. Carpenters, fitters, blacksmiths, and other artificers numbered in, bricklayers and labourers 122, persons in sundry employment 129, office clerks 34.

These figures show a large increase over those of twenty years before. At Spon Lane in 1850 the sheet glass manufacture gave employment to 451 persons, the crown glass to 203, the patent plate to 175, the ornamental to 48, the offices to 22, other occupations to 295—a total of 1,194. While more were then employed for the crown glass and the patent plate, the lighthouse and the rolled plate departments were not yet in being. Of the total 376 were youths under twenty-one and boys, 132 women and girls. At Oldbury then there were 277 males over twenty-one, 61 under that age, and 32 women and girls—370 in all; and at the Titford colliery 31 men and 11 boys.

A new manufacture attempted in the fifties was under a patent by Henry Adcock, for casting from melted basaltic stone hollow articles such as pipes, chimney-pots, and barrels.[1] His specification set out that slow annealing gave a hard stony substance, less slow one resembling marble, and yet more rapid an opaque glassy structure. That is to say, different rates of cooling produced different degrees of devitrification.

Henry Chance had principal charge of the experiments and has left full account of the process, both in its practical and in its interesting scientific aspects, in the paper read by him before the Chemical Society in 1868. The stone used was the well-known "Rowley Rag" of the vicinity. Experiments were carried on for some six years, at first for making pipes, afterwards for

---

1   English patent No. 13788 of October 23, 1851.

the production of solid materials for building. The melting, says Henry Chance, presented no difficulties, and those of the casting were surmounted, so that in the one case "after repeated trials pipes were produced completely reconverted into stone," and in the other "so far as the manufacture was concerned we were entirely successful." But in both the end was failure. With the pipes "the difficulty of uniformly obtaining thorough devitrification, when dealing with so small a mass, appeared to be insuperable," and the building stone, although here devitrification was "attained with the greatest certainty," could not be produced at sufficiently low cost.

William Chance would have stopped the experiments after two years. In a letter of October 28, 1853, he denounced the whole thing as an "entire failure." Whereas, he complained, when eighteen months before experience was had of the difficulty of making pipes it had been stated "that masonry to any extent could be made of such articles as would yield an enormous profit," it appeared now "that there was no more facility in making masonry than pipes, both seem to be up to the present time equally a failure." He saw no expectation of success; in further trials only money to be thrown away. He adjured his son James to take heed lest things should be done without his knowledge, believing his opinion to accord with his own, to drop the manufacture unless pipes could be made successfully.

His advice was not taken; endeavours were prosecuted for at least four years more. Adcock was desired to erect a furnace capable of burning a coating of stone colour on the articles produced and later to experiment on a practicable method for casting slag directly from blast or other furnaces. Pipes were tried again, but only one that was made could stand a strain of 70 lb.; some broke at the first pressures. Attempt was also made, with failure in both cases, to use the material for flattening stones for the lears and, in the glassy condition, for lagres. The stones cracked under the changes of temperature, the lagres retained heat too long and "scalded" the sheets flattened on them. Stonework was actually made for the Edgbaston Vestry Hall, but it was ordered, when that was done, to proceed with nothing but mantelpieces. Also sheets were made for roofing purposes and so used in the works early in 1855. But they could not stand the weather. Says Henry Chance in his paper: "A railway station in Scotland was covered with it; but the composition of the glass was such as to render it unable to resist sudden changes of temperature, and nearly every sheet was cracked during the alterations of a variable winter."

In the summer of 1856 Adcock was sent about his business and after that the manufacture was transferred from Oldbury, where hitherto it had been carried on, to Spon Lane, to be worked on plans of James Chance. These, however, were not for melting Rowley Rag, but for firing in close kilns at a bright red heat a mixture of sand with finely ground glass cullet.[1] "The body of the article" runs the account, "was cemented into a fine, hard and strong mass, but the surface was soft and crumbly" and no devices for indurating it succeeded. The articles to be made, and for which one Peter Thompson was commissioned to ascertain the probable demand and selling price in London, were "window-heads, steps, nagging, etc" The last reference in the Board minutes to the process being worked is of February 1857, when a shed was ordered near the House to shelter the stock. In June 1860 pot-arches were ordered to be erected on the site of the "old Basaltic furnace." Relics of the manufacture are the mantelpiece in the Board room at Spon Lane and a step at the old gate-house.

The following is Henry Chance's account of the experiments in detail:

An iron mould was made, with a core, and the stone being melted in a crucible was poured into the mould, which had previously been made red hot in an oven and was returned to the oven immediately after being filled and gradually cooled down. It was found that the iron mould (being so good a conductor) lost its heat so rapidly, that we could not convert the melted material into stone without heating the mould to such a degree as caused it to warp and spoiled it. It was therefore decided that iron was an unsuitable material for casting in. Mr. H. C. then suggested the possibility of using sand moulds for the purpose, and it was found that a mould made of sand inside a case of iron could be made red hot in an oven without injury, and that the melted metal poured into and annealed in such a mould was thoroughly converted into stone. The melting, at first made in a crucible, was after a few weeks' trial made in a reverberatory furnace, and no difficulty was found in melting it in this manner and running it thence into a mould placed in an oven below the furnace. When the experiments were conducted on a larger scale, it was found very inconvenient to have to run the metal direct from the furnace into the various ovens, and we therefore adopted the plan of running it into wrought iron ladles first, and thence through a hole in the bottom of the ladle, stopped by a moveable plug. When the metal was run into the ladle the portion next the outside of the ladle, and that on the surface,

---

1   Henry Chance took out a patent for this, No. 242 of 1856. A previous patent, No. 255 of 1855, in James Chance's name, covered apparatus for moulding and cutting off excess of metal from glass or other vitreous tubes.

formed a crust, which kept hot and fluid the metal within, and it was found that the best quality of stone was produced when the metal was cast as cold and stiff as possible, and that when it was cast immediately after filling the ladle the stone was rotten and badly converted. A difficulty occurred in casting pipes in sand moulds arising from the sand (in spite of having been coated with black lead &c.) adhering to the metal poured in, not allowing it to contract and thus causing horizontal cracks on the outer surface of the pipes. In order to obviate this a thin cylinder of wrought iron was placed inside the sand, and the cracks were thus avoided, but other imperfections were contracted owing to the scaling of the iron and other causes. In fine, although it was easy enough to cast a pipe, it was impossible to cast pipes sufficiently sound to stand the ordinary strain. This manufacture was therefore given up and we then proceeded to apply the material to the manufacture of various articles of stone for architectural purposes, such as windowheads, sills, coping stones, balusters, mantelpieces, &c. No difficulty was found in casting these articles, and by placing wrought iron rods in the moulds, so as to bring them in the middle of the casting, castings of any size or shape were prevented from breaking from the adhesion of the sand to the surface by the power these rods had, when contracting, to pull the casting with them. Stone of excellent quality was produced, and the ornaments on the castings were sufficiently clear and smooth for ordinary purposes, but not so for mantelpieces. The colder the metal when cast, the less sandbitten and the more smooth and clear was the result. This part of the manufacture, however, though successfully carried out as regards the mere making of the articles, failed for several reasons commercially. One great objection was the colour, a dark brown, too dark to be used without painting, a process which completely did away with all stone-like appearance. Another objection, and an objection to all artificial stone, was that every architect required different patterns, and as (the patterns being costly) the stone could be made cheaply only when a very large number were cast from the same pattern, it often happened that the cost of cut stone was less than that of cast stone. The intense hardness also of the stone, not allowing of its being cut in case of error, was another objection. In addition to this, cut plain stone being exceedingly cheap, and the expense of casting plain stone being quite as great as of casting ornamental stone, the trade in cast stone was limited to ornamental work and was therefore of a very confined character. Having after several years ascertained that we were manufacturing at a heavy loss, and that even allowing for all possible reductions there was no chance of lowering the cost to the selling price, we at length gave the matter up. Mr. James Chance proposed a plan for casting flags, steps, &c. on a large scale at a cheap rate, but it has never been carried out. The great difficulty in competing with cut stone in these ordinary articles lies in the excessive cheapness of the latter, and though the cast stone is far harder and more durable, people generally prefer to have a more perishable article, if at a less cost.

One remarkable fact elicited during our experiments was that if pieces of raw stone were placed in the mould and heated up with the mould and the metal poured round them the raw stone and the melted stone became firmly cemented together, and this composite was decidedly stronger than when made simply of melted stone.

An experiment was made by Mr. H. C. with a view to casting an artificial stone of a good colour through the same process of devitrification as that undergone by the basalt. Broken window glass was melted in the furnace both with and without the addition of limestone. When melted alone the glass produced an excellent casting, but was extremely difficult of devitrification and was in fact (excepting an outer skin of stone) still glass; when melted with limestone the devitrification became easy, but small particles of lime remained unmixed, and these, finding a place in the cast stone and absorbing moisture, expanded, and after a short exposure to the air blew the casting into little bits. The colour of the castings thus made was good, a pale greenish white, but the surface, though extremely sharp and clear, was so adhered to by the sand that it required a great amount of labour to remove it. The material also employed was far more expensive than basalt, being worth about £2 per ton.

CHAPTER V

1870 TO 1888

Save for the brief period of unhealthy inflation and excitement which followed upon the Franco-German war, this was a period of anxiety and trial for manufacturers generally, of competition ever growing keener, of continuous fall in prices, of reduction of wages and consequent industrial trouble. Chance Brothers & Co. had their full share of these difficulties, and suffered besides from embarking on two enterprises which promised well but proved disastrous—the purchases of the Nailsea Glass Works and of the Birmingham Plate Glass Works.

After the deaths of Lucas Chance's uncles, of Edward Homer in 1825 and of John Robert Lucas in 1828, the family continued to be represented at Nailsea by the son of the former, James Edward Homer, and by the son-in-law of the latter, Reginald Henry Bean (afterwards Rodbard), and his two sons. Of these and of the many other partners—Coathupes, Cookes, and others—there remained in 1862 only the younger Bean (Henry Lucas) and Isaac White, who had purchased the Coathupe interest in 1855. They then leased the works to Samuel Bowen, of West Bromwich, and John Powis, of London, and in 1867 sold them to the Hartleys. In 1869 Bowen was declared a bankrupt, and he and Powis surrendered their lease, whereon the Hartleys sought to dispose of the property. Chance Brothers & Co. thought well to seize the opportunity, and on May 19, 1870, bought the works for the sum of £14,000. Option was given to acquire the colliery, which supplied the works, on the terms on which the Hartleys held it, and accordingly we find a lease of coal and clay in the parishes of Nailsea and Wraxall granted to Chance Brothers & Co. on June 13, 1870, by Sir J. H. G. Smyth, Bart.

Furnaces at the works were three in number, two for sheet glass, with eight 65-inch pots, and one for rolled plate, with four. Orders issued to prepare one of the former, No. 2, for work forthwith, and Zachariah Squires was sent down to make pots at the rate of three a week. W. Stonier, of the ledger department at Spon Lane, was deputed to take charge of the office. Trouble came very soon, and principally from a cause that had not been foreseen, failure of the coal supply. Already in February 1871 it was found necessary to sink a new shaft. A few months later the works manager, William Brown, had to be discharged. Forster was offered the place, but eventually George F. Neale, a man who had gained his experience in America, was appointed—to have entire control of the manufacture at a salary of £400 a year.

For the year 1871 there was heavy loss, and though in 1872, with the rise in prices, profit was made, it was decided at the end of the year to suspend manufacture. The reasons are set forth in the following resolutions of the Board, of January and February 1873.

Resolved that the crown of the sheet furnace now at work having begun to drop very much the furnace be put out at once and as the stock of sheet glass is very large and the slack from the old pit has for some weeks past produced very bad glass, while the new shaft has failed as yet to produce any slack that can be used, the second sheet furnace be not lighted and the manufacture of sheet glass at the Works be suspended.

Resolved that trials be made at our Spon Lane Works of the Wareham sand and Nailsea limestone in use, as well as of our Nailsea slack, to ascertain how far these materials are adapted to the making of glass by Siemens' process.

Resolved that the manufacture of English sheet glass at our Nailsea Works be abandoned, the slack from our Colliery being unfit for a gas furnace (as proved by experiments here and at Messrs. Powell's bottle works at Bristol) as well as for the production of glass of good quality in a coal furnace.

Resolved that the four-pot rolled plate furnace at Nailsea be put out on Saturday fortnight and the Works closed as soon as practicable, it having been finally settled that the Nailsea slack is unfit for a gas furnace, and for producing good sheet glass in a coal furnace, and there being a limited supply of slack from the old pit, and no prospect of a supply from the new pit.

The resolutions were put in force with little delay. During the year all available plant, save what could be utilised at Spon Lane, was sold. Some of the employees (as Samuel Turner, afterwards cashier there) were transferred to Spon Lane; in March 1874 such workmen as remained were paid off and the works were finally closed.

It remained to see what could be done with the colliery. After an interview with Morgan, the manager, it was decided not to proceed with the shaft, where there seemed to be no prospect of success, but to get from the old seam all the slack available for sale. At first it could be sold only at a loss, and in May 1873 the coal-getting was stopped for a time. Next year Forster was able to report the hard rock of a fault pierced, and the rock overlying the coal seam reached ; the slack fetching 1*s.* 6*d.* a ton and the large coal improved in quality, with a ready sale. But that was all. In November 1875 it was " Resolved that we ascertain by working from No. 2 pit whether there is any coal in the adjoining ground, it being understood that if no coal can be found by this working we shall be in a position to state that we have exhausted all the means at our disposal of finding coal." Finally, in September 1876, James Chance was authorised to take such steps as he should deem expedient towards disposing of the colliery, and a year later the lease was surrendered.

Although the works were stopped, the property remained first on the firm's hands, and then on those of its three senior partners, for a number of years. It was valued in April 1876 at £3,250; thirteen acres of land at £750, the house at £500, thirty-four cottages at £2,000. Morgan was paid £1 a week for collecting the rents of these and for looking after the buildings and remaining plant. In 1885 Forster found all in good order and all the cottages but two occupied. In 1889 John Chance took over the sole ownership, and before long sold the property.

A fourth offer of the Birmingham Plate Glass Works had been made privately to Robert and John Chance in 1867, at the price of £70,000. The property was stated to comprise 16 acres of land, of which the half was field, six glass houses, forty kilns, and three engines. Stocks finished and unfinished were valued at £15,000 to £20,000, and production given at 15,000 to 18,000 feet of glass in the rough per week. The brothers were attracted by the proposal, but the difficulty lay in the direction. Henry Chance, in retirement for a time from active participation in manufacture and residing at St. Leonard's, declined for the present any such responsibility, and particularly that in question, and the matter dropped.

In November 1872, when prices were rising rapidly, there came through Mr. Gwilliam an offer of the late Major Cook's share in the works. From different reasons, the proposal was twice declined, but trade still booming, a third offer was accepted. In August 1873 it was agreed to purchase the seven acres covered by the works, with them, for the sum of £46,000. That done, it was resolved to increase the capital of the firm to £510.000, but to treat the works as a separate concern under their old title. Joseph Newey, of the export office at Spon Lane, was appointed secretary; Neale was brought from Nailsea to conduct the manufacture; and energetic steps were taken to put the plant into proper condition. It was decided to have gas furnaces, one with 20 and two with 16 pots, and to increase considerably the engine power and other equipment. C. W. Siemens offered to supply plans for the furnaces, with all the latest improvements, excepting his new continuous system, for a lump royalty of £250.[1] It was soon seen that the new plant could not be put in too quickly. Visiting the works in March 1874, after an interval of four months, John Chance was shocked by the large amount of polished glass that he found in stock, of sizes under 18 feet. He learnt that the large plates had frequently to be cut down for defects due to bad tables and imperfect machinery. And he found the glass indifferent in colour and of poor finish, faults that prevented a larger sale of the smaller sizes. Indeed, he noted sales to have averaged but little over £1,100 a week, far too small an amount to make the works pay. Figures for twenty-six weeks showed 23,581 feet of polished glass sent into the room, and only 14,327, barely 60 per cent., sent out. This he remarked to be due partly to deficiency in cutting power.

In this condition of affairs James and Henry Chance were authorized, with Dr. Hopkinson, to make such additions and alterations as they might think necessary to put everything into proper order. Neale had already been sent to Lancashire to make inquiries about casting tables. Information about furnaces was obtained from various places in England and abroad, and James H. Greaves, an engineer, was engaged to superintend the work. At the beginning of July it was resolved to suspend manufacture for a time, in view of the great difficulty of carrying on the works during the alterations. For fifteen months business was confined to the purchase and resale of plate glass made elsewhere.

The alterations decided upon included two new gas furnaces, each for twenty 36-inch pots, four additional kilns, bringing their number up to 42, three new casting tables, and new polishing machinery. Work was started from the first furnace in September 1875, the other was ready by the end of the year. But all this enterprise and outlay went unrewarded. The polishing

---

1   Letters of January 13 and 16, 1874.

machines worked badly owing to a defect in the engine and breakage was excessive. For this reason, much of the glass had to be sold unpolished. A large order for such for the Birmingham Arcade, for sizes too large to be supplied by rolled plate from Spon Lane, helped only for a time.

Changes in the management were tried. In June 1876 charge of the grinding and polishing machinery was transferred from Neale to W. Higginbottom. In September the former, whose engagement would expire at Christmas, was asked to resign, as matters were not "working comfortably." Elliot, who had come with Greaves to help in constructing the second gas furnace and had made an excellent impression, was installed in his place.[1] One Davis was brought in as foreman of the casters, and Hector Stingre, an expert in smoothing and polishing, strongly recommended, was engaged from Premontr to take charge of that department. Walter Clutterbuck was appointed "contre-maitre of the casting hall" at £5 a week. John Chance hoped now "that we shall begin to turn the corner." But Stingre, when he came in January 1877, had plenty of fault to find.

For profitable production it was necessary to accomplish six founds per week per furnace. But with the two at work together not more than ten founds per week could be got through, and seldom more than nine, "owing, it is said, to their being connected and pulling against each other." When No. 1 was out, No. 2 performed its task. Quality, also, compared badly with that of the glass from the old coal furnaces. The corner was not turned; while the losses for 1874 and 1875 had been £7,800 and £8,500 respectively, that for 1876 came near to £20,000. During the latter half of that year glass from the kilns fit for grinding fell short of requirements by a fifth. With prices falling in unprecedented fashion it was impossible to go on; in July 1877 the works were closed. John Chance recorded "our conviction that the management has been radically bad from beginning to end. We must not start again until some thorough understanding is come to as to the management."[2]

There never was inducement to start manufacturing again. Once more business was confined to the purchase and re-sale of glass, under the conduct of Newey and Higginbottom. James Chance took over successively

<hr />

1   Elliot, says John Chance, left in April 1877 "for a journey preparatory to going to Spon Lane as Constructor General." Higginbottom was sent in February to acquire information at plate glass works in Belgium.

2   The principal authority for the above is John Chance's journal, from which the quotations are taken.

the liabilities of his brothers and his cousins, and finally, in 1889, sold the works to the Credenda Seamless Tube Co., Limited.

At Spon Lane, besides the other troubles encountered, falling off in the quality of the sheet glass gave serious disquietude. Already in April 1871 strong representations had been made to Forster, and steps were threatened, should there be no improvement soon. The attention of C. W. Siemens, too, was called to the unsatisfactory results from the gas furnaces. One probable cause, it was already seen, lay in the deterioration of the Leighton sand. There was improvement, particularly in getting rid of "sandiver," a defect new with the gas furnaces, when a double screen for the sand was brought into use (July 1872). Forster received a bonus of £50, partly for its introduction.

Remedy was sought, not in obtaining better sand, but in change in the dimensions of the furnaces and in the number, and afterwards in the size, of the pots. In the first place, to see what diminished heating space would effect, No. 4, worked since February 1870 for sheet glass, had its width reduced and its crown lowered. The latter change was found not to answer, but the former was adopted shortly also for No. 2. Both furnaces, and No. 11, were placed, in October 1872, under Peter Rigby's charge. As to the pots, in 1870 all the nine large gas furnaces had eight of the 64- or 65-inch size.[1] The first experiment was to alter Nos. 7, 1, and 6 to take twelve of these, but after two or three years' trial the plan was found not to pay, and was abandoned. Reasons, given by John Chance, were the greater number of hands required to work out the same quantity of glass, the increased cost in wages, the greater quantity of blisters and consequently of coarse glass, owing to the necessity of working the metal hotter, and the irregularity of the working hours.[2] Nevertheless he considered the experiment

---

1   Seven were appropriated to sheet glass, No. 9 to rolled plate, No. 12 to crown glass. Of the others, Nos. 3 and 5 had been out of use for some years, while No. 8, the lighthouse furnace, had recently been altered to take two 42-inch covered pots for extra white glass, with three openings at the back for small casting pots.

2   A Board minute of October 13, 1874, states the following difficulties to have been found insuperable and counterbalanced by no advantages :

1) Inability to keep the blowers and gatherers at maximum production, since a 12-pot furnace could not be worked with less than 22 men, as against 14 with the others.

2) The long interval between the founds is very prejudicial to the proper working of the regenerative system."

3) There was no saving of fuel, either in founding or in working.

4) Attempts to diminish the number of men by employing two sets each journey, each set working six times a week, had failed, because the men's hands had not time to harden in the short intervals allowed.

satisfactory, the cost of conversion having been but small and the metal, after the first few weeks, worked out regularly. For quality, he says, No. 7 had done much better for the first months though "latterly she has sunk to her own level again," while No. 1 had not done well, and No. 6 very badly.

In 1876 came the idea, taken, perhaps, from the new furnaces installed at the Plate Glass Works, of having a much larger number of pots of a smaller size, 42 inches only in outside top diameter. Preliminary experiment in a little two-pot furnace, which had been built for the purpose of supplementing No. 8 for extra white but had proved a total failure, justified alteration of No. 7 to take fourteen of these, equivalent in yield to four of the 65-inch size, about 10,000 lb. per journey. It was fully expected, says John Chance (July 1876), to get 50 per cent, of best glass from the 42-inch pots, besides avoidance of large defects by having no more than one gatherer to each pot. The furnace was started in March 1877, and success was sufficient to warrant proposals for the similar alteration of Nos. 6 and 2. Calculating costs on the halved production of a furnace, John Chance found them to be 2s. per 100 feet in favour of the eight 65-inch pots, but on the other hand, that the glass from the 42-inch averaged 19s. 6d. per 100 feet in selling price, as against 16s. 3d.

The change was not adopted for No. 6; that furnace was rebuilt entirely for eight 65-inch pots, as formerly. For No. 2, "a narrow and unsuccessful furnace for some years," says John Chance, it was proposed to have eighteen of the small pots, fourteen within the cone and two more at each end. Actually, the furnace was rebuilt in 1878, with an iron bonnet in place of the cone, to take twenty-four, and room was found in addition for a 33-inch open pot at one end, used for various purposes, and a large "monkey" pot for green ship-lights at the other. But after a time these were found to interfere too much with the working of the end sheet pots, and were abandoned. Besides the ordinary sheet, such glasses were made in No. 2 as pot blue, sheet cathedral, "muffled," and the glass for flashing the opal, ruby and staining melted in adjoining "monkey" furnaces.

There were plans, also, for improving quality from the 65-inch pots. One such was to work all the eight at the same time, instead of four only. This, says Forster, answered well for working out the metal and diminution of large blisters, since but two gatherers worked at a pot instead of four. With sixteen blowers in Nos. 4 and 12 (working for a time for sheet glass, no crown being wanted) quality turned out good and work was completed in 7½ hours, but in No. 1, when there were but fourteen blowing holes and so the same number of blowers, 9 hours were required. Also were tried five journeys in the week, and this succeeded very well in No. 12, which warmed up more quickly than the other furnaces, founds being completed there in 22 hours, and the five journeys producing 93,000 lb. of very good glass.[1]

In September 1878 came the introduction of gas heating for the blowing holes. This was to be adopted, says John Chance, throughout the works, not so much in order to save fuel, as to get rid of blowing hole dust.

Meanwhile had been considered, and after four years' deliberation rejected, the new Siemens plan of melting glass in a tank instead of in a number of pots; the materials being filled in continuously at one end of the tank, and the molten glass worked out continuously at the other, without interruption of work. A principal advantage of this system is saving of the wear and tear upon the furnace consequent upon alternately raising the temperature for founding and lowering it for settling and working, a wear and tear yet greater with gas than with coal furnaces.[2] It was patented by Charles William Siemens in England in 1870 (No. 1513 of that year), while a second patent, No. 3478 of 1872, supplemented the earlier one in regard to floating bridges and rings, and protective cowls over the working holes. Tank furnaces themselves were not new; Bontemps had reported on some at work in Germany twenty years before; but for success with them was required combination with the Siemens regenerative system.

Siemens advertised his invention to British manufacturers in 1873. When Chance Brothers & Co. received a copy of his pamphlet they commissioned Professor Gore, of Birmingham, who had previously conducted for them scientific experiments on glass founding, to inspect on their behalf the furnace from which Frederick Siemens was making white Argand lamp glasses at Dresden. Nothing more was done until, on the invitation of C. W. Siemens (November 1874), Neale was sent to inspect a furnace at work at Jumet, in Belgium. It was his second visit to that country on the

---

1   Forster's note-books.
2   Henry Chance, writing on February 6, 1871, to R. S. Newall, of Gateshead, advised him that Siemens regenerative gas furnaces almost got rid of smoke, but went on : " Owing to the loss of heat during the settling and working of the glass, we do not obtain the saving of heat which we should do if the heat were continuous. If we were about to build fresh works, I think it is very doubtful whether we should not revert to the ordinary furnaces, the cost of maintaining the furnaces being very great."

firm's behalf, his previous one having been made in January 1873, when he was still manager at Nailsea. He found only a small furnace producing very inferior stringy glass, but attributed this to the bridge being greatly worn and the furnace built of very poor clay. He thought that the larger the furnace, the longer it would last, since corrosion of the walls would be retarded; one large enough to employ six blowers continuously he expected to produce very fair glass during about four months. Other information concerning nine furnaces at work for bottle glass in Germany and elsewhere abroad, besides that at Dresden mentioned, he was able to supply from communications of the Siemens brothers and other sources. While those first erected, he stated, had lasted but from three to eight months, he was sure that the later ones, constructed of proper materials and worked with care, would last twelve without any interruption of the working. He again held good results to depend on the size of the furnace, this time on the ground of advantage from increased surface and depth of the molten glass."

"Nothing" he said, "can be done with small furnaces; that complete success is dependent on the large size of the furnace has been sufficiently proved at my own works." He estimated the white glass furnace, with ten working holes and capacity for about 20 tons, to yield about 100 tons of glass per month, and of the bottle furnaces, holding from 30 to 50 tons, one with eight working holes to yield about 150, another, with twelve, about 240 tons per month, the depth of metal in all alike being something over two feet.[1]

There was little in all this to encourage. But in October 1875, after he had obtained protection for further improvements,[2] Siemens returned to the charge, intimating that Messrs. Pilkington had several tank furnaces in successful operation, and that he believed that glass of the best quality could be produced from them, that depending mainly on the skill of the glass-makers.

During the following year much information was collected about tank furnaces in use, and calculations were made of cost and production. Elliot, of the Plate Glass Works, put the cost of a furnace and gas-producers at £4,000, and the annual upkeep at not less than £1,500, while Peter Rigby made the latter figure £1,000. Henry Chance, assuming three tanks of

---

1   Neale's reports, January 27 and 28, 1875.
2   Patent No. 1551 of 1875.

the size projected to yield no more than two of the eight-pot furnaces, calculated saving on this basis at but 4 per cent. Economy, he opined, hinged on production. In January 1877 C. W. Siemens was approached about royalties and opportunities of viewing tank furnaces at work in Germany. He replied that he would be pleased to settle the royalties on the best footing, 4s. per ton of glass delivered into the warehouse, and that his brother Frederick would show his lamp-glass and bottle furnaces at Dresden and arrange for sheet-glass tanks in the vicinity to be seen. This, however, proved impracticable.[1] Finally, in November, after further correspondence with Siemens and an interview with his agent, it was resolved to decline moving in the matter at present.

The determining consideration seems to have been that of quality. With the heavy charges for carriage imposed by the inland position of the works, the firm could not compete with its rivals on the Mersey or the Tyne or in Belgium but by producing a high percentage of good glass, saleable at remunerative prices. That, with tank furnaces, did not at the time seem probable.

Though successful to a certain degree, the 42-inch pots introduced for sheet-glass in 1876 did not give full satisfaction. It was resolved to try a size which could just be worked out by a gatherer and a blower singly. Some, 49 inches in outside top diameter and 37 in height, calculated to hold half the contents of a 65-inch pot, were made and used, but it was preferred to have them of double the contents of a 42-inch pot, for which the said dimensions were 49 and 40 inches. No. 2 was started with twenty of these in July 1879. A great improvement was claimed in building the furnace 10 feet narrower in the middle than at the ends, the middle pots founding very well. Again small extra pots were inserted at the ends, and again with detriment to the working.[2] There followed in 1880 No. 3, with eight of the 49-inch pots, No. 1 with sixteen (a number reduced in the next year to twelve), No. 6 with ten, and No. 8 with four[3] No. 3 was a new furnace, so numbered, lately built for ten 42-inch pots in No. 12 house, with a view to supplement that furnace for crown glass, when required, and to increase the percentage of good quality. Excepting No. 1, the new furnaces proved more costly in working than those left with the 65-inch

1   Letters from C. W. Siemens, January 18 and 25, 1877.
2   As noted by James Chance in a letter of September 2, 1879.
3   Previously to this, No. 8 had had four 42-inch covered pots. During 1884 it had four 36-inch covered, and at other times four 42-inch open.

pots, as well as in building and repairs,[1] but nevertheless in 1881 No. 4 also was altered for ten of the 49-inch size, and so was No. 2, where with the greater number not more than seven journeys, sometimes but six, could be worked in a fortnight. No. 10 was ordered to follow suit, but it, in 1883, was converted instead into a 4-pot furnace for rolled plate, as said below.

In the same year No. 12 was altered to take fourteen 42-inch pots, like No. 7, and next year eight of these replaced again, though only for two years, the larger size in No. 3. That furnace was shut down in 1887, and in the same year No. 6 was made a third furnace for fourteen 42-inch pots.

Of flattening the sheet glass there is little in these years to remark beyond increase in the number of double-stone lears, partly with the object of equalising flattening power with production; the introduction of heating them by gas (March 1874) ; and the erection of two Belgian lears on the canal slope adjacent to the Spon Lane bridge. Lawton was still to the fore in improvements; in 1870 two double lears were built upon a plan of his, and in 1874 he received a present of £25 for his services in connexion with the gas heating and for finding a way to use the lagres a second time.

Industrial trouble had been foreshadowed already in 1868, when depression of trade had necessitated a general reduction of wages. This was followed in January 1871 by a further reduction of 16 per cent, for the unbound sheet glass men. In the autumn of that year, indeed, when the horizon was brightening, there was an advance, besides which, the hours of work were reduced from 59 to 54 hours per week and the Saturday half-holiday was instituted, it being ordered that work should cease for the future on Saturday afternoons at 2 instead of 4. In 1872 there were again advances, but by 1875 the time of fallacious prosperity was over. In June of that year the wages of the sheet glass men were lowered by 8 to 9 per cent., and in the autumn of 1877 again by 13 per cent., with corresponding reductions in other departments and stoppage of premiums. Short time was worked, for the first time, says John Chance, since 1855. In January 1878 it was deemed necessary to revert to payment of wages generally on a basis of 59 hours per week, and in September, for the sheet glass men, on the warehouse instead of on the cylinder room returns, a reduction of some 3 per cent. It was known that they intended to ask for an advance of 20 per

---

1   John Chance gives the figures, November 15, 1880.

cent., in concert with the men at other works—those at St. Helens had gone out on strike in August—but it was resolved to act in spite of this and to resist the demand strenuously. At a meeting on October 11 the men were informed that the firm had been losing money for the past two years, and still were losing, owing to the depressed state of the market and to underselling, and so far from meeting their request, had decided to adopt the St. Helens rates and system, and to reduce the tariff.[1] After consideration of the new proposals the men refused them unanimously, and when they persisted in their determination, all the sheet glass furnaces were shut down and preparation was made for a long strike. It lasted, however, but five weeks.

A couple of years later, prices still falling, it came under discussion whether the manufacture of sheet glass should not be given up altogether. This was negatived, but it was resolved to sell no more at present at the lowest prices, and in default of improvement to curtail production at the earliest time convenient. Effect was given to this decision in the summer of 1881 by reduction of the number of sheet-glass blowers to forty-five. Eight were accommodated in No. 3, ten each in Nos. 2 and 4, as ordered to be altered, ten in No. 6, and seven in No. 7. No. 1 was to be worked until Nos. 2 and 4 were ready, and sheet glass also to be made in No. 8, if possible, in case a half of No. 3 were wanted for crown glass.[2]

There were further demands for advances from various departments in the years 1881 and 1882, but only one of 5 per cent, to the rolled plate men was granted. Three years later, in October 1885, reduction was again held to be necessary. Economies were devised, to avoid a strike, bat they did not suffice; next year the partners agreed unanimously that should there be no rise in prices before the end of it, notice of a general reduction of wages must be issued. A confirmatory resolution was adopted in February 1887, but does not appear to have been acted on. The question was rather to the fore again, whether the sheet glass manufacture could be carried on at all.

That of rolled plate, on the other hand, was in the ascendant. For the year 1873 the make amounted to 5,400,000 lb., and had there been warehouse room, says John Chance, four journeys might have been worked in the week instead of three, and all the produce sold. In the

---

1   Particulars of the changes are given in a very interesting article in the Birmingham Daily Post of October 15, 1878.
2   John Chance, June 23 and July 7, 1881.

previous year, metal had been brought to No. 9 house from No. 11, to supplement, but this proved both awkward and expensive. The question of having a second rolled plate house was postponed until it should be seen whether the finding of coal at Nailsea would enable rolled plate to be made there again. This settled in the negative and the cone of No. 11 being pronounced unsafe, it was proposed in February 1874 to remove No. 8 house and build on its site, "thereby setting free the kilns of No. 9 house, now required for lighthouse glass, and enabling us at any time, by making use of these kilns, to make R.P. glass on one side of No. 11 furnace, and farther to build additional kilns, should we at any time decide to convert No. 9 into a 12-pot furnace." Further, to increase storage room and to enable the furnace to be worked without intermission, it was resolved to convert the pot-rooms over the packing-room into a store-room, and to erect, on the space adjoining No. 16 lear, a three-storey building, of which the two upper storeys should be used for pot-rooms, and the ground floor for storing crown and other glass.[1] Eventually, the old No. 11 house was pulled down and one for a new gas furnace with four 65-inch pots built in its place. Work was started in this in August 1875.

When a few years later "rolled cathedral," the glass, white or tinted, so well adapted for church windows, was coming into strong demand, No. 10 was made a third house for the department. In fact, the manufacture was now displacing that of sheet glass in importance at Spon Lane.

Connected with the rolled plate manufacture was a new process; one which accomplished that which Bessemer had failed to carry through and which Bontemps had pronounced impossible—lamination of glass between rollers. The first form of the machine employed was that patented in 1884 by Frederick Mason and John Conqueror (No. 13119). The molten glass was poured upon an inclined plate and passed between a pair of iron rolls, whence, by the inventors' design, the sheet formed was to be carried down another inclined plate to the annealing kiln.[2] The patent was acquired by William Edward Chance, who in June 1885 entered into negotiation with Chance Brothers & Co. for its trial and development. After prolonged experiments conducted by the firm, the secret of success was found and the early crude machine made workable; an agreement with

---

1 Board minutes of February 19, 1874. This new building was by the "Britannia" gate. It was valued in July 1875 at £2,100 and insured for £1,000, and the stock in it for £2,500 (Forster).

2 This was found not to work. Instead, the sheets are removed to the annealing kiln on carriages.

William Edward Chance was signed on February 24, 1887. Modifications of the process enabled not only plain sheets to be made, but also "figured "— impressed, that is, with a pattern upon one side. But the perfection of this manufacture came later.

Extension of the firm's landed property in these years appears to have been limited to the acquisitions of Yew Tree Place and Cottage, noticed in Chapter IV, and of the part of the canal bank next to Spon Lane bridge not yet obtained. This was bought in February 1873 for the sum of £50. In 1876 it was decided not to renew the lease of the Baggot Street Works, and in 1884 an offer of property behind the Schools, belonging to Edward Milward, was declined. The firm's holdings in West Bromwich were disposed of, and though there were recurrent negotiations for purchase of the "Royal Oak" and "Britannia" public-houses, the one in Spon Lane, the other on the Oldbury road, nothing resulted from them. An offer of £800 for the "Britannia" and two cottages adjoining, made in 1870, was declined. The "Royal Oak," at the entrance to the works, was wanted in order that it and the adjoining cottages might be pulled down and the entrance widened. A succession of resolutions to buy appear in the Board minute-books, and negotiations on the subject have continued to the present time.

One of the West Bromwich properties adjoined the premises of Messrs. Salter, and stretched from High Street across to St. Thomas' Street; the other, purchased originally from Messrs. Grigg and Johnson, and covering nearly half an acre, lay at the corner of Spon Lane and Union Street. The buildings on the former property had been leased in 1850, by George Silvester, partly to Samuel Bowen and partly to John Payne. Chance Brothers & Co. held a mortgage on them, effected previously to Bowen's bankruptcy in 1869, after which it was resolved "to obtain a release of the equity of redemption of Bowen's premises at West Bromwich from the bankrupt's assignees, so as to become absolute owners of the property."[1] That done, the firm disposed of it in 1880 to the then occupants, Messrs. Dunn & Broughall, glass and lead merchants, for the sum of £2,300. The Union Street property was put up to auction in September 1877, and sold for £2,000 to a Mr. Simcox.

Of building in the works there is little to record, besides what has been mentioned. Although the erection of new gas works was contemplated in 1875, the existing supply being inadequate for winter requirements and so

---

1   Board minute of December 30, 1869.

impure that Staffordshire gas had to be used in many parts of the works, at a cost of £250 a year, nothing more was done than to put up near the plate works a gasholder for storage, brought from Nailsea.

It remains to write of those who conducted or who served the firm during the period under consideration.

The partnership formed in 1870 between James, Robert, Edward, John, and Henry Chance remained unchanged until the end of 1879. Only, from the beginning of that year, the Oldbury works were constituted a separate firm under the style of Chance Brothers, partly in order that Alexander Chance, who had managed them for ten years with such brilliant success, might become a partner in them without participation at Spon Lane. The separation was hardly more than nominal, the other five partners remaining the same. James Chance from this time no longer took any active part in the management, and Henry Chance not much, while Robert Chance gave up the commercial direction to his brother John.

From January 1, 1880, Edward Chance was replaced as a partner by James Chance's second son, George Ferguson, who, after a term of initiatory work between the times of his leaving Harrow and going up to Cambridge, had come to Spon Lane in permanence after taking his degree in 1878. Other junior members of the family at work there were the two sons of Robert Chance, both also Cambridge men. The elder, Arthur Lucas, began in July 1876, but before long retired to take charge of his father's clay works at Himley. The career of the younger, Walter Peyton, was sadly cut short by his death in January 1880.

A fourth accession was that of Kenneth Alan Macaulay, nephew of James Chance, an Etonian and afterwards of King's College, London. He worked for two years, from July 1877, in the manufacturing departments, and after a temporary absence returned on Robert Chance's retirement and Walter Chance's death to assist in the offices. He was admitted a partner in Chance Brothers & Co. in January 1883.

Later comers were James Frederick Chance, third son of James, and Edward Ferguson Chance, second son of Edward. The former, another Etonian, after taking a respectable place in the Classical Tripos at Cambridge, had studied science at South Kensington and Strasburg University. He took up work at Spon Lane and Oldbury in 1883, and next year was admitted a partner in both firms. But he retired from participation in the work in 1888. Edward Ferguson Chance, another Cambridge man,

JAMES TIMMINS CHANCE
at the age of forty
from an oil painting by J. C. Horsley R.A.
(To face page 135)

previously at Harrow, began in 1884. His first charge was that of the coal supply both at Spon Lane and Oldbury. Then he took over the charge of the rolled plate manufacture. George Chance, Macaulay, and he were the first Managing Directors of the Limited Company formed in 1889.

Edward Forster continued to be the trusted chief manager of the works until his death on October 12, 1887, at the age of 73. His loss was deeply felt. He was a thorough gentleman, whose kindly disposition endeared him to all. A notable gathering at his funeral, at which the firm was fitly represented, testified to the respect and affection in which he was held. A stained glass window was erected to his memory in St. Paul's Church.

Earlier losses of moment were those of the office heads in London and at Spon Lane—John Macartney and Franklin Taylor. The latter resigned his position in the autumn of 1884, while still comparatively young. He was accorded a pension of £50, but after a year intimated that he did not need it—in fact, that he had resigned early in order to have time and energy for reading and other pursuits. He died after about five years spent quietly at Southsea. Macartney, recorded as a "most faithful servant for 45 years," retired in 1875, at the age of 77, with a pension of £100 a year. He was replaced by a remarkable man, William Stevens, who already for some years had transacted the principal work and had received the larger salary. Beginning by driving a horse on the canal, he was engaged for the patent plate at its start as a lad in the grinding room, and had risen to be a foreman in the rolled plate department, when Lucas Chance picked him out to sell some glass in Manchester. This proved to be his real vocation. He was sent to the London office, by his singular uprightness quickly acquired the perfect confidence of the customers, and passed a long life of keen commercial competition with conspicuous success and absolutely trusted by all. When at the end of 1891 the London office was given up, he received a pension of £200 a year, having completed a service of fifty years eighteen months before.

Of Job Genner and George Crowther notice will be taken in later chapters.

After Edward Forster, the most notable man in the manufacturing departments was Peter Rigby, difficult to get on with or to trust, but of a practical ability that made him James Chance's right-hand man at the lighthouse and patent plate works. The value attached to his services is shown by rises of his salary, in April 1871 to £300 a year, and in October 1872, after, as said, he had been given charge of certain of the glasshouses, to £450. But a year later he was requested to resign his post at the lighthouse works, since Dr. Hopkinson, now in charge of them, found it impossible to work with him. When the reductions of 1877 came, it was pointed out to him that his salary had not been touched, although two departments had been taken from him. Thereon he gave notice to leave, but on reconsideration agreed to accept his £450 for himself and his son jointly. But he neither stayed, nor lived, much longer.

The reductions affected also Job Lawton, Samuel Withers, and John Neale. Lawton was a man of like type to Rigby, of great practical ability, but not readily to be trusted. Time and again he had received reward for services rendered, as has been noticed, but now his time was come; he had notice to leave at the end of March 1878, and died on April 15.

Withers had had notice to leave in 1871, the same year that his brother Joseph departed. It did not take effect; he continued until 1878 to look after the coal supply both for Spon Lane and Oldbury, and after that, with a reduced salary, for Spon Lane only. He retained this charge until circumstances rendered it necessary to dispense with his services in 1887.

Neale was informed of the necessity of reducing the number of hands in his department, so as to bring the wages list down to the level of 1871. It was shown him that previously to the present reduction the wages paid had amounted to £30 a week more than at that time, when as much sheet glass was made as at present, the number of men employed having been increased, and that of boys decreased; and it was intimated that if he could not effect a reduction of expenses the Board would find someone to take his place.[1] At the same time his salary was reduced to £300. Failing to give satisfaction in the matter in question, he had notice to leave in August 1878.

On Edward Forster's death his son, Edward John, who had assisted him for a number of years, succeeded him as chief manager, with a salary raised to £400 a year. He had received frequent recognition of good service, but retained his new position only for five years, retiring in 1892 with a pension of £100. He died, as the result of an accident, in May 1904.

Of the departmental managers specially to be held in remembrance is the veteran James Simcox, who passed away at the great age of 90 on November 23, 1916. Aided in his latter years by his son, Albert Lewis, he continued in partial charge of the rolled plate department, where he had

---

1   Board minutes of January 2 and February 11, 1878.

served so well, long after old age had imposed its disabilities; and to the last he loved to revisit the scene of his life's labours.

Walter Maltby, the works' engineer, retired in 1895 after a service of nearly 50 years, with a pension of £52 a year and special presents. Other principal men were Forrester Lloyd, of the mixing department; James Stokes, who on Lawton's departure added the management of the flattening to his own charge in the cylinder rooms; John Sheldon, head builder; William Ingram, foreman of the sheet glass makers; Joshua Spooner, of the boiler shops ; Joseph Field, of the carpenters ; James Field, his son;[1] Job Parish, skilled in the manufacture of optical glass; Mangematin and Gilbert in the lighthouse works; Emmanuel Sorrill, at the patent plate; Samuel Taylor in the cylinder department; Daniel Smith in that of the coal supply; David Churchley, foreman bricklayer; Solomon Cutler in the wages office; John Silk; Arthur Bates, storekeeper; and in the various warehouses David Bates, W. J. Biddlestone, E. J. Cotterell, Enoch Hall, Thomas Roberts, and George Stone, who succeeded Solomon Cutler as foreman of the sheet glass cutters. Nor must be forgotten William Grigg, contractor for the hauling, still with us, though having handed over his business to his sons. He entered the Schools on the first day that they were opened, in 1845. On his eightieth birthday, October 21, 1916, the firm had the pleasure of handing him a gold watch with an inscription commemorating his long connexion with them, approaching 70 years. James Simcox had a like presentation in 1900, and Forrester Lloyd on his retirement in 1912 after 47 years' service.

Of the potmakers, Robert Squires and William Robson had by 1870 ceased work. The former died in 1884. His brother Zachariah, who died in 1902, kept on until his fifty-second year of service, 1890. John and James Johnson retired in 1876 and 1878 respectively, Thomas Squires and the younger William Robson in 1883, Joseph Squires and John Carter in 1889, Joseph Carter in 1891, Arthur Squires in 1893. The present pot maker, Robert Humphries, began work in 1889.

Founders employed throughout the period 1870-1888 were Edward Grayson, Joseph Harwood, Thomas Launchbury, Joseph Mason, and William Westwood, and from 1871 John Lloyd junior. John Jakeways, Stephen Leming, J. Swain, D. Turley, and E. Whitehead retired in 1870, and J. Evans in 1871. Others employed for this important work for several

---

1   Now principal manager.

years on from 1870 were Charles Nolan, J. Stansfield, Josiah Weston, and
J. Wheeler, from 1871 Thomas Care, and from 1872 Matthew Brown.
Later comers to whom it was entrusted within the period were James Care,
Edward Price, George Potter senior, Frederick Lloyd, Levi Payne, Samuel
Lunn, M. Freeth, George Potter junior, J. Stevens, and S. Bradshaw.
Furnace-keepers were W. Ingram, Charles Laban, senior and junior, Wil-
liam Gittins, John Deeley, Henry Sims, Thomas Hadley, William Brown,
Elijah Cox, John Smith, Samuel Burden.

CHAPTER VI

CHANCE BROTHERS & CO. LIMITED

As from January 1, 1889, the firm of Chance Brothers & Co. was converted, for reasons of private convenience, into a Company with limited liability. The first members of the Board, besides Robert Chance, were John Chance, Chairman, Henry Chance, Deputy Chairman, George Chance, K. A. Macaulay and Edward Chance, Managing Directors. James Chance retired, handing over his share in the concern to his two younger sons. After the death of Robert Chance, in 1897, he was replaced by his son Arthur. Previously in the same year Henry J. S. Stobart had joined the Board with special charge of the Lighthouse and Optical departments. John Chance dying in November 1900, Henry Chance was appointed Chairman and George Chance Deputy Chairman. When Henry Chance died but two months later George Chance filled his place, Macaulay succeeding him as Deputy Chairman.

James Chance died on January 11, 1902, at the age of 87. Eighteen months before, in recognition of his national lighthouse service and of his great public benefactions—principal among them his gift and endowment of West Smethwick Park and his establishment of the Chance School of Engineering at the Birmingham University, in each case at the cost of £50,000—he had been created a baronet. Full obituary notices appeared in the newspapers at the time.

A feature of the constitution of the Company was permission to approved employees to place money with it on deposit, at interest equal, between a guaranteed 3 per cent, and a maximum of 10 per cent., to the shareholders' dividends. By this arrangement, establishing a profit-sharing scheme on a sound basis, the depositors have a direct interest in results, inducement to save money, certainty of some return on their investments, no liability, and security for their capital so long as the Company is solvent.

An important condition is that no one but the secretary and the cashier, who keep and audit the accounts in confidence, have knowledge of the amounts of individual credits. All that the directors know is the total sum deposited, privacy being ensured by entry of the particulars in the cash-books under numbers. Within twelve months from the institution of the scheme fifty men availed themselves of the privilege offered. By 1915 the number accepted amounted to 121, and the total interest paid to over £12,000. Then the Board, taking the view that investment in War Loan was more fitting, decided to suspend acceptance of applications. Moreover many of the depositors, reluctantly and from a sense of duty, withdrew their money for War investment. After the cessation of hostilities the working of the scheme was resumed. At the end of 1918 the number of depositors was 50, the sum of their balances, reduced by the withdrawals, £8,837, and the interest paid, since the commencement, £14,150. In the first months of 1919 about 50 more were enrolled, including some of the employees at Glasgow.

In 1903 the Articles of Association were altered so as to allow some of the staff and others to hold shares in the Company. The first of these new shareholders were Messrs. Genner, Crowther, Field, and Lindsay Forster.

Job Genner was among the first to profit by Talbot's teaching at the Schools and took care to utilise afterwards the advantages of the Evening Continuation Classes. Entering the offices as a boy, he soon proved his capacity in the heavy work of the Export department, and advanced stead-ily, so that on Franklin Taylor's retirement in 1884 he was marked out for his successor. Auditor of the new Company for some years to 1911, he was then appointed Secretary, a post which he retained until his retirement, with a pension, towards the end of 1913. In recognition of the way in which he had performed his duties during nearly 62 years, he was presented on leaving by the Directors with a silver tea-and-coffee service, suitably inscribed, and by his colleagues on the staff with a Sheraton cabinet. He died on July 17, 1919, aged 80.

George Crowther also began in the offices as a boy and also rapidly distinguished himself, rising to be for many years principal traveller for the firm and head of the Sales department. His services are noticed more particularly in Chapter XIV. Similar recognition was accorded to him, in the form of a gold watch, on his retirement in July 1915 with 59 years' service to his credit.

James Field, another in the service of the firm from boyhood, has been principal manager for the Company since 1903, when he succeeded Henry Talbot. He has had charge of all building and furnace work, and his services in designing and carrying out improvements have been many and important.

A. Lindsay Forster came as chief engineer in 1901, replacing Albert Jackson, who in his turn had succeeded Marshall, noticed in the chapter on the Lighthouse Works. Among his many services that of his installation of the electric power plant is recorded below. Since 1913 he has managed the Company's works at Glasgow. Besides these four, several others of the staff now hold shares.

Henry Talbot was the fifth son of the Schools head master. He succeeded E. J. Forster as principal manager in 1892, having assisted him since 1887. After eight years his health began to render him unable to perform his duties properly, and finally he retired in the summer of 1903, accepting a less onerous post elsewhere.

The long and trusted services of James Simcox and Forrester Lloyd have been noticed. Among other prominent men, no longer with the Company, may specially be called to mind in the works Joseph Edward Scott, and in the offices Samuel Turner, the cashier, who died, greatly regretted, in May 1905.

Scott came in 1887 to bring into operation at Spon Lane the continuous method of flattening glass, already established by him elsewhere. He also gave great help with the installation of tank furnaces and in other ways. But after ten years' good service his health broke down, and irremediably.

As time went on, it became necessary to look out for young men who might qualify to be Managing Directors. Firstly, in 1900, came on probation Arthur Frederick Dobbs, a grandson of James Chance, and a Scholar of King's College, Cambridge. But in spite of his ability and industry his services were not retained. In 1903 came from Oxford Walter Lucas Chance, an Etonian, eldest son of Arthur and great-grandson of the founder of the firm, and from Cambridge Clinton Frederick Chance, youngest son of Alexander. The latter worked with singular zeal and energy, but left after two and a half years, transferring his services to Chance & Hunt Limited. Walter Chance began work in the offices, and at the end of the year was appointed Secretary to the Company and an assistant to the Managing Directors. A year later were delegated to him general charge of the offices and special charge of the sales. He has been a Director of the Company since 1909.

Another was John Robert Wharton, of Eton and Cambridge, son of Admiral Sir John Wharton, K.C.B., for many years Hydrographer to the Admiralty. Coming in 1906 he was given charge of the optical glass, and then, on Clinton Chance's departure, of the sheet glass makers. Later he was transferred to the lighthouse works. He joined the Board in January 1914.

Next, in July 1908, were obtained the valuable services of Captain Wilfrid Lionel Foster, D.S.O., R.A.(now Major and a C.B.E.), second of the famous cricketing brothers and one who had gained his distinction in the Somaliland campaign of 1903. He began as assistant to Edward Chance in the manufacturing, soon proved his worth, and became a Director at the beginning of 1910.

Later comers were G. Trevor Williams and Richard Evelyn Threlfall, the latter a son of that leading man in the world of science, Sir Richard Threlfall, K.B.E., F.R.S. Williams, a victim of the War, was endowed with exceptional ability and scientific perception. Having gained a Pemberton Fellowship at Armstrong College, Newcastle-on-Tyne, and the degrees of Master of Science of Durham and Bachelor of Science of London, after two years' practical experience at the Parsons electrical works he had gone to Cambridge for research work, and would doubtless have been elected a Fellow of the Royal Society had he lived. His loss to the Company was a heavy one.

The latest development in the management of the Works has been the appointment of "Departmental Directors," in the words of the final resolution on the subject (March 19, 1918) "to have charge of particular branches or departments of our business, and their remuneration, duties and powers to be decided from time to time by the Board, but no Departmental Director shall be regarded or treated as a Director under the Companies' Acts or our Articles of Association." Those first appointed were Messrs. Field, Lindsay Forster, Jeboult, and Bassett—the second as "Glasgow Director." Jeboult will be noticed as chief man under Stobart at the lighthouse works; A. E. Bassett rendered valuable service in the rolled plate department before occupancy of his present post as assistant to Walter Chance in the Commercial department. A further resolution appointed the Managing Directors and their assistants, and the Departmental Directors, to be a Managing Committee, and laid down preliminary rules for its proceedings.

The other principal men in the offices at present are W. P. Fielden, who replaced Genner as Secretary to the Company in 1914, Arthur T. Stephens, Arthur Laban, Samuel Grice, F. Maltby, J. W. Daughtery, and George Lewis, cashier from 1905, and at Glasgow, John Aitchison. Most of them have been in the service of the firm from boyhood.

The latest change to record (1919) is K. A. Macaulay's resignation of his directorship, after more than forty years of invaluable service.

Prospects, when the new Company started, were brightening. In 1890 the wages of the sheet glass men could be advanced by ten, and those of the rolled plate men by five per cent., and general hours of work were again reduced from 59 to 54 per week, on the principal condition that work in the warehouses should be continued up to 6 p.m., if necessary, without payment for overtime. The advance in prices, however, which made this possible, proved but temporary. In June 1891 a return to former rates of wages and a decision to pay them again on the warehouse returns caused the sheet glass men to strike, and they stayed out for some months. The rolled plate men, on the other hand, after demur, accepted the reduction. In spite of this it was found necessary to discharge a set and a half of them. Reduction of cost was obligatory, and to this end it was decided at last to try what could be done with a continuous tank furnace.

With the sheet glass what hampered most was the continued falling off in quality, principally due, no doubt, to the deterioration of the Leighton sand. The Hanmer mine there had furnished very good coarse-grained material, such as was required for the pot furnaces, but by 1878 it had been given up, and others opened in its place failed to satisfy regularly. When tank furnaces were introduced, the fine-grained sand necessary for them was obtained from Belgium, and for a number of years the supply came almost entirely thence or from Lynn, the Company's Leighton holdings having been disposed of in 1893.[1]

The tank furnace for the rolled plate was finished and lighted in July 1892. Results giving satisfaction, it was resolved to try one for the sheet glass also. This was built on the site of the original No. 1 furnace, and was started on June 22, 1893. Decision to build a third, on the site of No.

---

1   The "old mine" so called, was given to William Reeve in recognition of his family's long services as contractors for working the sand.

2 pot-furnace, followed in November. It was on a new plan devised by Scott—namely, in two parts connected by channels of fireclay; the object being to give easier access to the bridge for repairs.[1] But the invention did not survive trial; the tank was altered to the ordinary form.

It was soon found that one tank was sufficient for the sheet glass, indeed, the men were informed in October 1895 that even one could be worked only if the unbound men accepted a reduction of ten per cent, in wages. To this they consented. The second sheet tank, No. 3, was turned over to the rolled plate manufacture, for which enlargement of No. 1, on a plan of Scott's, had been required in the previous year, in consequence of the steadily increasing demand. In August 1897 the wages of the rolled plate men could be advanced again by 10 per cent, when working three shifts, and by 5 per cent, when working two — rates maintained for four years.

The Mason and Conqueror machine, noticed in the last chapter, had, by the time that the Company was formed, been so developed under the care of George Chance that little remained of the original but the principle.

Rolled glass was being turned out from it regularly and successfully. In 1890 Edward Chance, now in charge of the process, perfected the machine by the addition of a second pair of rolls, specially for the production of "figured" glass. The first pair formed the sheet, the second, in the words of the patent, "further operated on and improved it and when desirable impressed it with a pattern on one side." [2] This manufacture quickly became, and has continued to be, of the very first importance. A number of successful designs were introduced and registered, and licences for them and for the process were granted to other manufacturers, A specially artistic type is the glass known as "Flemish," of undulating surface instead of impressed with a pattern. To quote from a catalogue:

Flemish glass achieved immediate success with those best qualified to judge, and its use and popularity have been continuously extending ever since. It is a very translucent

---

1   Patent No. 6264 of 1894. The claims were for " separating a short distance apart the melting and working chambers of tank furnaces " and
connecting them by one or more tubular fireclay channels. It was decided in February 1898 not to keep the patent up.
2   Patent No. 785 of 1890, granted to Chance Brothers & Co. Limited, and Edward Ferguson Chance.

glass with a smooth undulating surface, to which dust particles do not readily adhere and which is consequently very easily kept clean, and though not transparent in the ordinary sense it has the effect in many positions of actually transmitting more light than clear glass. Its general character is one of dignity and refinement. In appearance it is attractive and restful to the eye, nor can any other glass give such a distinguished tone and finish to high-class work.

The glass is universally recommended and employed by architects, and though mostly found in private houses and offices, is also in common use for the better class of public buildings.

The double-roll machine was also applied to the manufacture of "wired" glass, in which wire-mesh is embedded in the sheet. This had come up for notice in 1894, in connexion with a patent by Frank Shuman, of Philadelphia, but the matter was not then prosecuted.[1] In 1905, however, patents were taken out for making the glass on the double-roll machine in various ways.[2] But there was not sufficient inducement from the point of view of profit to pursue the manufacture on a working scale until quite recent times, when experiments have been renewed.

In connexion with the double-roll machines must specially be recalled the services of Edward Beetlestone, a man who, taken from the lighthouse works to attend to them, after the success with the "figured" glass had been achieved, speedily made himself master of their intricacies, and is now, with A. L. Simcox, a manager in the rolled plate department. Noticeable in particular are his patented three-sided "teeming-plate" and wiring roll with non-continuous ribs.[3]

By the year 1903 a second tank furnace was again required for the sheet glass. This, No. 4, was built on the site of the old No. 1 blowing holes, with its new blowing holes on the site of the old No. 6 pot-furnace.

A special feature of these years was the extraordinary increase in the demand for extra white glass, principally for the figured and for spectacle glass. In 1890 No. 3 pot-furnace, after a brief trial with eight covered pots, was furnished with four for the manufacture. In 1898 another pot-furnace, No. 4, was given over specially for the figured, and two years later came under consideration a tank for the same purpose. This was built in 1905

---

1   It was not thought fit to complete a patent applied for, No. 12997 of 1894.
2   No. 12495 in the names of George Chance and the Company, and No. 12688 in those of Edward Chance, Lindsay Forster, and the Company, both of 1905.
3   Patents Nos. 6533 of 1905 and 112099 of 1917

on the site of No. 7 pot-furnace, whose number it retained. In 1907 the increasing American demand for spectacle glass brought No. 12 furnace, with four covered pots, into use for that, and the extra white warehouse was ordered to be extended at a cost of £1,500. Next year was discussed erection of yet another furnace for the glass. Eventually, in 1910, was tried in No. 4 house a small so-called "tub" tank, where the glass was founded and worked alternately, as with pots. The experiment failed from early appearance and rapid increase of clay stones. Instead was tried a furnace for open pots, with special arrangements for a clear flame above the pot-level, so as not to affect the colour of the glass. This gave very satisfactory results, and next year (1912) the principle was adopted in No. 3. In 1913 a "tub" tank was tried again in No. 6 house, with the same result as before. Here also was then introduced the clear flame over open pots, and in 1918 in No. 12.

Before No. 12 furnace was turned over to the extra white, it had been used for ten years, with six 49-inch pots, almost exclusively for another important and successful manufacture, that of "muffled" glass, impressed with patterns by blowing into moulds. This, placed on the market by a Berlin firm about the year 1884, met a demand for something different from rolled cathedral for ornamental glazing. Chance Brothers & Co. at once undertook the manufacture, and in course of time, by improved make and by facilities for quicker delivery in England, succeeded in capturing practically the whole home trade, their production approaching for some time a million feet a year. From 1911, however, demand began to fall off owing to the competition of figured, and on the outbreak of the war the manufacture had to yield to its special requirements.

Of the other pot-furnaces No. 5, used chiefly for the lighthouse glass, was rebuilt in 1904 for heating by gas. The 42-inch pots were now obsolete; in 1903 all in stock but eight had been ordered to be broken down.

A new manufacture undertaken in 1902 was that of "opal tiles" by another name "venturene," intended principally for the interior lining of walls in place of the usual tiles of earthenware, and well suited for the purpose both from the decorative and the sanitary points of view. They were first cut from rolled plates of opal glass, afterwards pressed from glass melted in a special furnace devised by Oscar Andries. Several patents were applied for in the years 1901 and 1903 in the names of Edward Chance or Rosenhain, the Company's scientific adviser, but neither were they completed nor a later one of 1907 (No. 27714). The tiles were turned out successfully and at no excessive outlay, but low prices made the venture unremunerative. In 1909 Rees, the Company's chemist, proposed production of a similar article from semi-founded glass, a process entailing much greater expenditure and work. Captain Foster, on his arrival, took charge of the manufacture of these "vitreous tiles" and made it a success. An illustrated catalogue showed examples of more than a hundred standard tints and a large variety of ways of employment. As stated therein, the non-absorbent nature of the tiles suited them for outdoor as well as indoor use. Cut into small fragments, they found application as "mosaic" But again the very low selling prices could not be met. On the outbreak of the War the manufacture was suspended, and in default of prospect of profit, and from other causes, has not yet been resumed.

Against this, early in 1915, came a request from Government for a new manufacture to be undertaken, that of the heat-resisting glass globes essential to present systems of lighting by gas, oil, or electricity. Previously to the War the whole of the supply of these came from abroad, mainly from Germany and Austria. The undertaking was a difficult one, the glass differing in composition, founding qualities, and working methods from any previously made at Spon Lane, excepting certain of the optical glasses. When the manufacture of what were principally wanted, globes of large size for outdoor lighting, had with great trouble been perfected, suddenly the lighting restrictions practically put an end to their employment. Attention was then turned to production of the smaller articles of like kind required for indoor lighting. The manufacture of these was also mastered, so that the disappointing set-back turned to advantage; the Company is now in a position to supply glass proper for all purposes where it may be subjected to marked variations of temperature. Quality is reported in every respect equal to that of the best continental makes. Given the maintenance of possible selling conditions, the manufacture promises to become an important branch of the Company's business.

Continuous lears were introduced with the help of J. E. Scott, as said, in 1887, the first being built for the sheet glass, by conversion of No. 3 lear in March of that year.[1] Certain difficulties encountered having been successfully overcome, the use of these continuous lears was speedily extended

---

1   One form of continuous flattening had been tried about 1871, but the glass turned out had proved too hard.

both for the sheet glass and the rolled plate. To suit the thicker substance of the latter George Chance devised the use of two tunnels alongside each other, in order that the sheets might be fed into either alternately, and so the time for their cooling be doubled. Another arrangement proposed was to have two sets of travelling stones in a tunnel of double width.[1] In the course of years the continuous system has proved a complete success. For the rolled plate the double tunnel is in settled use at Spon Lane, while the sheet glass lears have been improved by lengthening and in other ways.

On the other hand a substitute devised by Scott for lagres, a cast-iron slab coated with "purimachos" or other fire-resisting cement,[2] turned out a failure. Much was expected from it; saving was estimated at £400 a year, Scott received a present of £200, and the invention was protected also in France, Belgium, and the United States. But on investigation of prospects abroad, it was resolved not to prosecute the matter there. Finally, in 1902, the patent was allowed to drop.

In February 1902 was considered installation of an electric power plant. Lindsay Forster submitted proposals, and was authorized to obtain tenders. Various schemes having been discussed and preliminary resolutions passed, a first plant was set up, driven by a 150-h.p. engine. A second followed in 1903. The power was almost all at first used for the saw and clay mills and the patent plate, and one result in that department was the disuse and sale of the fine old beam-engine that had been steadily at work, often both by night and day, since 1840. Before long, the use of electric power became general throughout the works, and mechanical could be substituted for hand power for various purposes. It was applied in the course of the years 1905 to 1907 to drawing the glass out of the rolled plate kilns, to operating the casting tables, to mixing the materials, to carriage by telpherage, and to conveyors for the gas-producers. Blowing by compressed air was introduced for the "muffled" glass in 1904 and successfully applied to the sheet glass cylinders some years later. Practically all this important work was designed and well carried out by Lindsay Forster.

The principal landed development of these years was turning to account the Company's property at the corner of Spon Lane and George Street, still unbuilt upon save for some small tenements let to various persons. In 1891 it was offered to the town at 4s. a yard for the purpose of

---

1   Patent No. 12212 of 1890.
2   3 Patent No. 5899 of 1894.

a recreation ground, but the offer was declined for the reason that laying out would cost £500. It was resolved to decline to accept less than £800, but, if the Local Board should purchase, to contribute £200 towards the cost of laying out, provided that the land were used for no other purpose without the Company's consent. But with the opening of West Smethwick Park the want of a recreation ground here lapsed. A temporary lease at £10 10s. a year was effected in 1897 to Mrs. Sarah Clarke, and in December of the same year the property, saving a small piece, was let for building on leases of 999 years to William Joseph Cox.

In 1916 and 1917 more land was acquired for extension of the light-house works, a small piece from Messrs. Glossop and Smallwood for £60, and a larger one from the trustees of P. A. Fraser for £1,346 5s. Other recent purchases have been a strip of land, with dwelling houses, adjoining the south-eastern corner of the works—this in November 1916 from Miss Clara Marsh for £500—and 846 square yards behind the Schools.

Away from Spon Lane there was an acquisition of great importance— glassworks at Glasgow. Negotiations of 1895 with the Glasgow Plate Glass Company for the formation of a joint Limited Liability Company fell through, but twelve years later Chance Brothers & Co. purchased the works, and in 1908 transformed them into a Limited Liability Company. The Glasgow vendors were Messrs. Brogan and Mallock, and the services of the latter were retained to direct the works. In 1911 the new Company was taken over by Chance Brothers & Co., who, as a first step, decided upon alterations and additions estimated to cost £11,000. The works are now carried on under the direction of A. Lindsay Forster.

Lastly are to be mentioned the arrangements which in 1912 resulted in the transfer of shares of the Company to the St.-Gobain Company of France, the oldest and still the premier plate glass manufacturing company of the world. The close connexion between the two Companies of late years has been of the greatest mutual benefit to them.

Of alterations in the works, principal items were in 1894 the demolition of the old cottages known as "Scotch Row," country homes of Lucas Chance's workmen seventy years before, and in 1898 provision of a new engine and machinery for the mixing department. The old saw-mill was disestablished in 1903. In 1905 the offices were extended and re-arranged, next year the old stables near the House were cleared away, and in 1911 came the new gate-house. Extension of the buildings in its vicinity followed, and particularly, from 1914, of the optical department, with establishment of the Research Laboratory, in consequence of the requirements of the War. In 1916 were built a garage and a fire-engine house on the site of the old playground.

A novelty of 1905 was the "Suggestion Scheme" under which suggestions of improvements are received from the workmen and remunerated according to the value of them adjudged. A Committee appointed in December 1905 included Clinton Chance, as Chairman, and Messrs. Field, Forster, and Robson. In February 1907 the number of members of the Committee was raised to nine, Edward Chance being Chairman and Stobart Deputy Chairman. On October 6, 1908, it was resolved that the Committee should be asked to report improvements, proved of value after trial, with a view to further remuneration being considered by the Board. With natural changes in membership it continues its work.

A reform in the hours of labour was the institution of the "One Break Day"—that is to say, abolition of the breakfast interval, the men breakfasting before they came to work. The scheme was under consideration in 1901, when the question of a general reduction of wages was deferred until experience should be had of its working. In February 1904 three proposals were set before the men : hours of work to be (a) from 7.0 to 12.0 and from 1.0 to 5.30 ; (b) from 7.30 to 12.30 and from 1.30 to 6.0 ; or (c) as the former in summer and the latter in winter. The system, it was stated, "is in force in many Works, and has proved satisfactory to both employers and employed. It is essential, in the face of keen competition, to make use of all improvements which work for the mutual advantage of our Company and our workpeople." Facilities for breakfast at the works, before starting work, were offered to those who might desire them, any sort of meal during working hours being prohibited. For women and girls special arrangements were to be made, after the hours for men and boys had been fixed. Voting on the subject, confined to men of 21 years and upwards in the Company's employ for not less than a year, went in favour of "A" scheme, which came into force in April and was maintained for nearly fourteen years.

From January 1919 the hours of work have been reduced to 47. Two schemes for this were set before the work people; by the one, work to be started at 7.30 in summer and at 8.0 in winter, by the other at 7.30 all the year round. The notice issued contained the following passages:

In considering this question, it is well to emphasise the fact that a shorter working week has been agreed to by employers on assurances given by the men's representatives that (1) output will not be decreased, and (2) there shall be 47 hours of actual work. Therefore, in deciding the hour at which work should start in the morning, it is essential to bear in mind that it should be one which will permit of everyone having breakfast before starting, as no eating can be allowed during working hours.

There are decided advantages in not starting work at an earlier hour than 8 a.m., at any rate so far as the winter months are concerned, and it may be well to indicate these.

1. In America, where somewhat similar hours are worked, 8 a.m. is the hour generally adopted for starting, and is the one which has proved to be the most suitable.

2. From experience of a one-break system over a period of fourteen years there is reason to believe that many of those employed in these works have habitually started work without first having had a proper breakfast and have, therefore, either had to work five hours without a proper meal or else consume food during working hours, which is contrary to the rules.

3. It is essential, if trade is to be secured to keep the Works in this country fully employed, and if any diminution of output owing to the shorter hours is to be avoided, that lost time should be reduced to a minimum. This was difficult to prevent when the starting-time was 7 a.m., and it is believed the same difficulty (though perhaps to a less extent) would continue through the winter months if work started at 7.30. We shall be glad to give facilities for Breakfast at the Works, before starting work, for those who desire it. Whoever wishes for this must give notice as soon as possible to his Manager or Foreman, in order that the necessary arrangements may be made.

In spite of the above recommendation, a large majority of the work people, 512 to 104, voted for always starting work at 7.30 a.m. It continues until noon and is resumed, excepting on Saturdays, from 1.0 to 5.0 p.m.

CHAPTER VII

THE LIGHTHOUSE WORKS[1]

THE luminary in a modern lighthouse is caged in a complex structure of glass lenses and prisms, which refract or reflect the diverging rays and concentrate them upon the sea, where they will be of service to the mariner. Application of the refractive powers of glass to this purpose was the work of the great French mathematician and physicist, Augustin Fresnel, early in the last century, and the splendid "dioptric" instruments of to-day are developments from his original constructions.

Scientific knowledge and technical skill of a high order are required to construct a dioptric light. The glass-founder, in the first place, must exercise his highest art to produce a colourless glass as free as possible from striae and other flaws. Roughly cast in moulds, the lenses and prisms must have their surfaces accurately ground to particular curvatures, whose calculation is the province of the skilled mathematician. When they have received their final forms, and have been finely polished, they must be fitted into their places with the most scrupulous nicety, in order that the rays falling upon each may be transmitted exactly in the direction required. Lastly, the plans of the engineer must be controlled by what is practicable with so difficult a material, and by the masterful dictates of economy. He cannot employ always the same design, but must have regard to particular local conditions, which may require him to devise novel forms and arrangements to suit them. The resources of physical, mathematical and mechanical science alike are taxed in the production of these beautiful instruments.

Glass, says Thomas Stevenson, the leading expert of his time in lighthouse engineering, is the best material for lighthouse apparatus, and the fewer agents that are employed, the less the waste of light. "That apparatus is the most perfect, which sends the greatest number of rays at the same

---

1   Much in this chapter is taken from the author's *The Lighthouse Work of Sir James Chance, Baronet* (London: 1902), to which reference may be made for details.

time to the eye of the observer, provided the light, when revolving, is allowed to remain sufficiently long in view for the sailor to take a compass bearing" To obtain the maximum effect of a flame of given intensity it is necessary "to transmit the whole of its rays in the required direction with the least possible loss from absorption or irregular scattering." [1]

Fresnel standardised his optical apparatus in six "orders," decreasing in dimension from 920 millimetres focal distance for the first order, to 150 for the sixth. In modern times have been added the great "hyper-radial" order, of 1330 millimetres focal distance; the "meso-radial," of 1125, seldom used; the small third, of 375, and others smaller than the sixth for ship-lights, buoys, and other purposes.

The necessity for distinction among lights obliges also variation in their character. They may be fixed, when the rays are condensed in the vertical plane only, or revolving, when there is condensation also in the horizontal plane, or in "azimuth," and beams of intense power are made to sweep round the horizon in succession, or partly fixed and partly revolving. Again, fixed lights may be varied in character by sudden interruption of their rays at given intervals, and revolving lights by changes in the period and duration of their flashes. Modern contrivance has so greatly amplified this means of distinction among revolving lights, that fixed lights now are seldom employed but of the smaller orders for harbours and river estuaries. In Fresnel's time dioptric apparatus was made only in his native country. But soon after his death the manufacture was attempted by Messrs. Cookson & Co., of South Shields, instructed by Leonor Fresnel, brother of Augustin. The methods of making their annular or "polyzonal" lenses, of which they constructed the first in 1831, were crude, and they were grievously hampered by the excise restrictions in force, but nevertheless they supplied apparatus for a number of lighthouses in the British Isles. In 1860 it was reported to the Royal Commission on Lighthouses, then at work, that twelve British and one Irish were provided with refracting apparatus made by them. However, Messrs. R. W. Swinburne & Co., who bought their works in 1845, did not continue the manufacture. Mr. Swinburne informed James Chance in 1864,

from the difficulty and expense attending the infancy of the manufacture Messrs. Cookson & Co. were never adequately remunerated. The matter was taken up with

---

1   *Lighthouse Illumination*, 1871, p. 2.

great zeal and ability by a junior member of the firm, and, as the oldest plate-glass firm in England, they felt themselves impelled to attempt the establishment of a manufacture, for so patriotic an object, that might prove worthy of this great maritime and manufacturing country.

In January 1845, the year in which the excise restrictions were removed, the manufacture of dioptric apparatus was suggested to James Chance by Sir David Brewster. Acknowledging receipt of certain "plates of ground glass" of such perfect surface as he "could scarcely have conceived it possible to execute" he referred to Messrs. Cookson's failure to produce "built-up polyzonal lenses" for Scotch lighthouses to the satisfaction of Mr. Alan Stevenson, who had been obliged, in consequence, to order from Paris, and went on : "I cannot doubt from what I saw at your Works that you could overcome the difficulties which had been insurmountable by them, and that the branch of manufacturing lenses would be both an interesting and a profitable addition to your establishment." The suggestion could not in any case have been followed up at that time of pressure, but doubtless lay in germination. It was revived by Bontemps, on his coming to Spon Lane in 1848. As a first step, then, inquiry was made at the Trinity House as to prospects of success. Robert Chance paid a visit there in October 1849. He was received with the habitual courtesy of the House, but the information given as to probable demand was not encouraging. Nevertheless a second visit in December, on which occasion he was introduced to Mr. W. C. Wilkins, who practically monopolised at that time the manufacture in England of everything connected with the fitting of lighthouses, excepting the glass, had for its result a decision to make the venture. An agreement was made with Wilkins, and Bontemps obtained for the firm the services of a French expert, Tabouret, who had been for thirty years in the lighthouse department of the "Ponts et Chaussees" and had worked for Augustin Fresnel himself. Engaged from January 1, 1850, he brought with him as assistants Mangematin and his own son, agreeing to pay the latter himself.[1] In December he was desired to construct a revolving apparatus of the first order to be shown in the lighthouse under construction for the Great Exhibition of 1851.

This was a combination of Fresnel's revolving drum of eight refracting

---

1   The terms arranged were one-third of the net profits with a minimum of £62, 10s. per quarter, the agreement to be terminable without previous notice at the end of 1851 or 1853.

annular lenses with fixed reflecting or "catadioptric" zones, thirteen above and six below. Revolution of the drum being completed in four minutes, the effect was that of a fixed light waxing to maximum intensity every thirty seconds. "The workmanship" the Jury reported, "was not characterised by any degree of finish—a fact in its favour, as any great degree of finish, or adoption of ornament, would involve an increased outlay of capital without compensating advantages." The glass, purposely made very hard to resist corrosion by the atmosphere, they pronounced "dark, and of a greenish colour"; but they did not consider this of importance, "as it is known that such tinge does not affect the transmission of light." As regarded striae, they considered the glass to be equal in quality to that of a French apparatus made by Letourneau, of Paris, and shown by Wilkins. This differed from Tabouret's light in giving "short eclipses," as the term went. Instead of eight annular lenses the revolving drum had four, alternating with segments of a fixed-light belt; and vertical refracting prisms placed outside the fixed reflecting zones condensed the rays from them in azimuth. Ornamental finish given the Jury considered actually objectionable.[1]

Previously, in February, Wilkins had offered Chance Brothers & Co. an order for another first order light, on condition of approval of a segment of it by the Trinity House, but the offer was declined on the ground of desire to make a first impression by the apparatus for the Exhibition.

In these first attempts the firm enjoyed the expert advice of Messrs. Alan and Thomas Stevenson, engineers to the Northern Lighthouse Board, the latter the inventor of almost all improvements made in the optical apparatus since Fresnel's death, and now setting up the first great sea-light on his new "holophotal" system at Ronaldshay in the Orkneys. Robert Chance records much useful information about the Trinity House given by Alan Stevenson on the occasion of a visit to Spoil Lane in 1851. Subsequently he advised against manufacturing for stock in anticipation of demand, there being "too many peculiarities in each case of a given lighthouse to make it probable that your stock should be suitable to the wants of any customer." Yet, he said, "if the Trinity Board were to improve all their fixed lights, you might then have orders enough." Similarly, Thomas Stevenson remarked on the continual changes in form and design, instancing his own inventions. Were apparatus made for stock, he advised that it should be for fixed rather

---

1   Reports of the Juries, Exhibition of 1851, p. 272. The two apparatus are figured in Stevenson's *Lighthouse Construction and Illumination*, 1881, p. 79.

than for revolving lights. Yet, "should your glass and manufacture equal the Parisian, there is little fear of a demand." One sentence of his letter asserted a principle on which James Chance also acted, when he took up the manufacture: "I have never taken out a patent, and never intend to do it, so that of all my inventions the public derive the benefit."[1]

The good start was not followed up. A segment sent to the Trinity House for examination in January 1852 was condemned as bad in colour and too full of striae. The Board was anxious to encourage, says Robert Chance, but deemed good colour essential. Thereon, as said elsewhere, a special furnace, No. 8, was built for melting the glass in covered pots.

Yet, though the firm had assistance from Sir David Brewster, both in correspondence and by personal inspection, and was supplied with a tracing for a holophotal quadrant by Messrs. Stevenson, another segment sent to the Trinity House in February 1853 was again found unacceptable. Thereon it was resolved "that all the present operations connected with the lighthouse department be suspended at the earliest possible period in consequence of the signal failure on the part of M. Tabouret to produce any lighthouse to the satisfaction of the Trinity Board" His agreement was cancelled, and he left the works on April 16.

The next two years were occupied in mastering the details of the work and in gaining experience, management being entrusted to Mangematin and Masselin.[2] Seven apparatus were completed, all for fixed lights; two of them of the third order, two of the fourth, and three of the fifth. Two at least, one of the third and one of the fourth order, were made for Wilkins, and appear to have been put up respectively at Broadhaven and at Spit Bank, in Cork Harbour. The other of the third order went to Beeves Rock (River Shannon), the other of the fourth to Samphire Island in Tralee Bay, and one of the fifth to the Levant or the Black Sea. On one of them Professor Faraday, as scientific adviser to the Trinity House, reported on March 14, 1854:

> Having this day examined one division of a Catadioptric Apparatus constructed by Mr. Chance, of Birmingham, and compared it with one of French construction, which the Corporation possess, mounted in the Comparative Frame, I am of opinion that, in the colour of the glass, the working of the various pieces, and the fitting of the whole

---

1   Letters to the firm from Alan and Thomas Stevenson, April 1851.
2   Bontemps's cousin, see Chapter VIII.

together, the former is equal to the latter; and, from the effect of the light upon the screen, I believe that one would not be distinguishable from the other when seen at sea.

Transmitting his report to the firm, Mr. Herbert, secretary to the Trinity House, wrote:

The observations made by many of the Elder Brethren, who were present on the occasion, induce them to concur with Dr. Faraday in the opinion which he has expressed.

The success obtained with these lesser apparatus justified venture on one of the first order for the Paris Exhibition of 1855. It had to compete there with exhibits by Lepaute and by Sautter et Cie, Letourneau's successors. They obtained medals of honour, and Chance Brothers & Co. one of the first class, awarded, said the Jury, in consideration of their efforts to introduce the manufacture into England. But Sir David Brewster reported: "In the quality of the glass, either in reference to colour or to freedom from striae, and in the fitting up of the various parts, this apparatus seemed to be very little, if at all, inferior to its rivals"[1]

After this the firm set seriously to work to develop the manufacture. A prospectus issued emphasised the necessity of distinguishing neighbouring coast-lights from one another, and figured and described the apparatus that could be supplied; apparatus, that is to say, fixed, flashing or revolving, of any of the six orders and constructed either on Stevenson's holophotal system or on Fresnel's. The flashing lights were of the type of Letourneau's of 1851, of which a drawing was appended. Also were shown so-called holophotal reflector lights (a combination of glass agents with metallic mirrors), a dioptric apparatus for lightships, and a patent dioptric signal lantern. During the next four years were completed and sent to various parts of the world nine lights of the first order, five of the second, two of the third, eleven of the fourth, and a number of smaller ones. Those of the first two orders were erected at Rathlin Island (two), Bardsey Island, Inisheer in Galway Bay, Lundy Island, Whitby (two), Whalsey Skerries in the Shetlands, Cape Schanck (Victoria), Rhu Val (Islay), Pencarrow in New Zealand, Seskar in the Baltic, the Cani Rocks on the north coast of Africa, and Race Rocks (Vancouver). At the works was built a brick tower,

---

1   *Rapport du Jury Mixte International,* p. 455; Reports on the Universal Exhibition, i. 291.

eighty feet high, on which to place the lights for examination of the effect produced at the height at which they would ultimately stand. Robert Chance continued to supervise the manufacture, James Chance helping and in particular performing the mathematical calculations, as is shown by a resolution of 1857 to engage a clerk, who understood logarithms, to assist him in them. The work was greatly improved, and the glass no longer open to the reproach of bad colour as compared with the French. When in 1859 the firm made overtures to the Spanish Government, the engineer, Senor Lucio del Valle, was sent to inspect. As the result of a thorough examination of all that was being done he had no hesitation in recommending an order for three lights and iron towers for the mouth of the Ebro. Letters and certificates from the Messrs. Stevenson, and the report of Professor Faraday, he stated, gave

> assurance of the most formal manner that Messrs. Chance's apparatus are made of materials of good quality, that the curvatures of the lenses and prisms are correct, that the polish is perfect, that they give a great light, and, in fine, unite all the qualities necessary for them to be considered apparatus of complete finish, equal in everything to the lights produced from the French works.

It is easy to deduce from my observations:
 1 That the catadioptric lights made by these gentlemen are not inferior to the French lights as regards the optical part.
 2. Nor do they yield in anything as regards the mechanical part; they have even some advantages over the French lights in points of detail.[1]

Meanwhile, in May 1856, Mangematin being found unable to conduct the department with due regard to prompt execution of the orders, the entire responsibility had been handed over to Masselin. An office was provided for him at the lighthouse works, in order that he might exercise closer control. Yet, just a year later, he was requested to resign the management, in consequence of inefficiency of administration displayed in connexion with the Lundy Island light. He retained the position of engineer, but the practical work, excepting in the fitting shop, passed into the hands of Peter Rigby, previously in charge of the grinding. In the fitting shop William Maltby ruled for a time, with Mangematin under him. Head of the office

---

1   Translated from Del Valle's report to the Spanish Government.

from 1854 was the future distinguished manager, James Kenward.

It was during this period that public attention was directed to the unsatisfactory condition of the lighting of the British coasts. In 1858 a Royal Commission was appointed to investigate the truth of the complaints, an event in the history of lighthouse illumination second only in importance to the introduction of dioptric apparatus. After preliminary work the Commissioners, assisted by such distinguished men of science as Airy, Astronomer Royal, and Faraday, personally inspected most of the lighthouses in Great Britain and Ireland and many in France and on the north coast of Spain, visiting also works where lighthouse apparatus of various kinds was constructed. The condition in which they found many of the lights examined amply justified their appointment. Even in the best cases, they reported, a large proportion of the light was wasted. Sometimes a part of it was thrown too high, sometimes it shone upon the land. They found fault alike with the lamps in use, excepting in Scotland, the quality of the glass, the construction of the apparatus, and the neglect of proper adjustment of the lenses, in particular to the "dip" of the horizon. That is to say, the beams were habitually directed to the horizon, or above it, leaving the sea between it and the shore, where the light was wanted, dark.[1]

The Commissioners paid their first visit to Spon Lane on December 23, 1859. Under the guidance of James Chance and Masselin they made a thorough inspection of the processes of manufacture and of the arrangements for testing the optical accuracy of the lenses and prisms.

They remarked on the "very superior quality" of the glass and on the machinery for grinding the surfaces accurately, of a superior description to any yet seen."[2] They discussed and allowed the disadvantages of the custom of ordering different parts of an apparatus in different quarters, and of neither giving to the manufacturer of the glass information as to the nature and size of the luminary to be used, nor allowing him to make the adjustments himself on the spot. They were informed that Chance

---

1   See the *Report of the Commissioners*, 1861, i. pp. ix, x, xiv.

2   Fresnel, said James Chance in the paper that he read before the Institution of Civil Engineers on May 7, 1867, "contrived expressly a system of grinding the glass rings by combining a cross-stroke with rotation, thus translating his geometrical conceptions into corresponding mechanism." The Astronomer Royal said, in the discussion on this paper, "that which struck him most was the cross-stroke in the polishing; when there was a ring-lens to be made, the cross-curvature was not given by grinding in a bowl, but by the cross-work of the polisher, and by some small adjustment of the mechanism, which Mr. Chance had arranged, there was a power of altering the degree of curvature which would be given by that cross-stroke. Upon that everything depended."

Brothers & Co. were not even allowed to tender for the metal framework, though obliged to construct such for their own use in adjusting the glass. It was clear that, under this system, when the glass came to be set up in a lighthouse in a different framework and by other hands, the pains taken by the contractor to ensure accuracy of adjustment might be thrown away. Or, as happened in some cases, an unsuitable lamp might be provided, or the bars of the lantern be so placed as to intercept some of the light. Nor could allowance be made for the "dip" of the horizon. "Nothing," said James Chance in 1867, "could be more unscientific than the system which was, until a recent date, frequently practised by the lighthouse authorities of this country."

The importance of this question caused the Commissioners to make it at once the object of their special inquiry. Early in 1860 they circulated among a few leading men of science and experts in lighthouse apparatus a set of questions intended to elicit opinion on the propriety of giving to the manufacturer of dioptric apparatus, to guide him in making his adjustments, information as to the height of the light above the sea and the horizontal arc required to be illuminated. In a letter of March 7, 1860, Admiral Baillie-Hamilton, Chairman of the Commission, invited James Chance's "individual and special" attention to the points of inquiry, expressing desire that the Astronomer Royal should meet him and compare views. In another letter the Admiral hoped that the two would not confine themselves to a simple answer to the questions, but would go further and "suggest other and perhaps more important data as necessary to be furnished to the manufacturer." Again, on March 24, he wrote that the Commissioners "particularly desire to have Mr. James Chance's answers—such as he may be disposed to give—to the questions," and any additional observations or suggestions that he might be inclined to make.

Airy gave two days at the beginning of April to thorough examination of a large apparatus in course of construction at Spon Lane for the Government of Victoria, presumably that of the first order sent to Gabo Island. He found the individual prisms to be all properly curved and well adjusted, and could not say that one was better than another. "Each panel of prisms that I observed appeared excellent." No light-frame, he believed, had ever been examined so well before, and he commented on the "general excellence" of Messrs. Chance's methods, the necessity for a final examination when all the panels were united, and the importance of not being bound by the fixed rule adopted by the engineer.

> I examined carefully (in the day) the mathematical process on which is founded the experimental process by which the curvature of the curved reflecting side is examined. It appears quite correct. Subsequently I saw the testing of one of the external rings of a lens in the long gallery. This was going on as a matter of daily manufacture, and was not put up for my edification. It was excellent. I had no idea that a ring could be ground to do its duty with so much accuracy.

Among various questions discussed on this occasion was that of the proper height of the lamp-stand. Faults observed disappeared when the lamp, "adjusted by the engineer's usual rule" was raised five-sixteenths of an inch from the customary English height to the full French height.

During the next two months the Commissioners, with Airy, examined again, in the light of their experience, various British and French lights, and among them the two of Messrs. Chance's construction at Whitby. Again the Astronomer Royal was full of praise for the quality of the glass and the accurate formation of the prisms, pronouncing the whole "a most beautiful piece of work, possible only where the maker is a man of science and also a practical man." Only in the southern lighthouse he thought the details of the form of the reflecting prisms bad, and his impression was that they were of little use. Adjustments to the horizon he found to be all wrong.

> My impression is that in the north lighthouse three-fourths of the light is absolutely thrown away, and in the south lighthouse nine-tenths of the light is absolutely thrown away. . . . When, with a ruler, I covered the part of the flame which merely gave light to the sky, it was absurd to see how little was left for the useful part. ... It really gave me a feeling of melancholy to see the results of such exquisite workmanship entirely annihilated by subsequent faults in the mounting and adjustment.

He expressed the hope that, while the state of these lights must be made public, this should be done in such a manner as to throw no blame on Messrs. Chance, whose workmanship, as shown in the glass, was admirable, or upon the engineer's work in the framing and mounting, which appeared to be of the highest order. The necessary statement should be made "in such a shape as would prevent the commission of any injustice or the excitement of any painful feeling."

Further, Airy expressed a wish that James Chance should be called in to assist personally in readjustment of the Whitby lights. He wrote to him on June 28: "I very much wish that I could induce you to look at the Whitby lights. I think that it would lead to an extensive and beneficial revolution in lighthouses." And to Admiral Hamilton:

> I enclose a letter, which I have just received from Mr. Chance. It is clear, I think, that by judicious co-operation with him we may do much to improve the lighthouses. . . . The Whitby light is the most flagrant instance of mismanagement. The constructor of every part of the Whitby apparatus is at hand. The said constructor is willing to go heartily into the improvement of the Whitby light. Therefore, leave all others and rest on it.

Next day Admiral Hamilton, Dr. Gladstone, F.R.S., one of the Commissioners, the Astronomer Royal, and James Chance met at Blackheath. Airy stated that he and the last-named were agreed upon the best method of remedying the defects at Whitby. As the result, a letter was prepared inviting the Elder Brethren of the Trinity House to meet the Commissioners and others at the North Foreland and Whitby lighthouses some time in the month of August.

James Chance was now fired by the attraction of scientific work undertaken in the service of the country. During the next twelve years of his life he made the subject his absorbing occupation, to his own honour, indeed, and that of his firm, but to the detriment of its more profitable undertakings, from which his attention was necessarily diverted. Heavily burdened as he was with private business, and greatly interested in various public matters, he yet gave up to the construction and improvement of lighthouse apparatus nights as well as days at the works, and occupied himself at home for hours together in solving the novel and intricate problems which presented themselves, and in working out the elaborate calculations required for each new design. He personally supervised every detail of the work; with each new light the reputation of his firm for excellence and accuracy of work increased through his exertions; and he was acknowledged an authority on the subject second to none.

The conferences at the North Foreland and at Whitby took place on August 2 and 9 respectively. They were attended by representatives of the Trinity House, with Faraday, of the Northern Lighthouse Board (Thomas

Stevenson), and of the Irish Board. Sautter, who had made the North Foreland apparatus, was present on both occasions, but James Chance, with Masselin, only at Whitby. When the opinion was advanced there, that in spite of the defects observed from within the lantern the lights would nevertheless be seen properly from the sea, observation was made thence. The result was to prove that in both lights the lower reflecting prisms were all but useless, while in the case of the southern light impression was given that no rays were received from the central refracting zone at all. Indeed, the well-adjusted upper reflectors of the northern light were judged to be equal in efficiency to the whole of the southern apparatus. Summarising under fifteen heads the defects of the Whitby lights, the Commissioners thought proper to add to their report the following note:

> It is due to Mr. James Chance to state that the orders given to him are simply to construct a certain well-known apparatus of a given size. Up to the time of the commencement of our inquiries he had not directed his mathematical researches into investigations connected with the scientific questions bearing on the subject. Mr. Chance was never informed of the height of a proposed lighthouse; and that very inferior description of lamp, the fountain, was ordered of another firm, leaving him no option in the matter.

Professor Faraday, they said, and the Elder Brethren of the Trinity House, had always disclaimed to be considered opticians. They had depended on Fresnel's calculations, and supposed that adjustment after his rules was applicable to any height of flame and to any elevation of the lighthouse.

Full technical discussions of the questions at issue followed in correspondence of James Chance with Faraday and Airy. The former visited Spon Lane and worked with him for two days on the determination of focal points. In one of two large lights under construction for Russia he found that James Chance had, of his own judgment and experience, adjusted the prisms to unusual foci, with "very excellent" effect. In fact, those calculated for the lower reflecting prisms coincided practically with his own determinations. But for the upper prisms, he said, their conclusions differed considerably, and the matter required further investigation. On this report the Trinity House requested Messrs. Chance to put up at their works a panel for the purpose of experiment. That done, Faraday came again to make the desired observations in September.

Meanwhile the firm had formally requested permission from the Trinity

House to make alterations at Whitby south lighthouse at once. The method of procedure having been arranged between James Chance and Faraday, they proceeded thither, and the work was finished early in October. Faraday reported:

> All the time we were at Whitby (eight or nine days) Mr. Chance and myself were occupied in learning, practising new methods of adjustment and correction and using new instruments; and I cannot say too much in thanking Mr. Chance for the earnest and intelligent manner in which he has wrought with me in the experiments, working and thinking every point out. The method of adjustment is now so perfect, that the authorities can hardly require more accuracy than the manufacturer can ensure. The Trinity House may direct at its pleasure that the light of one part of an apparatus shall be thrown chiefly in one direction, as the sea horizon, and that of another part in another relative direction, as nearer to the coast; and I have no doubt that if the electric light, or any other of the compressed intense illuminations, be hereafter adopted, the principles and methods of adjustment now devised and carried into practice will prove of very great and special advantage.

The method adopted was that of "internal observation"—that is, looking at the horizon through each lens and prism in turn from within the apparatus. It was not new, but disused, and had been revived in the course of the work of the Commission.[1] Airy, who had employed it when examining the Cap d'Ailly light, sent to James Chance his hearty congratulations. Referring to his own previous trials of other methods, he said : "Last came the simple notion of merely looking with the eyes. Simplicity always comes last."

The method was perfected by James Chance at Whitby by a device which rendered it unnecessary to observe the horizon itself, and enabled the final adjustments to be made at the manufactory. The horizon being obscured during several days by haze, it occurred to him to fix a vertical staff upon the cliff and to mark the line of the horizon on it, graduated to correspond with the successive zones of the apparatus above and below the middle refracting belt. But this could be done equally well, by calcula-

---

1   The usual practice at Spon Lane had been "to place a white ball or a minute gas flame in an assumed conjugate focus of a lens or prism, and the eye of an observer in the other conjugate focus at a short distance outside. The whole apparatus was tested in like manner, and the difference between the conjugate focus for the distance and the focus for parallel rays was calculated." At the works of Messrs. Sautter et Cie. the upper reflecting prisms were set "by looking from the outside along a spirit-level at the centre of each prism in turn, and at the reflected image of a red ball suspended in the centre of the apparatus, and reflected by the prism." (*Report of the Commission*, ii. 627.)

tion, at the works; on trial there, the method was found to ensure perfect accuracy, and it has been in use ever since.

On examination of the result by the Commissioners and by a deputation from the Trinity House it was seen that whereas, before, the northern unaltered light had been somewhat superior to the southern, now hardly any difference was perceptible at the greatest distance of observation, while at a few miles' distance it was manifestly inferior. Thereon the Trinity House instructed Messrs. Chance to proceed with permanent alterations at the south lighthouse at once. James Chance was able to send in his report on them to Faraday on November 17, whereon the latter wrote:

> I have not seen the lighthouse since the adjustments were made, but they were made by Mr. James Chance himself, and I have the fullest trust in him. Everything thus far confirms me in the opinion that what the Trinity House has done in this case has been done well; that every future case can be considered in relation to the adjustments necessary for it from the very beginning; and that the adjustment can be carried out with certainty.

There ensued a long discussion on proper focal heights, in which the Messrs. Stevenson took an important but dissentient part. In the course of it, in January 1861, James Chance sent in to the Commissioners an elaborate paper dealing with the whole question of adjustment of dioptric apparatus.[1] Admiral Hamilton acknowledged its receipt in the following very complimentary letter :

> I was reluctant to leave this office last night without having written to thank you, and to express my *admiration* of the paper you have supplied us with.
>
> If the time and labours of the Commission had had no other end, it would have been sufficiently answered in their having led to the earnest application of your talents and your time to a subject of the very last importance as regards the science of lighthouse illumination—to be mastered as that subject has been by you.
>
> Scientific men may be more minutely conscious than myself of all the value of your work, and at any rate it will stir the minds and mettle of many of them; but as even I am able to understand every axiom as well as the whole theory in your clear and complete treatise, I can yield to none in appreciating its merits, and in the feeling of satisfaction at its being thus given to the world.

---

1   See *The Lighthouse Work of Sir James Chance*, pp. 29-43.

And in the discussion on James Chance's paper of 1867, cited, Admiral Hamilton

> could not say how much the Royal Commissioners were indebted to him. . . . The Trinity House, and others who were interested in the maritime concerns of the country, were aware how much was owing to him, and he himself considered himself fortunate in having been at the head of an inquiry in which the services of such a man as Mr. Chance could be made available.

One of the principal matters in debate had been the best type of lamp. Three were in use: the fountain, supplied with oil by gravitation; the pump, in which the oil was forced into the burner by pumps actuated by clockwork; and the pressure or moderator, where the oil was forced up out of a cylinder by a weighted piston. The last two, when properly worked, gave the free overflow admitted to be necessary, but the first did not. Yet the Commissioners found this lamp in universal use in England and Ireland, and its inefficiency, especially in contrast with the French and Scotch pump lamps, struck them forcibly. At Whitby south lighthouse, for instance, they found that the proportion of oil overflowing to that burnt was as 1 to 4, whereas in Scotland it was about 3 to 2, and in France 3 or 4 to 1. As the result, the effective flame at Whitby attained but one-third of the height for which the optical apparatus had been designed, and if the reservoir that supplied the oil were raised, they said, "the flame immediately rises, the pipes quickly become hot, the specific gravity of the oil in the rising branch is diminished, the influx of oil is increased with great rapidity, and the flame becomes extravagantly high, smoky and unmanageable." Moreover, for the sake of economy, the lightkeepers were instructed to keep down the consumption of oil as much as possible, and this, the Commissioners discovered, had led to the rejection of the pump lamp by the Trinity House; used on this principle it had been found unsuitable.[1]

Faraday recommended a pair of lamps for the Whitby south lighthouse, such as would give a plentiful overflow, and Messrs. Chance were commissioned to supply them as part of the alterations. They supplied accordingly two pump lamps, which gave great satisfaction, but James Chance, in his report on the alterations, stated that he proposed to send ultimately a pair of pressure lamps preferred by him as simpler in construction. These were of a new type contrived by Masselin, the principal improvement be-

---

1   Report, i. 65.

ing that the weights were placed outside the cylinder and underneath its centre, instead of within the piston. Rigidity was thus obtained, and the weights could quickly and easily be altered. The whole construction, also, was very solid, which that of the old lamps was not. The new ones, after thorough testing, were approved by the Trinity House as the best;[1] two of them were ordered for Whitby, and another for St. Catherine's in the Isle of Wight, where the keepers promptly asked for blue spectacles. In the course of the winter, Masselin contrived further improvements, which he explained in a paper read before the Society of Mechanical Engineers on April 24, 1862. In 1869 James Chance was able to say: "We have supplied our pressure lamps to all parts of the world, in many cases to places quite isolated, where no repair would be possible, and we have never yet heard of failure." In fact no further improvement was necessary until the substitution of mineral oil for colza obliged new methods for its combustion.

The next apparatus to be treated on the new principles was one of the first order under construction for the Smalls Rock, near Milford Haven, whose completion had been delayed pending the experiments on focal points. When it was ready, in January 1861, James Chance expressed his earnest desire that some of the Trinity House Brethren should come down to inspect it, "because the adjustment of the lenses is so entirely novel." He further stated:

As any inferences drawn from practical results at sea would be valueless, unless the final erection at its destination is performed with perfect accuracy, I would suggest to your Board the expediency of making our firm responsible for the final erection. As the matter now stands, the responsibility of our firm closes with the examination by Professor Faraday.

The inspection took place on January 28. Faraday reported on it, after explaining the adjustments made:

The apparatus has been put together by Mr. James Chance with these adjustments, and being in a proper place I had the focimeter set upon the burner, and a true sea-horizon

---

1   Captain Nisbet wrote on April 25, 1861: "The result of the experiment last week has surprised us all. . . . We are all most favourably impressed with your lamp, but the small consumption of oil is rather astonishing in comparison with other lamps." And it was stated officially: "The superiority of the action of the lamp over the others is so evident, that the Light Committee strongly recommend the principle to be adopted wherever a new pressure lamp may next be required."

mark placed in the distance. The whole was so true that the ray proceeding to the eye through the middle of each piece of glass passed by the focimeter at the point desired. The greatest departure was but 2 mm, and very few of these occurred. Further, the manner in which, as the apparatus revolved or the eye was moved about, the object at the horizon passed laterally from one panel to another, or vertically from one rib to another, showed the perfection of the adjustment of each individual piece by the harmony and consistency of the whole, though there were above 300 pieces of glass associated together. At night the lamp was lighted and observed from the distance; the results accorded perfectly with anticipations. As the head was raised or lowered, each piece of glass showed its maximum effect at the right place, its light coming in or going out as it should do in relation to the distant horizon; and I think that, as far as regards the system of adjustment, the power of carrying it into effect, and finally of examining its correct application, everything is proved to be practicable, and has here been realized. The essential points now are to supply a good lamp, and to provide that it be kept in good order. In relation to colour and striae the glass was very good.

After this came a formal request from the Trinity House (May 22, 1861) that James Chance would join in an examination of all the dioptric apparatus under its charge and in the execution of any changes that might be necessary. He willingly assented, and as a first step accompanied Captains Baily and Nisbet on a week's tour of inspection of the Skerries and other lights on the Welsh coast.

But the first light chosen for correction was that at the North Foreland. On request from the Trinity House for its examination and for an estimate of the cost of alteration, Masselin was sent thither to report. He pronounced the state of the light to be almost as bad as could be. James Chance undertook the readjustments personally, and the work was completed on August 8. The next week was occupied by like alterations at the South Foreland. Then were taken in hand the lights at St. Catherine's, at Whitby North, and at St. Ann's, Milford Haven; in 1862 the low light at Orfordness, those at the Skerries and Bardsey Island, and the high light on Lundy Island; and in 1864-5 the high light at Spurn Head, the two at Trevose Head, and those at the Needles, the Eddystone, and the Bishop Rock. All these, excepting the Eddystone, were of the first order. Ten Irish lights also were readjusted, including those at Fastnet, Kinsale Old Head, Ballycottin, Youghal, Minehead, and Dungarvan. The best that was possible was done with all, but while one or two, as St. Ann's and the Eddystone, required little more than levelling, in many cases the faults due to wrong curvatures, to defects in the glass, and to want of solidity and accuracy in the fitting together of the panels, could only have been properly remedied by complete reconstruction. Of twenty-seven lights treated, twenty were of French manufacture.

New apparatus shown at the London Exhibition of 1862 included a first-order holophotal revolving apparatus, afterwards erected at Innistrahull, on the north coast of Ireland, two of the sixth order made for the Messrs. Stevenson, one of the fourth order, revolving, used to exhibit the magneto-electric light of Professor Holmes and afterwards sent to Demerara, and the first practical form of the totally reflecting dioptric mirror, an instrument remarkable in that to a person standing behind it all is dark, though nothing intervenes between him and the source of light but a screen of clear glass.

This mirror is composed of the doubly reflecting prisms invented by Thomas Stevenson in 1850 and first formed by him into rings by revolution of their sections round a horizontal axis. He had actually constructed such a mirror of beehive form, having a parabolic conoid at the centre. Approached by him about the construction of one for the Exhibition, James Chance proposed the simpler principle of generating the rings round a vertical axis. Advantages thus obtained were that only segments of them were required, that the awkward conoid was got rid of, and that the image of the flame was not reversed in reflection, as in the former case, but exactly super-imposed on the original, so that the back rays were reflected in the manner best calculated to augment the intensity of the front ones. Stevenson allowed this arrangement to be much the better, and explained that it had been his first idea, but that practical difficulties had determined the other construction. He wrote:

I am very glad indeed that you proposed the horizontal zones, for I now quite agree with you in considering them preferable to the vertical. The reasons you adduce are, to my mind, conclusively in favour of the horizontal.[1]

And again, after the mirror was finished:

Though I have not yet been able to see the mirror, I find from all the inquiries I have made that it is most satisfactory. It certainly does great credit to your firm, but spe-

---

1   The horizontal zones being, of course, those generated round the vertical axis, and *vice versa*.

cially to yourself. There is no need whatever for doing anything with it. I am very glad that you did not waste time in any attempt to avoid the stepped or slightly irregular internal surface, which is neither an eyesore nor theoretically any objection … It is a very beautiful specimen of most accurate workmanship and does you every credit.

The difficulties encountered in constructing the mirror of hemispherical form led James Chance, before it was finished, to devise for others a simpler form approaching to the cylindrical and occupying much less space. To quote his paper of 1867:

> During the progress of this instrument the idea occurred to the Author of separating the zones, and also of dividing them into segments, like the ordinary reflecting zones of a dioptric light; by this means it became practicable to increase considerably the radius of the mirror, and thereby to render it applicable to the largest sea-light, without overstepping the limits of the angular breadths of the zones, and yet without being compelled to resort to glass of high refractive power. The separation of the zones also rendered it feasible to avoid giving to the aggregate structure a spherical shape, which would have encroached most inconveniently upon the space required for the service of the lamp.

Or, as he wrote to Stevenson in the same year (June 1):

> What rendered the mirror a lighthouse instrument was the separating the zones and dividing them into segments. . . . The 1862 instrument gave me so much trouble in consequence of the joints, that I was driven to the separate arrangement, and the segmental plan followed and the consequent dispensing with flint glass arising from the additional practicable diameter combined with the segmental division. My own reason for adopting the horizontal rings was mainly in the first instance to avoid the conoid.

Mirrors in their new form, with further improvements, were applied in the following years to the lights at Cape Saunders and Tairoa's Head in New Zealand, at Double Island in the Bay of Bengal, at Buddonness (River Tay), at Ushenish in the Hebrides, and elsewhere.[1] The last-named and another made for the Trinity House were shown at the Paris Exhibition of 1867, when it was found that they increased the power of a first-order light by about one-third.

At that Exhibition the firm showed also, besides a number of small lights

---

1   See *The Lighthouse Work of Sir James Chance*, pp. 77-82.

intended for the River Hooghly and elsewhere, two apparatus of the first order, one fixed and one revolving, a duplicate of the remarkable third-order light recently made for Messrs. Stevenson for Buddonness, which found a home in the Edinburgh Industrial Museum, and another of the third order for use with Professor Holmes's electric arc, also fitted with a dioptric mirror. This was afterwards set up by the Trinity House for experimental purposes at its wharf at Blackwall, and was subsequently utilised in the construction of one of the Lizard lights, of which below. Results were triumphant. Although inferior to the French apparatus in finish and in the quality of the glass, photometric tests of the lights showed a clear advantage in efficiency. With those of the first order the superiority amounted to 5½ per cent, for a refracting panel, and to nearly 7¾ per cent, for a complete revolving panel. A gold medal was awarded for the exhibit. The men employed for the installation were Edwin and William Stokes, William Kerr, and young Mangematin, working under the superintendence of David Henderson, the firm's chief draughtsman, and of Peter Rigby.

The number of lights of the first to the fourth orders made at Spon Lane under the direction and mostly from the designs of James Chance, during the years 1860 to 1871, exceeded 180. Of the first order there were 52, of the second 29 and of the third 31. As the most notable of them may be remarked the revolving red light for the Hanois Rocks, Guernsey; the first-order fixed light at Great Orme's Head, where vertical refracting prisms threw into a red beam, over an arc of 90 in a certain direction, rays which otherwise would have fallen uselessly upon the land; the very important first-order fixed light at Europa Point, Gibraltar, similarly throwing a strong red beam over the shoal near the Pearl Rock; lights of the fourth order for the entry to the River Dart and for Somes Island, New Zealand, showing white light for the fairway channels and red and green to illuminate the dangers on either side; the third-order Buddonness light, remarkable, says Thomas Stevenson, "at the time it was made for containing every kind of dioptric agent then known," including his new right-angled expanding prisms, arranged in a half-cone above the apparatus; the Lochindaal and the Stornoway "apparent" light, with their "back prisms"; the red and white revolving lights for the Wolf Rock and Flamborough Head; the revolving electric light at Souter Point, with its peculiar arrangements for distribution of the rays and its secondary reflected beam below; and the two great electric lights at the South Foreland. The apparatus at Europa

Point, with its huge vertical condensing prisms, the present author saw in perfect preservation more than forty years after its erection.

The Board of Trade had first accepted for this Gibraltar light the somewhat lower tender of Messrs. Sautter, to the great annoyance of James Chance's friends at the Trinity House. Captain Nisbet wrote to him on April 6, 1863:

> I cannot tell you how disappointed I was, and also several others, that your tender for the Gibraltar lens was not taken, as we should then have had the advantage of your latest improvements. The Board of Trade, who unfortunately hold the purse strings, insisted on the lowest tender being accepted, although the Brethren protested against it, and a paltry £100 made all the difference in the eyes of the people at the West End, and M. Sautter has it.

And Captain Arrow on August 21:

> I wish to goodness you had had it for many reasons, and especially because it is at a place where there are more visitors of all nationalities in a week than any other lighthouse sees in a week of years. It was specially on that account I advocated the pressing the Board of Trade to take your tender as the English maker, but the Elders overruled me.

In consequence, when invited to alter the French light at the Eddystone, whose condition, in Captain Arrow's opinion, was "enough to make your hair stand on end," James Chance wrote (September 11):

> It seems rather inconsistent that we should be concerned in any way in altering an apparatus made at Paris, and yet that, when a light is wanted for an important post like Gibraltar, the work should be entrusted to a Parisian maker in consideration of a paltry difference of price, but regardless of the superiority of the English apparatus and also regardless of the very great discouragement to the English manufacture itself. Also, whether it is decided to have a new lens at the Eddystone or to alter the old one, it is wiser for our firm to have nothing to do with that decision.

However, in the end the Board of Trade, intimating that the concession was not to be taken as a precedent, allowed the apparatus to be made at Spon Lane. The work was begun at the end of 1863 and completed at Gibraltar in November 1864. It proved to be what the Elder Brethren had desired, the best of its kind. Sir James (then Mr.) Douglass spoke of it in the discussion on James Chance's paper of 1867 as a signal instance "of care in design, great perfection in the material of the glass portions, and optical accuracy in construction." The work, he said,

> was so accurately performed, that when the light was tested by a Committee of the Elder Brethren of the Trinity House the line of demarcation between the white and red lights at the Pearl Rocks was found to be identical with that determined on, and not the slightest alteration was required in the adjustment of the apparatus.

And Captain Arrow spoke on the same occasion to like effect. Nautical men generally accorded the light the highest praise, and it contributed greatly to enhance Messrs. Chance's reputation.

The Wolf Rock light was first shown on January 1, 1870. Captain Lethbridge, of H.M.S. Simoom, who saw it from the sea next day, reported to the Admiralty:

> When abreast of the Lizard light we sighted the light of the Wolf Rocks at a distance of 23 miles . . . The night was fine and clear . . . Both the white and red lights are very good, and throughout my experience of Channel work I have never seen any red light to equal the red light on the Wolf Rock in brilliancy. I was so struck with the superiority of this light that I sent down for the officers to come up and see it. I think the Hydrographer of the Navy would be glad to know what a success this light is, setting aside the importance of it.

And Mr. Douglass, in his account of the light, termed the optical apparatus "probably the most perfect for the purpose that has yet been constructed." Like praise was accorded to the electric lights subsequently erected at Souter Point and the South Foreland. Of the former Mr. Douglass wrote in his paper of 1879 on "The Electric Light applied to Lighthouse Illumination":

> The Author desires to express his admiration of the excellent fulfilment by Mr. Chance of the optical requirements for both the upper and lower lights. Every portion of the apparatus acts in the most efficient manner. The beams have their vertical and horizontal limits accurately and sharply defined; and those of the upper light, with a uniform intensity at full power of 230 times that of the electric luminary, or about 700,000 candles, are probably the most effective that have yet been sent from a lighthouse.

In fact, in every case the work was executed to the perfect satisfaction of the authorities. Captain Arrow wrote on July 21, 1868:

> Your last new lights have been thoroughly appreciated, and wherever changes have been made they have been the admiration of seamen.

In the case of the South Foreland lights, first shown on January 1, 1872, the full power of the high light, in the most illuminated plane, was estimated at 152,000 candles, and that of the low light at 131,000, being 20 and 90 times respectively the power of those, the one dioptric, the other catoptric, previously in use.[1]

These lights were among the last constructed under James Chance's personal direction. His firm secured in 1872 the services of Dr. John Hopkinson, the brilliant Senior Wrangler of the previous year, and he, after a short time, took over the management of the Lighthouse Works. The next apparatus for the electric light ordered by the Trinity House, those for the Lizard, were designed by him. He said about them in the discussion on James Chance's paper of 1879 :

> The end to be obtained at the Lizard was closely similar to that of the two Foreland lights. He happened to be well acquainted with the details of them, and with the reasons which determined those details, for when he first became interested in the optics of lighthouses Mr. Chance had placed in his hands the calculations concerning those apparatus, to give him an insight into the methods of design which he had successfully practised. He had also examined the finished apparatus, when Mr. Chance had initiated him into the system of trial adjustment and testing which he had introduced and constantly used. The South Foreland lights had been a great success, and he was naturally anxious to suggest improvements on a good model. He therefore gave some time to reconsidering the whole question, but finally had to confess himself baffled, and to admit that there was no room for display of any originality, but the best he could do was to copy the South Foreland, making only such small alterations in detail as the slight difference in the circumstances demanded.

---

1   Mr. Douglass in his paper of 1879. Dr. Hopkinson, in his account of the Macquarie and Tino Island lights rendered to the Institution of Civil Engineers on December 7, 1886, said: "The practice of appropriating certain elements of the apparatus to different distances on the sea was first introduced by Mr. James T. Chance in the lights for the South Foreland exhibited in 1872."

James Chance's paper mentioned, on "Dioptric Apparatus in Light-houses for the Electric Light," was read before the Institution of Civil Engineers on April 22, 1879. For his previous paper on "Optical Apparatus used in Lighthouses," read on May 7, 1867, he had been awarded the Telford Gold Medal and Premium of the Institution, and elected an Associate.

Of James Chance's helpers at Spon Lane, Masselin retired at his own wish at the end of 1862, although his agreement had been renewed for five years from January 1, 1860.[1] After that time James Chance was principally aided in the engineering work by David Henderson, but certain action on his part obliged dispensation with his services in 1867. This brought Peter Rigby completely to the front. He received frequent recognition of his services in the form of presents, and his salary was raised in 1871 to £300 a year. Other bonuses were accorded from time to time to Mangematin and to William Brown. Kenward, as head of the office, continued to prove the reliance that could be placed upon him.

Room for extension of the works was provided by the purchases of adjacent land in 1859 and 1860, elsewhere mentioned. The road between the fitting shop and the two-storey store-room was ordered to be covered over to provide a place for erecting first-order lanterns; the grinding shed and the fitting shop to be enlarged; and an office for the draughtsmen and clerks to be erected within the precincts. Also it was proposed to erect on the seven-storey building on the old side, in place of the belfry, a wooden shed for the reception of such lights as it might be desired to examine from a distance. In 1867 came the further purchase of land from Benjamin Darby, which enabled the new erecting room to be built at its south end.

Dr. Hopkinson ceased regular attendance at Spon Lane in 1877 and soon removed to London, leaving the direction of the lighthouse works to Kenward, but continuing to act as the firm's scientific adviser and to furnish designs for the lights until 1894, when he assumed the position merely of consultative engineer. Four years later came the tragedy by which he, in the plenitude of his activity and fame as the foremost electrical engineer in England, perished with his son and daughter in the Alps.

James Kenward, sole manager of the lighthouse works from 1878, was no ordinary man. Capable of the duties of the position to which he had

---

1   A particular stipulation of this agreement was that Masselin should not be required to have any communication with Peter Rigby excepting in the presence of a partner.

risen by pure merit, in his leisure hours he was a poet, an archaeologist, a scholar deeply read in Celtic lore and literature. Volumes of verse and a chain of essays testify to his inspiration and his attainments. He remained in control until 1895, when he was relieved of his responsibility at the works by Henry J. S. Stobart. It was agreed that he should transfer his residence to London, there to act, principally, as negotiator for the firm with the authorities and with Colonial and foreign governments. This he continued to do, with conspicuous ability, until in 1902 his advanced age obliged his retirement. He received a suitable pension and leaving presents, one of them a silver salver with an appropriate inscription. On his death in July 1906 the firm's letter of condolence to his widow recalled the many years during which, in the words of the Board minute, "his ability, uprightness and considerate behaviour had earned the esteem of all with whom he was brought into contact."

Principal draughtsman in Kenward's time was Alfred Perry, a man of genial presence and worth not soon to be forgotten by those who knew him. He died in 1893, and was succeeded in his office by W. F. A. Richey. Peter Rigby had been removed from the manufacturing in 1873, when Dr. Hopkinson had found it impossible to work with him. Mangematin died in 1889 at the age of 76. On his retirement some years previously he had been replaced as foreman of the fitting shops by William Kerr, a leading workman for many years. In 1893, in view of Kenward's advancing years, it was deemed advisable to have in control of the shops a trained engineer, and Jenner G. Marshall was appointed. Next year thorough reorganization of the works was undertaken. A number of resolutions appear: to give Marshall more effective control; to abolish piece-work; to allow the leading hands[1] one per cent, of the profits for the credit of their deposit accounts; to reduce the number of men; to re-arrange wages; to close the general fitting shop and transfer all the work and some of the machines to the lighthouse shops proper; and to pension off the worthy old Frenchman Gilbert, handing over the grinding and focussing processes to James Smith and George Thomas Lane, respectively, in his place. Some of these changes were carried out, some not, or were but temporary; the "leading hands" arrangement, for instance, was abandoned as unsatisfactory in 1898.

Dr. Hopkinson's first large lights were the two for the Lizard, men-

---

1   William Kerr, John Henry Tickle, William Allan Hughes, John Beech, Albert Pendrous, David Parish, William Rathbone, and John Hill.

tioned, fixed and of the third order. Another apparatus for the electric arc, small but of importance, which followed, was that supplied for the signal-light in the Clock Tower of the Houses of Parliament. A six-sided holophotal revolving light, of special construction and intended for the South Stack Rocks at Holyhead, was shown at the London Exhibition of 1874. When it was put up, the Harbourmaster at Holyhead reported: "The captains of the steamers sailing in and out of this harbour tell me that the new South Stack light is truly splendid. They are delighted with it." Of another of the first order erected at Heligoland the Governor of the Island wrote (May 1, 1876):

> The Hamburg Naval Authorities, and the Heads of the Pilot Department at Cuxhaven, have strongly expressed to me their appreciation of the new Light which has just been placed by the Trinity House at Heligoland, and I also hear from the Heligoland steamer, as also from various Merchant Captains passing to the Elbe and Weser, that the new Light is a subject of admiration and gratitude to every one.

In 1874 Dr. Hopkinson introduced his great "group-flashing" innovation. By this invention the single beams of a revolving light are divided into groups of two or more rapidly succeeding flashes, by whose variations in number and period, combined with the usual variation of the beams themselves, a vast facility of distinction among sea-lights is afforded. Before long the use of coloured media for the purpose of characterization, objectionable always on account of the great amount of light absorbed, was almost completely obviated. As Mr. Kenward wrote in 1896:

> The revolving lights on a coast can now be unmistakeably distinguished by their single, double, triple or quadruple flash. In each case sufficient intensity is preserved for long ranges, and abundant opportunity allowed for taking a bearing from the group flashes.

The first lights of this character made at Spon Lane were for the Little Basses, Ceylon, and for the Gulf of Mexico, in 1875, for the Casquets, shown at the Brussels Exhibition of 1876, and for Vlaake Hoek, Java, in 1877. The first was double, the others triple-flashing. Excepting the Gulf of Mexico apparatus they were of the first order. The Casquets light "enabled the Trinity Corporation to dispense with two of the three lights hitherto employed, and to show from one tower a half-minute light in triple flashes, each lasting two seconds, each interval between them three seconds, and the long interval between the groups eighteen seconds."[1]

These lights, and many others of the kind that followed, were holophotal. Before long it was found possible so to extend the vertical angle subtended by the refracting portion of the apparatus as to dispense with the reflecting prisms, when so desired. Says Kenward:

A remarkable variation of the usual elements of a dioptric sea-light dates from 1879 or 1880. Lower prisms for sea-lights had, at the suggestion of the writer in 1874, been suppressed on several occasions; and for port-lights Messrs. Chance had dispensed with all prisms, and raised the lenses to a vertical angle of 80°. But now it was determined to produce a first-order apparatus with refractors only, extending the vertical angle to 920 from 560 or 570, the old normal height. This was attained by Messrs. Chance by means of dense flint glass in the superior and inferior limits. The power of the lenses, always counting for 75 per cent, of that of the complete light, was thus considerably augmented, while the cost and bulk were reduced, though doubtless at the expense of symmetry. The first order lights, Anvil Point (Dorset), the Eddystone, and the Minicoy (Indian Sea) were constructed on this principle at Birmingham (1880-83).[2]

Of this Eddystone light, twelve-sided and double-flashing, we read:

As scientific adviser to the Trinity House Professor Tyndall examined the apparatus at Messrs. Chance's works and pronounced it to be in every way a most excellent piece of workmanship as regards the glass, and a perfect and beautiful specimen of optical adjustment.[3]

One advantage of these "lens-lights" was that their form enabled them to be piled, if desired, one upon another, thus doubling, trebling, or even quadrupling the power of the light. To quote Kenward further :

The plan of superposed lenses was first suggested, in 1859, by Mr. J. W. D. Brown, of Lewisham, and first practically set forth, in 1872, by Mr. John R. Wigham, an engineer of conspicuous ability, in connection with his large gas flames for Irish lighthouses; and it has been since fully approved and adopted by the Trinity House. The great lights of Galley Head, Howth Bailey, and Rockabill attest the excellence of this arrangement of lenses, and the Eddystone biform is not less successful.

---

1   Kenward, *A Review of Lighthouse Work and Economy in the United Kingdom during the Past Fifty Years* (1837-1887), p. 11.

2   Ibid, p, 13.

3   Price Edwards, *The Eddystone Lighthouse*, pp. 22, 28.

Next to be noticed as of Dr. Hopkinson's design are the three great electric apparatus made for South Head, Macquarie, Sydney, 16 panels, revolving (1880), for Tino Island, Spezia, 24 panels, triple group-flashing (1884), and for St. Catherine's in the Isle of Wight, single-flashing (1888). The Macquarie light, of the first order and of seven to eight million candle power, could well claim, when it was erected there in 1882, to be the most powerful in the world. It presented a novelty for electric revolving lights in that the condensing effect was produced by means of a single agent, the vertical prisms usual for the purpose being dispensed with, and so absorption of the rays lessened. Kenward wrote to James Chance in 1883:

> The Macquarie light is always spoken of in the very highest terms. It is said to be one of the constant attractions of Sydney, and the theme of praise from all mariners. It is below the horizon at 60 miles, but at this distance the reflection on the sky is intensely visible.

The Tino light, of the second order, was observed from a mountain behind Savona at 73 miles distance, and was judged brighter at that distance than the noted "Lanterna" at Genoa at 24 miles. Kenward (in the same letter) was

> inclined to think its record better than any other as yet. When the direct beam of a light can be seen in moonlight from a station 73 miles distant, and in a " quasi-opaque " atmosphere from a station 30 miles distant, it would seem difficult to find, and needless to wish for, anything more potent.

For the St. Catherine's light—the "queen of British lights" in his expression—also of the second order—the apparatus was composed of refracting panels only, sixteen in number. The intensity of the white beam (there were two red arcs) was estimated at between six and seven millions of candles.[1]

A fourth electric light of the foremost rank was constructed to the designs of the Messrs. Stevenson for the Isle of May in 1885. In this apparatus, says David Stevenson, the condensing principle—the principle which

> consists in darkening certain sectors by diverting the light from them and throwing it into the adjoining sectors so as to reinforce their light—was carried further than

---

1   Kenward, *Our Lighthouses of the English Channel*, British Association, 1899.

in any previously constructed, a quadruple group-flashing effect being produced by vertical condensing prisms of first-order radii revolving round a fixed-light apparatus of the second order.

The power of the beam was estimated at three million candles with one electric machine in use, and at six million with both.[1]

A further development in flashing lights was so to condense the beam in amplitude that it affected the eye of the mariner just as might a flash of lightning; hence the name for this French invention, "feu eclair." By the physiological effect of the sudden flash its practical power is augmented in a very high degree. Kenward wrote in 1897:

> The principle of luminous compression and instantaneous integral perception has been well known to European physiologists of the present century, but the application of it to lighthouse construction is due to Mr. Bourdelles, the distinguished head of the French Lighthouse Administration, under whose direction several striking specimens of the feu eclair system have been constructed in recent years. To attain—at least where oil is the illuminant—the utmost advantage of the " lightning light " in respect to intensity of beam three conditions are necessary, (1) optical panels large in size and few in number, (2) a burner of adequate volume and power, (3) rotation of the apparatus at a high speed.

It was decided in 1895 to construct on this principle a first order oil light for Cape Leeuwin, in Western Australia, which had lain under consideration for ten years. To quote Kenward's description:

> The light is composed of a "bivalve" optical apparatus, the first of this size and this arrangement yet made. . . . The two panels, separated by a space of 114°, are mounted on a carriage made to revolve by suitable clockwork in such a manner that the flash from each of them lasts one-fifth of a second, with an intervening eclipse of 4 4/5 seconds, the period being thus 5 seconds, and the time of revolution 10 seconds. This represents double the time of a usual French "feu eclair" arrangement hitherto adopted. It has in England been judged prudent not to adopt for the present the minimum duration of flash, but gradually to accustom the mariner to the reduction from two, three or four seconds to possibly one-tenth of a second.

The necessary speed of rotation was obtained by flotation of the apparatus on a bath of mercury. The power of the flame of the six-wick burner was

---

1   David A. Stevenson in his account of the light read before the Institution of Mechanical Engineers on August 3, 1887.

about 850 candles in the focus. To continue the quotation:

> The intensity of the beam through the lens cannot now be exactly stated, but it may be provisionally estimated as of at least a quarter of a million of candles. This high value results altogether from the "feu eclair" principle. It does not express the actual photometric intensity of the panel at rest, but rather the light which enters the eye during a fraction of a second from the same panel rotated at great speed, and of which the eye has "integral perception." The concentration of the beam into one lightning-like flash, where nothing is wasted, and the very rapid alternation of light and dark, are conditions clearly excelling those of the slowly-moving panels, where light from being unduly prolonged is wasted, and where the tedious succession of beam and eclipse is far less impressive. Physiological principles are indeed the basis of the "feu eclair" system.

A largely extended experience must, however, be invoked ere the measure of advantage to the mariner arising from this beautiful expedient can be adequately determined. Meantime it is noteworthy that, by using the electric light, powers of 20 millions of candles have been realised by the French authorities with small optical apparatus, and that in the great light to be erected on the Breton coast— the phare d'Eckmühl—the beam directed to the horizon will attain, it is expected, the almost fabulous figure of 40 millions of candles.

Commander P. Harris, of the P. & O. steamship China, after seeing the Cape Leeuwin light for the first time at a distance of 24 miles through mist and in full moonlight, reported:

> Given a dark clear night its rays, I feel confident, would be seen from this vessel's deck at least 30 miles away. It is without exception the finest oil light I have seen. The quick flash is grand, it compels attention, and once seen will leave no uncertainty as to where is the Cape.

A third-order light on the same principle by Chance Brothers & Co. was shown at the Fisheries and Nautical Exhibition of 1897. It was said in its description:

> The clockwork and the mercury-float carriage, by means of which the effect of weight and friction is minimized and rapid rotation ensured, have been raised by the constructors to a much greater degree of efficiency than in anterior types. A revolution of the Light is completed in 5 seconds. The duration of the flash is from 1/9th to 1/10th of a second. The mean apparent power of the flash to an observer's eye at sea is estimated as equal at least to 100,000 candles. These figures represent double the

intensity of many established revolving lights of the first order on the old system. The beam would be brightly visible in ordinary weather at 30 miles, at a suitable elevation. But it must be remembered that the physiological value of a feu eclair, depending largely on rapidity of rotation, may be largely reduced when other characteristics, demanding smaller panels and slower movements, are in question. All that can at present be fairly maintained is that without displacing well-known and well-tried forms the feu eclair is an important and valuable addition to the resources of the Lighthouse Engineer.

Power of distinction similar to that conferred by "group-flashing" upon revolving lights was obtained for fixed lights by the "occulting" system, under which the rays are suddenly cut off at regulated intervals whether by revolving shades, by a dropped screen, or in the case of gas-flames by cutting off the gas. The two systems have been described, the former as darkness interrupted by light, the latter as light interrupted by darkness. The first occulting apparatus, one of the second order, was made for St. Tudwal's South Island in Cardigan Bay in 1876. Three more followed in 1878, one of the same order for Point Lynas, Anglesey, and two of the fourth and fifth orders for Malmo, in Sweden; while in 1880 two of the first order were constructed for China. But the principal use of these occulting lights is for harbours and pier-heads. At the International Maritime Congress of 1893 it was stated that they had solved this problem, at least for Italian harbours. After trial of two near Messina, said Signor Domenico Lo Gatto, "their success both from the technical and from the economical point of view was complete, and they were quickly adopted at Genoa, Naples, Cagliari, Catania," and many other ports. In some cases, as of the lights for Sunderland North East Pier (1888), Barry Dock (1889), and Tynemouth South Pier (1895), lights respectively of the fifth, fourth and third orders, the occulting mechanism was actuated automatically by clockwork.

When Kenward entered on his sole administration mineral had displaced vegetable oil as an illuminant. After Captain H. H. Doty had constructed a multiple-wick burner for it, giving a steady and brilliant flame, its use had been authorized for the Pentland Skerries (1871) and for Flamborough Head (1872). After that its employment quickly became general. James Douglass's early four-wick burner gave an intensity of 415 candle-units, as against the 230 of the four-wick vegetable oil burner of Fresnel. In the course of years he arrived at a nine-wick burner with a power of 1,785 candles. Meanwhile Wigham had introduced the use of gas, with yet greater results. His burner of 108 jets, arranged in concentric circles, gave a flame of 2,433 candle power, to be outdone by Douglass with a ten-ring burner giving an equivalent of 2,619 candles.

With the mass of ex-focal rays sent out by these huge flames, veritable furnaces, it was beyond the power of the largest usual optical instruments to deal. To meet the difficulty the Messrs. Stevenson devised in 1884 their "hyper-radial" apparatus of 1330 millimetres focal distance, nearly half as great again as that of the first-order apparatus of Fresnel. Lights of this size were constructed at Spon Lane in 1887 and 1888 for Bishop Rock and Round Island in the Scillys, for Bull Rock in Ireland, and for two lighthouses in Ceylon, and in 1895 for Spurn Head, in Yorkshire. The three first were "biform"—that is to say, one apparatus was placed upon another—and all but the first were single-flashing. After the electric lights that at the Bull Rock, with its maximum intensity of 333,500 candles, was the most powerful in the United Kingdom.[1]

At the other end of the scale were several varieties of lights of smaller radius than the 150 millimetres of the sixth order. A special application of these small apparatus was for ships' signal lights. It is as important to mariners to be warned in time of the approach of other ships, in order to avoid collision, as to be guided on their way by coast and harbour lights. And with the velocities of modern travel the distance at which such warning must be given is measured in miles. To do this the moulded, often roughly moulded, lenses of ordinary ship-lights are utterly insufficient. But with proper dioptric lenses, especially if used with the electric light, not only is the range of visibility vastly increased but also the red and green side-lights can be intensified to equal the white mast-light. Without such intensification the white light is seen long before the red, and the red before the green, owing to absorption of rays by the coloured media.

Dioptric side-lights, made at Spon Lane, were first used by Thomas Stevenson for the Northern Lighthouse Board's tender Pharos as far back as 1866. They were small Fresnel lenses of but three inches focal distance, supplemented by a hemispherical mirror of silvered copper and straight azimuthal condensing prisms. After this first experiment the fights were much improved; they were made larger, and the somewhat cumbrous condensing prisms were dispensed with. A circular issued by Chance

---

1   Kenward, *Lighthouse Apparatus and Lighthouse Administration* in 1894, a paper read at the meeting of the British Association in that year.

Brothers & Co. in September 1883 stated:

> Our attention has repeatedly been directed to the need of better Signal Lights for Steamships and Sailing Vessels than are usually provided under existing statutory regulations, and we have been asked, as the only makers of Dioptric Lights in the Kingdom, to introduce a type of apparatus which may efficiently help to prevent disastrous collisions like those that have made notable the navigation of the past five years.
>
> We have accordingly designed a set of Shiplights on the general principles applied to the great Coast Lights of our construction. They are truly dioptric, formed not of moulded or pressed glass but of pure optical glass accurately curved, ground and polished, having a focal distance of 125 millimetres, or 4·92 inches, and comprising a cylindric belt with five lens-rings—six pieces in all—the height being about 6·5 inches. The optical glass, framed in gun-metal, is mounted in a polished cylindrical lantern of copper and gun-metal, about 27 inches in height and 12 inches in diameter, having a domical top. . . . The apparatus presents every facility for cleaning and lighting, is symmetrical in shape, solid in construction, and little liable to injury.

> The set of lights for a Steamship is composed of four apparatus: a Red Port Light, of 112°, a Green Starboard Light, of 112½°, a White Foremast Light, of 225°, and a White Anchor Light, of 360°.

> In the place of petroleum we have now resorted to Electricity for a far greater degree of intensity. . . . The Edison Electric Light Company has designed for us an Incandescent lamp whose initial intensity is not less than 100 candles, so that a beam four times as powerful as that from the duplex burner is at the command of the ship.

> It is probable that when fuller public knowledge shall have been acquired of the resources offered by science in this direction, the choice of a particular vessel by travellers will be materially influenced by the circumstance of whether or not she be protected by Signal Lights of the greatest attainable power.

> Commending these facts to the consideration of Government Authorities, Shipowners, and Shipbuilders, we are prepared to receive their orders for our Lights.

Nevertheless it took years to persuade shipowners of the advantages of the new lights and of the necessity of their use, in spite of their greater cost. The Peninsular and Oriental Company were the first to adopt them.[1]

---

1   For particulars on this subject see Ken ward's papers on "Recent Improvements in Ship Signal-Lights" and on "Ship-Lights and Collisions," read respectively before the Birmingham Philosophical Society on April 24, 1884, and at the International Maritime Congress of 1893. On the former occasion, speaking of tests of the efficiency of the lights made at Spithead, he said: "The atmosphere was clear. One typical result was that at three miles the red light of the true lens was distinctly better than the white light of the best pseudo-lens; and another that at five miles the white light of the true

Another important branch of manufacture, developed on a large scale, was that of mirrors and diverging lenses for Admiralty and other search-lights. By the year 1895 could be counted up as supplied 562 "Chance" mirrors, 41 mirrors of the Mangin type, and 1,264 divergers. By the present time, and especially since the outbreak of the War, these figures have been vastly increased.

Of other matters Kenward wrote at the close of his administration:

> Accuracy and finish of the optical glass, focal adjustment to suit every condition of height of flame above the sea, and distribution of beam vertically or in azimuth, are not the only conditions studied at Spon Lane. It has been the great care of Messrs. Chance, side by side with the improvements in their lamps and burners, to remodel the revolving carriages and clockwork of all Sea and Harbour Lights ; and what would seem to be the highest improvement attainable in these parts in regard to stability of bearing, precision of motion, and excellence of material, has been especially their work. The lanterns also, of which there are six sizes and three or four types of construction, have been brought up to the essential requirements of modern Lights.

The Spon Lane Lighthouse Establishment has been greatly and often extended in area, and in working plant and resources. The methods of grinding and polishing the optical glass, which is, as it has always been, cast specially for Lighthouse purposes in the general Works, have not been changed materially; but the number and variety of machines have been largely increased. Similarly in the mechanical shops there is now a complete outillage for turning, drilling, shaping, planing, wheel-cutting, screwing, &c, applicable not only to the old designs and processes but to the newer branches of Lighthouse business added since 1865. Convenient areas and appliances for the construction and erection of Towers of cast and wrought iron of special designs have also been provided. The work-people employed are, by the nature of the business, few in relation to its importance, seldom exceeding 200, but the skill and intelligence of a great proportion of them are necessarily of a high order in view of the precise and highly finished work they are engaged in, where the knowledge and practice of Engineering and of Optics are habitually required.

---

lens exceeded by 2 to 1 that of the best pseudo-lens. All the other pseudo-lenses were invisible at five miles. Different experiments have given similar results, the Trinity House tests showing 2·6 to 1 as the ratio of the intensity of the true lens to that of the pseudo-lens."

On Kenward's retirement in 1895 direction of the lighthouse works was entrusted to their present head, Henry J. S. Stobart. William Kerr was relieved of his position as foreman and given a pension of £50 a year, with the promise of any work that he might reasonably be expected to perform. Two years later Marshall was replaced as engineer by W. H. Robson. He remained for a year in charge of the general engineering department of the works and of the patent plate, but then retired. Robson retained his post until, in 1908, his health broke down.

W. A. Jeboult, for a number of years now Stobart's principal assistant at the lighthouse works, began as head of the drawing office in 1906. To him are due many important improvements, particularly in mechanical design. Richey also, employed since 1884, has rendered valuable and steady service.

In 1898 Chance Brothers & Co. succeeded in effecting an arrangement with the lighthouse authorities, which placed their manufacture on a more assured footing. In past times had been pressed often enough upon the Board of Trade and the Treasury, that if the firm abandoned the manufacture Great Britain would be dependent for her coast and other lights solely upon the French makers, who would then be enabled to raise their prices to any height they pleased. And moreover, that in case of war with France, the supply would be cut off altogether. In spite of this, the said authorities had held faithful to the fetish of taking the lowest tender, regardless of consequences and often without regard to quality. It was not at large profits that Chance Brothers & Co. aimed; they have always regarded their lighthouse work as a national undertaking, and have carried it on in not a few years at a loss; what they wanted, was to be reasonably assured of continuous employment of their workmen and plant. The French makers had this assurance, for the French Government, alive to the danger of dependence on a foreign nation, have always kept their orders at home. The Trinity House had constantly, since James Chance took up the lighthouse work and gained their entire confidence, favoured Spon Lane, and the Irish Board in less degree, but the Scottish Commissioners had mostly had resort to France.

Already in 1867 Captain Arrow had proposed an arrangement on a fixed contract basis. Some twenty years later, such an arrangement was carried out; a continuous supply at prices regulated by cost. In 1894 negotiation on the subject was opened with the two other Boards, and this was carried to a successful conclusion after four years.

Now therefore it was justifiable to lay out capital in extension and reconstruction. In the first place purchase of machinery to the value of £1,500 was authorized in April 1898, and plans were prepared for providing the erecting and fitting shops with overhead travellers. Next year it was resolved to expend up to £10,000 in complete rebuilding, and the further purchase of machinery. The contract with Messrs. Barnsley & Sons for the work was sealed in July. Further extensions and other large purchases of machinery followed in the next three years; in 1904 electric power was applied; and in 1908, and again in 1913, there was further large alteration and extension.

Under Stobart's direction the principle of aiming at perfection in construction has been maintained. While no special novelty in optical contrivance calls for notice, every genuine improvement, mechanical or otherwise, has been adopted. Everything has been done to maintain work at the high standard of James Chance, Hopkinson, and Kenward, and the great expansion of sales has borne witness to the success achieved. At the Franco-British Exhibition of 1908 a "Grand Prix" was awarded for the Company's exhibit.

In 1904 was introduced the "Chance" incandescent oil burner, which soon proved itself superior to all others. A modern author writes of it:

> One of the best known, most powerful, and most scientifically perfect is the Chance incandescent light. This burner is used in many of the most powerful lights of the world and has given complete satisfaction. The mantle varies in size with the size and type of the light, ranging from 35 to 85 millimetres in diameter, the latter, in conjunction with a hyper-radial apparatus, producing a light exceeding a million candle-power.

And this at the rate of only 2½ pints of oil consumed per hour.[1] Although the Trinity House preferred their own type, the "Chance" burner was adopted by the Scottish and Irish Boards and by most Colonial and many foreign authorities. It has obviated to a large extent employment of the gigantic flames of Wigham and Douglass.[2]

---

1   Frederick A. Talbot, *Lightships and Lighthouses*, p. 48. For particulars see *A Few Notes on Modern Lighthouse Practice*, issued by Chance Brothers & Co. in 1910, pp. 26-30, 58-67.

2   Patent No. 3421 of 1903, for "Improvements relating to Incandescent Vapour Burners," was granted to John Hendy, foreman, and Henry John Scott Stobart, Director, of Chance Brothers & Co. Limited.

In 1909 was developed the use of the oxy-acetylene flame for search-lights and signalling lamps. A long series of trials were undertaken for the War Office, with a view to superseding the "Begbie" signalling lamp. But the officer in charge of the Signalling School was removed elsewhere, and the War Office, apparently, lost interest in the subject. Exhaustive tests, however, at Petrograd against like instruments of German make resulted in decisive victory for the Chance apparatus, particularly on the score of reliability, and in large orders for the Russian army. Easy of transport, the lights are of service wherever powerful illumination is required locally, as in night attacks, loading and unloading vessels at night, lighting up railway wreckage and salvage work generally. On board ship they are of special value as absolutely independent of accidents that may suddenly cut the electric current off. Features are their extraordinary power for signalling purposes and the ease with which they may be applied thereto; a "Scott" shutter can be fixed before the glass in a few moments, and with the help of telescopes signals be read over a distance of from 20 to 30 miles in daylight, and from 60 to 80 miles at night. Four standard sizes are made, rising from 10 inches aperture with a weight for the whole equipment of 96 lb., to 20 inches with a weight of 217 lb. Special forms are an instrument for aeroplanes of 16 inches aperture and less than 60 lb. in weight, and a miniature signalling lamp for short distances, easily carried by one man.

Recent development in these acetylene lights is on the lines of their performing their work unwatched. In this a very keen competitor is the Swedish Aga Company, who have a most useful contrivance in their sun-valve, which turns off the gas during the day-time automatically. With it an instrument recently patented by Mr. Lamplough may, it is hoped, compete successfully.

Nor in their mechanical work are Chance Brothers & Co. to be beaten. To quote, for example, Mr. F. A. Talbot again:

> The weight and clockwork system perfected by Messrs. Chance is regarded as one of the best in the service. The rotation is perfect and even, owing to the govern-ing system incorporated, while the steel wire carrying the weight is preferable to the chain, which is subject to wear and is noisy in action. In the Chance clockwork gear the weight is just sufficient to start the apparatus from a state of rest, the advantage of such a method being that, should the apparatus be stopped in its revolution from any

outward incident, it is able to restart itself.[1]

Chance Brothers & Co. make not only the optical apparatus and the lanterns that protect them, but also the iron or steel structures for their support, where such replace the usual towers of stone. They supply also every form of that subsidiary means of warning to the mariner, the fog-signal, whether it be siren, explosive signal, reed-horn, whistle or bell.

When the War broke out, part of the lighthouse shops was given up to the manufacture of high quality machine tools. In this direction very important contracts for milling machines were carried out for Messrs. Alfred Herbert Limited, reputed to produce the best tools of this description in the United Kingdom. On the other hand, part of the old patent plate works, now disused, was taken in to provide room for new grinding machinery, specially for the manufacture of small lenses for buoys. And a further portion of the Blakeley Hall Estate was purchased, as said elsewhere, with a view to extension. For the manufacture of lighthouse apparatus had to be carried on, in spite of the War, to satisfy the world's needs.

Lights sent out to almost every part of the world since 1895 can but be enumerated. Chief among them, perhaps, are the great hyper-radial lights made in 1906 for that exceedingly dangerous promontory, Cape Race in Newfoundland, and in 1908 for Manora Point, Karachi; the two most powerful oil-lights in the world. The glass alone at Cape Race weighs three tons, the whole apparatus seven. But the huge mass revolves on its bath of 950 lb. of mercury easily in half a minute, and its million candle-power beams are seen at a distance of 19 miles every 7½ seconds. Other hyper-radial apparatus have been supplied for Oigh Sgier (Hyskier) Rock (1903) and Rhu Rhea (1909) in Scotland, and for the subsidiary light at Orford.

Of first-order revolving lights, 58 in number, may be noticed the North Foreland (1900), the Bass Rock (1901), the biform Fastnet (1903), the Cap de Couedie, in South Australia (1908), the Chilang, China, triple-flashing (1910), the Dhu Heartach, double-flashing (1913), and the biform Souter Point (1915). In the case of Cap de Couedie two dangers had to be marked in a subsidiary manner: the one, 8¾ miles distant, by a red beam; the other, 1¾ miles, by a green. The installation has been pronounced highly successful and very economical.[2]

---

1   *Lightships and Lighthouses*, p. 43.
2   Ibid. p. 55. The light is figured and described in Chance Brothers & Co.'s *Modern Lighthouse Practice*, pp. 33-35.

Lesser revolving lights, nearly all of the second to the fourth orders, have numbered well over 200. Of fixed lights, on the other hand, superseded for high powers now that means of distinction among revolving lights are so vastly increased, few have been made larger than of the fourth order, but of that order, again, more than 200. A large number of them are furnished with occulting mechanism. Reckoned all together, there have been sent out from Spon Lane since the year 1855, in approximate figures, 11 hyper-radial lights, 205 of the first order, 120 of the second, 170 of the third, 27 of the small third, and 520 of the fourth, a total of some 1,050; while smaller apparatus, mirrors and divergers, and various types of fog-signals, are to be counted by hundreds. Apart from France and Germany, it is hard to find a country in the world on the lights of whose coasts the name of Chance is not engraved.

APPENDIX

Notes by James Kenward

June 25, 1886.

Canada.—The clear atmosphere of Canada renders needless high powers of light. Therefore no first-order light, and no revolving light, has been sent thither. A very large number of our apparatus of 125 mm shiplight section have been supplied for the bays and rivers of the Dominion. Some white lights of these are visible to the very limit of their horizon, 7 to 10 miles, some coloured lights 4 to 7 miles. For similar reasons, and also because of the exceeding cheapness of oils, reflector lights are still largely in use.

Australia.—In Australia, owing to differences in climate and atmosphere, to the more rapid growth of commerce, and perhaps to an honourable rivalry between the various Governments as to endowing the coasts with the most advanced specimens of optical and mechanical design, the choice has fallen on powerful revolving lights. The equipment of the seaboard of 8,000 miles is by no means complete. Several large lights must yet be put up. The great Electric Light of Sydney Heads (Macquarie) is simply the most powerful in the whole world. It has been found too powerful, and is now being worked with the lower power of arc. The proposal to erect another Electric Light on Cape Leeuwin, the great land fall of the Continent on the S.W., has been abandoned, and we shall probably supply a first-class revolving, with burners of the 7-wick type.

In *Queensland* a large use of condensing agents has been made in the small lights, as well as a very wise choice of Sea-lights, under the skilled administration of Captain Heath, R.N.

In *Western Australia* a good example of economical adaptation is the Point Moore Light, iron tower, second order revolving, all specially designed. Another is the King's Island Light in Bass Straits, iron open-work tower, first order revolving and lantern, a work that has excited the admiration of all who know it.

Fiji will shortly have its first lighthouses, both revolving and of the most modern type. For Na Solo, the larger one, we have recommended and are preparing a cast-iron tower of the Point Moore description.

Straits Settlements.—Among the fine Lights sent to this Colony is a third order group-flashing. This valuable form of apparatus, designed by Dr. Hopkinson, was first constructed by us. It has since become the most popular form in the world. The French makers copied our work at the very earliest moment, and it is probably known in foreign countries as a French design. "Sic vos non vobis" has been the rule. Colombo, Minicoy, Little Basses, Watling Island, are among the examples of most striking importance of our work. The first group-flash in Australia will be the Cape Everard, now being made for Victoria.

Red Sea.—Our first successes here have been our only ones. The concessions for lighting both sides have been obtained through diplomatic astuteness by the French makers, quite contrary to the wishes and efforts of our Government.

India.—Besides the group-flashing Lights glanced at the Eastern Grove is noteworthy, with its condensing prisms, and the Minicoy lens Light, with extended refracting angles of Eddystone type. The Oyster Reef and Krishna Shoal Lights have been destroyed by cyclones. The occulting Light at Perim is a unique instance of this new arrangement in a fourth order, and is very successful. The four apparatus for Hooghly floating Lights deserve notice. I have a similar one now at the Liverpool Exhibition, with improvements in the Lantern.

CHAPTER VIII

OPTICAL GLASS

THE glass from which are fashioned the lenses of telescopes, field-glasses, microscopes, photographic cameras and other scientific instruments is made in a different way from any other. It is neither gathered nor poured, but left, when melted and refined, to solidify in the pot. The block that results, or fragments of it, are moulded by reheating into forms suitable for the completing work of the optician. The manufacture is of very great nicety and difficulty, demanding the utmost care and patience. Messrs. Chance Brothers & Co. undertook it seventy years since, and remained until quite recently the sole makers in the United Kingdom. As with sheet glass and other enterprise, they owed inception of this manufacture to Georges Bontemps.

The function of the finished lenses is to refract the rays of light that traverse them. That they may perform this function perfectly the glass must be free from striae and other defects that may distort the rays in their passage. It cannot, like window glass, be limited to one simple composition; different purposes require glasses of different refractive and dispersive powers. In modern times the exact scientific work of Abbe and Schott and their collaborators and successors at Jena, applied to an exhaustive variety of compositions, has furnished the optician with material of any such particular powers as he may want. In 1757, when John Dollond presented to the world his achromatic lens, he had but two glasses from which to shape its twin components, the ordinary crown and flint. Hence the survival of these names for the half dozen standard types of optical glass made at Spon Lane and elsewhere a century later, the "flint" glasses being those into whose composition lead enters, but not lime.

It was a Swiss mechanic, Pierre Louis Guinand, who after years of patient experiment towards the end of the eighteenth century discovered how to produce the homogeneous glass required. He equalised the density of his molten mass throughout by long continued stirring, allowed the whole to cool slowly undisturbed, then broke the pot away and obtained a block from which, when broken up, perfectly homogeneous fragments could be selected. In 1799 he was able to show Lalande at Paris some flawless discs from 4 to 6 inches in diameter.[1]

Six years later, the celebrated optician Fraunhofer persuaded Guinand to come to work with him at Munich. They were able to produce from time to time discs suitable for working into object glasses, one or two of as much as 13 to 15 inches in diameter, but they never made the process reliable. For many years yet opticians had to depend for their regular needs on the rare discovery of suitable fragments of ordinary glass. We have the testimony of Sir George Airy that about 1828, when he was at Cambridge, they "had to wait for years for a 4-inch object glass"[2] Faraday stated in his Bakerian lecture of 1829, the paper in which he detailed the work of the Committee of the Royal Society appointed to investigate the subject:

> Mr. Dollond, one of the first of our opticians, has not been able to obtain a disc of flint glass four and a half inches in diameter, fit for a telescope, within the last five years, or a suitable disc of five inches in diameter within the last ten years.

After eulogising the labours of Guinand and Fraunhofer, he went on to say:

> Both these men, according to the best evidence we can obtain, have produced and left some perfect glass in large pieces; but whether it is that the knowledge they acquired was altogether practical and personal, a matter of minute experience, and not of a nature to be communicated ; or whether other circumstances were connected with it; it is certain that the public are not in possession of any instruction, relative to the method of making a homogeneous glass fit for optical purposes, beyond what was possessed before their time; and in this country it seems doubtful whether they ever attained a method of making such glass with certainty and at pleasure, or have left any satisfactory instructions on the subject behind them.

In spite of this, Guinand's process was even now being brought to give regular results. His elder son was working upon it with Daguet in Switzerland, his younger, Henri, in conjunction with Bontemps and the optician Lerebours in France. In the autumn of 1828 were exhibited to the

---

1   It is related that Guinand began by grinding spectacles for his own use, went on to construct telescopes by fixing lenses in pasteboard tubes, and then embarked on his experiments in melting the glass.

2   Letter to James Chance, July 10, 1880.

"Academie des Sciences" a number of successful discs made at Choisy-le-Roi, among them one of 12 and another of 6 inches in diameter. From that time, says Bontemps, production went on regularly, if for a period without great activity.[1] In 1836 Henri Guinand produced a disc of flint glass 14 inches in diameter, which Bontemps was able to match with its crown glass complement in 1843. The objective constructed from these by Lerebours became in 1849 the property of the Observatory of Paris.[2] At Munich, Merz and Mahler, the successors of Fraunhofer and Utzschneider, were able to produce two somewhat larger, one of which went in 1840 to the Russian Observatory of Polkovo [Pulkova], the other in 1843 to Harvard University.

It was natural, in view of their friendly relations, that Lucas Chance should be apprised of Bontemps' work, natural also to him to seek new outlets for his energy. Negotiation for the manufacture of optical glass at Spon Lane opened in May 1837 through the same mutual friend, A. Claudet, who had played a like part in connexion with the sheet glass. Proposing terms on which Chance Brothers & Co. could be instructed, Claudet intimated that he had himself a small interest in the undertaking, as agreed with Bontemps. It was arranged that the latter should receive for his instruction a premium of 3,000 francs, the sum that he had himself paid to Henri Guinand for the like service, and that of the net profits five-twelfths should be allotted to him, five to the firm, and two to Claudet; the 3,000 francs to be paid only when the firm had realised that amount of profit to itself. In September it was resolved to erect a small furnace and its appurtenances on the south side of No. 4 house, and in March 1838 Lucas Chance took out an English patent for Guinand's process.[3]

Some trial was made, but the exigencies of the new patent plate and of other work took precedence. Nor was success likely to be obtained on written instructions alone. In June 1840 Lucas Chance invited Bontemps to come over "to perfect the optical glass," hoping also, he told his brother, to glean advice from him about flattening sheet glass. He intimated willingness to show him the patent plate works. Bontemps accepted the invitation and proposed to send in advance an expert workman named Halle.

William Chance objected strongly. In the absence of his brother on business affairs in Ireland and the north he informed Bontemps of his "most

1   *Guide du Verrier*, pp. 652-3.
2   *Rapports du Jury, Paris Exhibition of 1855*, p. 403 ; Reports on the Paris Universal Exhibition, ii. 284 (Sir David Brewster). Cf. the *Guide du Verrier*, pp. 663-4.
3   No. 7596 of March 19, 1838.

decided opposition" to any more experiment at present, and requested him to stop Halle's departure. He gave as the reason "the vital importance to our concern, that while we are carrying on our present important operations our attention and that of our managers should not be distracted in the slightest degree by any new operations."

The correspondence that ensued between the brothers was not a little heated. Lucas Chance held that little of the time of Withers, the proper person to conduct the process, would be taken up in learning it, indeed, that it would be no bad thing for him to be up at night occasionally with Bontemps, for he could then look after the work in the crown houses, always bad when no manager was about. He was anxious to be in closest intimacy with Bontemps, to have him, indeed "considered one of us"; he being "the only man for optical glass," the best maker in France of shades and sheet glass, conversant with every mode of flattening in France, Belgium, and Germany, and "a scientific and practical man," who understood every branch of glass-making "far better than any one we have, or all of them put together." He called to mind in illustration of the consequences of his brother's principle "of doing every thing within ourselves" the rejection of Bontemps' liberal proposals in connexion with his combined Houtard-Hutter lear and the total failure of the attempt to set up such a lear without him.

William Chance was not to be moved, at least in regard to the present, when there were "so many things under hand of ten times the importance" of optical glass. He was strong on the necessity of concentration on perfection of the patent plate, and would have Withers and his sons devote their whole attention to supplying good glass for that. Moreover he was opposed to letting Bontemps see "the patent plate polishing and other improvements." His views, supported by James Chance, prevailed; inception of the undertaking had to await the coming of Bontemps to Spon Lane, eight years later. That Lucas Chance kept the matter in mind is shown by drawings of optical pots, which his son Robert brought from Choisy in 1844. In the meantime, thick ordinary crown glass continued to be supplied to Messrs. Dollond for optical purposes, and afterwards, at least up to 1867, to other opticians, whose purposes for small and cheap telescopes it served.

Immediately on Bontemps' arrival, in June 1848, a new plant was ordered. He began with the "hard crown" and "dense flint" in demand for telescopes, and the "soft crown" and "light flint" for photographic cameras. With the last, for which two parts of red lead were used to three of sand, London opticians were supplied already in the spring of 1849. The "hard crown" differed but little in composition from ordinary crown glass, save that potash was substituted for the soda. For the "soft crown" equal quantities of lime and red lead were used, 9·66 parts of each to 100 of sand. The first "dense flint" was made from equal quantities of sand and red lead; but very soon Bontemps increased the proportion of the latter by 5 per cent., in order, it would seem, to match in density Daguet's Swiss flint, which he found in general use in Germany, when travelling there for the firm in 1850. At all events, he was able to show German opticians a "new dense flint" on his second tour in 1852. A fifth glass of his make was an "extra dense flint" for special use in microscopes, a glass for which he had found demand in Germany, especially at Berlin. The proportion of red lead to sand in this case was as 128 to 100.[1]

At a later time, after 1867, was introduced a "double extra dense flint," for which the proportion of red lead to sand was as high as 9 to 5. The densities of these six glasses rose in gradation from 2·485 to 4·45. With some modification in the quantities of alkali they remained the standards at Spon Lane until the introduction of the Jena novelties, certain special glasses being made also for particular opticians. It may be noted that the hard crown was known as "German" and the soft crown as "American."

One man interested in the undertaking was James Nasmyth. On January 21, 1849, having heard that optical glass was to be made at Spon Lane, he explained in a long letter a method for its manufacture proposed by him before the Astronomical Society in 1846. His plan was to bring the glass to the highest degree of liquidity possible, to keep it in that state for two or three days, to allow it then to cool slowly undisturbed, and when cold to saw the block horizontally into parallel slices. He stated that a large experiment of the method had completely succeeded. However, on receiving from James Chance Bontemps' comments and explanation of the Guinand process, he wrote (February 14): "I need not say how delighted I am to know that so highly intelligent and experienced a person, as Mr. Bontemps is known to be, has brought his high intelligence and experience into close contact with such liberal and intelligent parties as you and your partners." Having read Bontemps' remarks he felt obliged "to yield all my ideas on the subject as unworthy of notice."

---

1   *Guide du Verrier*, pp. 659—65, and Bontemps' letters.

In April 1849 Bontemps was able to report orders from London opticians to the value of £250, and more to follow, when they had disposed of their stocks of Swiss discs and Pellatt's plates. He was

> happy to mention their unanimous approbation of the glass they received. Mr. R. L. Chance, Jr. heard by himself that they never got better glass, and one of the most competent, Mr. A. Ross, says that our flint glass is even superior to the Swiss flint, not being altered so easily by the atmosphere. . . For summary I have the conviction that we are going on in the right way. The glass is approved, and the customers expressed their satisfaction not to be obliged to bring their materials from abroad, having an equal or perhaps better article in their own country. As for the British flint plate, they confess that our light flint is superior by far: and consequently we must expect by and by to supply all the consumption.[1]

In the autumn of the same year, specimens of the glass were shown at the "Birmingham Exposition of Arts and Manufactures" held in connexion with the meeting of the British Association.

Travelling the Continent on the firm's business in 1850, Bontemps was able to see all the principal opticians in Germany, excepting Voigtländer (whom he did not know to have left Vienna for Brunswick) and Schweizer of Augsburg. He had best success at Vienna with Ploessel, "the most celebrated optician in Germany," Prokesch and Waibl. Merz of Munich he found to be making for his own use only. Altogether, he obtained orders for discs and plates to the value of nearly £200, with promise of more, should the glass prove satisfactory. At Berlin "not a town of luxury, neither a very commercial place," he found demand only for very dense flint for microscopes, and that in small quantity. As the special result of his journey of some 3,700 miles he claimed to have made the acquaintance "of all the more important opticians in Germany, who most likely will increase the demand for optical glass, in a great proportion in your works." Visiting the country again in 1852 he found the Viennese to be well satisfied with the quality of the glass supplied to them, and obtained a large order from Voigtländer.[2]

At the Great Exhibition of 1851 the firm had a notable show of optical glass. There were discs and plates of light flint (one of the former 20 inches in diameter) "adapted for the construction of object-glasses for

---

1   Bontemps' *Report on my Excursion in London*, April 1849.
2   Bontemps' reports, 1850 and 1852.

Daguerreotype and Talbotype apparatus and cameras," a 9-inch achromatic lens worked by W. Simms from discs made at Spon Lane, and others of 8 to 4 inches, of which the crown components came thence. But the great triumph was a large 29-inch disc of dense flint, 2¼ inches thick and weighing about 200 lb. It stood examination well; excepting in one place near the surface, and a few "utterly inoffensive" threads, no appearance of striae whatever could be discerned in any part, nor any bubbles of consequence. It was shown that in working the disc into a concave or a concavo-convex lens most of the single fault would be got rid of, or that a perfect disc 22 inches in diameter could be cut from it or one of 25 inches obtained by moulding. The Jury had no hesitation in recommending a Council Medal. Since, they reported, "the great object-glasses of Pulkowa [Pulkova] and of New Cambridge [Harvard] in the United States" did not exceed 16 inches, the present achievement constituted an enormous advance.[1]

The disc was shown again at Paris in 1855 in company with one of equal size of crown glass, which Bontemps had hoped in 1851 to obtain by using larger pots[2] and had at last been able to produce. Foucault, he says, pronounced this to be the most perfect and pure piece of glass that he had ever seen.[3] Sir David Brewster states in his report that he did his best to persuade the British Government to buy the pair and to "construct with them the greatest achromatic telescope that was ever contemplated by the most sanguine astronomer" but could rouse no interest; and that when success promised in raising private funds for the purchase he received the mortifying intelligence that Le Verrier had the discs in his possession under a provisional arrangement with the French Government to buy them at a high price, should they prove fit for the construction of a telescope.[4] They were bought for £1,000 each, but only the crown disc, it would seem, was used. A modern author states definitely that the flint component of the great achromatic object-glass of the Observatory of Paris, subsequently constructed, was made by Feil of that city, Henri Guinand's grandson. Presumably this was the 28·3-inch disc shown by him there in 1867.[5]

Following on the success of 1851, twelve large and six small optical pots

---

1   Reports of the Juries, pp. 269-70. The disc was polished by Ross.

2   *Examen historique et critique*, p. 116.

3   *Guide du Verrier*, p. 665.

4   Reports on the Paris Universal Exhibition, ii. 287.

5   Henrivaux, *La Verrerie au XXe Siècle*, Paris, 1911, p. 239; *Rapports du Jury International*, Paris, 1867, P- 451 ; cf. the *Guide du Verrier*, p. 665.

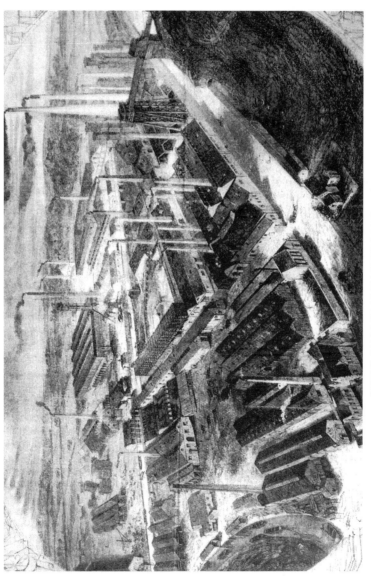

VIEW OF THE ALKALI WORKS 1862
(To face page 198)

were ordered always to be kept in stock, and the manufacture proceeded regularly. On Bontemps' proposition his cousin, Arnaud Masselin, was engaged to conduct it, as also to manage the Coloured and Ornamental departments. He was to have so much of Bontemps' share of the optical profits as they should agree upon between themselves. Power to retain his services after 1854, if so desired, was made an essential condition,[1] and the arrangement was duly renewed for five years from January 1, 1855. Bontemps, returning to France, undertook to aid the firm "in the extension of their optical trade and generally to give them any assistance and information in his power in relation to their various manufactures." In return for this, and for his undertaking to assist no other manufacturer, he was to receive a quarter of the profits of the department. Before this time the plant had been removed to the new side of the works. Net profits for 1853 had amounted to over £1,900, and Bontemps' premium of 3,000 francs had been paid.

Examples of the firm's work shown at Paris in 1855 included, besides the two large discs mentioned, a pair of 20-inch for photographic purposes, a pair of 16-inch for a telescope, and a great variety of smaller ones. The Jury stated that the exhibit would have gained a medal of the first class, had not that honour been accorded for the great lighthouse apparatus and the "beaux cristaux." Bontemps in his report expressed much dissatisfaction that no medal was given for the two great glasses.[2] In November of the same year another 29-inch crown disc was ordered to be made, and twelve months later Masselin was pressed to produce a 24-inch crown and a 29-inch flint disc of the best quality.

The next opportunity to show results to the world was afforded at the London Exhibition of 1862. Besides a pair of 25-inch discs, subsequently worked by Messrs. T. Cooke & Sons, of York, into an object-glass for Mr. R. S. Newall, of Gateshead, which eventually, through his son, Professor Newall, became the property of Cambridge University, there was an objec-

---

1   " If M. Masselin comes here, he will be a learner, and will acquire a thorough acquaintance with the various branches of our business, and it is therefore necessary for us to have the power of retaining him after the 31st Decr. 1854. ... . I can assure you that nothing would have induced me to entertain the admission of M. Masselin into our concern, except the desire to comply with your own wishes, for I consider that our present deficiencies consist not in the want of additional administrators, but in our not availing ourselves properly of those which we have" James Chance to Bontemps at Paris, August 9, 1851). Bontemps agreed.

2   *Reports on the Paris Universal Exhibition*, ii. 287, 402; *Rapports du Jury Mixte International*, p. 404.

tive made from the firm's glass mounted in an equatorial telescope of 20 inches aperture by J. Buckingham[1] stated to be the largest yet in existence. The discs, we learn, had been made to the order of the French Government, but rejected on account of a flaw in one of them. Buckingham satisfied himself that the flaw could all but be eliminated in shaping the lens, and found his expectation completely justified. Besides these large specimens were shown " many others of smaller dimensions, all of which appear to be of first-rate quality," and "an immense block of very pure glass, capable, it is presumed, of being moulded into a much larger disc than any of those exhibited." [2]

At the Paris Exhibition of 1867 the firm again showed discs of great purity and dimensions, in the Jury's estimate, but on this occasion they were outdone by Feil, whose exhibit included the great flint disc noticed.[3]

Other large glasses were supplied to Alvan Clark & Sons, the noted opticians of Boston, Mass. One pair, sent in 1860, was worked by them into an object-glass, mounted at the University of Chicago in 1863. They reported on it: " The 18-inch is the best definer we have ever seen. . . . The discovery of the companion of Sirius has established the character of the glass and won for us the Lalande prize." A pair of 26-1-inch discs was supplied to them in 1871, and another of the same size in 1874. From the former they made the object-glass for a large equatorial telescope for the Naval Observatory at Washington, an instrument pronounced to be their masterpiece. It enabled Encke's comet to be observed at a distance of 182 million miles.[4]

Negotiation, however, for the huge discs wanted for the Lick Observatory in California came to nothing. They were to be of not less than 36 inches, nor of more than 42 inches diameter in the rough. Arrangement of terms is recorded in a Spon Lane Board minute of February 15, 1875, but eventually Alvan Clark preferred to get them made by Feil of Paris, as provided with larger furnaces and pots. There is further record in a minute of ten years later, January 12, 1885, of willingness to make a 36-inch disc for him.

Of exhibits at Philadelphia in 1876 the largest was a 24-inch crown disc, afterwards cut up as bad. Others were pairs of 19, 16 and 12 inches.

1   [This telescope by Buckingham was sent to Calton Hill Observatory, Edinburgh]
2   Reports of the Juries, 1862, Class XIII, pp. 5, 27.
3   *Rapports du Jury International*, Paris 1867, pp. 450-1.
4   *Scientific American*.

Another 19-inch disc, of dense flint, was cut down and sent to Alvan Clark & Sons to replace one of 16 inches supplied to them in 1879, which, with its crown glass companion, had been shown in that year at South Kensington.

The difficulty and cost of producing these large discs are illustrated by the prices charged for them. In 1880 the quotation for small ones of the best quality was 4s. to £4 per dozen, as they advanced in size from 1 to 2¾ inches in diameter, but a single 6-inch disc was priced at £5 17s., one of 12 inches at £34, of 18 inches at £170, of 24 inches at £470, and of 30 inches at no less than £1,000. Whether a profit were made on these great sizes was much a matter of hazard; early success meant a large gain, repetition of failures heavy loss. Their production, with its slow processes of moulding and annealing, was in any case an affair of months. If first attempts were not successful, the months might lengthen into years. And all through the long ordeal there was constant liability to accident. In one case, for instance, imprisoned air blew a hole through a very fine crown disc in the final moulding. In another a magnificent one of flint was lost through having been set flat instead of vertically; in cooling its weight prevented proper contraction, and it broke through the middle. And even if the end were safely reached, a very slight fault, only then apparent, might yet render the disc unserviceable.

In later years there was little call for objectives of great dimension. The reflecting principle, to make the specula for which presented no great difficulty, began to be preferred to the refracting for large telescopes. The latter were of use only in a clear atmosphere, and the larger the objective the clearer the atmosphere must be. Hence the selection of a remote mountain-top in California for the Lick Observatory. Until quite modern times the only other discs of great size made at Spon Lane appear to have been a 28-inch pair for Greenwich Observatory. The contract for them was concluded in October 1887. The flint disc was obtained at the first attempt, but it took some four or five years to obtain the crown. A 15-inch glass was supplied through Messrs. T. Cooke & Sons for Teramo in Italy, and in 1911 a pair of 16-inch discs were sent to Sir Howard Grubb. Not long afterwards a 24-inch pair were put in hand for the late Sir David Gill, Astronomer Royal at the Cape, but the outbreak of the War prevented execution. Just before that date was produced a 28-inch disc of crown glass, which was shown at the British Scientific Products Exhibitions in

1918 and 1919.[1]

It was not on these exceptional productions that the manufacture depended for success, but on regular satisfaction of the ordinary requirements of opticians. As a total in tons the demand amounted to nothing great, but since, after Daguet stopped work and before the Jena works were started, the only makers of optical glass in the world were Chance Brothers & Co. and Feil of Paris, there was enough to keep both firms busy. The work proceeded at Spon Lane steadily and without much incident.

The agreement with Bontemps and Masselin was renewed in 1860 for five years, the latter undertaking also to manage any other departments for which he might be competent and have time. His salary was raised to £500, of which £150 was on account of his optical management. But at the end of 1862 he left Spon Lane of his own will. Charge of the optical glass was then committed to William Edward Chance, who retained it until he also retired ten years later. The founder under him was C. Halle, presumably Bontemps' man noticed above. He was engaged up to the end of 1867, having been promised £100 then, provided that he had taught one or more men to carry on the work efficiently. For services at the Paris Exhibition of 1867 he received a present of £20. Job Parish was his successor, and he, assisted in his latter years by his son, William Arthur Parish, carried on the founding until 1901.

After William Edward Chance's departure Henry Chance, and then Dr. Hopkinson, had the direction. When the last-named removed to London, Ernest Talbot, third son of the schools' headmaster, acted for a time as his representative, but shortly went also to London to assist him in his engineering work. After that George Chance supervised, until in 1895 Stobart relieved him of the charge. At length, now, it was decided to attempt manufacture of the new glasses, which for some ten years past had been sent out from Jena.

That factory, devoted to the production of glasses for optical and other scientific purposes, owes its existence to the devoted labours of Professor Abbe, purely a man of science, and of Dr. Schott, possessed also of some practical knowledge of glass making. Abbe was moved in the first instance by sense of the need of improvement of objectives for microscopes; need, namely, of getting rid of the "secondary spectrum" the fault of the ordinary achromatic lens, and of greater variability of relations between the refractive

---

1   A photograph in *Nature*, March 27, 1919.

and dispersive powers of its components. He discussed these requirements, and the means for fulfilling them, in a paper read in 1876. Four years later Schott came forward and concerted with him an exhaustive investigation of what optical properties might be imparted to glasses by the introduction into them of substances other than the ordinary. He set to work to produce a complete series of laboratory specimens, which Abbe and his assistant Dr. Riedel submitted to spectroscopic examination. At last, in 1884, they and the two Zeiss brothers ventured at their own expense, and with little or no prospect of pecuniary reward, on an initial factory at Jena. Soon afterwards the Prussian Government, enlightened in such matters, came to their aid with subsidies, and further by prescribing the exclusive use of Jena glasses for all instruments under their control. The works could then be established on a real commercial footing. The first trade catalogue, of 1886, listed utilisable glasses of forty-four different compositions, nineteen of them essentially new. Schott's laboratory specimens had exceeded the thousand.

In the mere employment of unusual ingredients there was nothing novel. Fraunhofer and others in Germany had made optical glasses of peculiar composition early in the century. Maes of Clichy had shown specimens containing boric acid, zinc oxide, baryta and magnesia at Paris in 1849, and had gained a Council Medal for them at the Great Exhibition of 1851. Thallic oxide is mentioned in the reports of the Paris Exhibition of 1867 as giving special refractive and dispersive powers. Vernon Harcourt in England had spent a quarter of a century producing on a minute scale glasses containing some thirty different oxides, among them boric, phosphoric and titanic acids. After Professor Stokes joined him in his researches in 1862, they had worked specially to find a perfectly achromatic combination, and in 1874 Stokes was able to show a small object-glass which, though very imperfect, achieved their aim.[1] He had been aided in his experiments by Dr. Hopkinson. That the secondary spectrum *could* be overcome, Abbe knew before he took up his work. What he and Schott did was to determine and tabulate systematically the optical properties of all possible compositions. With silicate glasses alone they found that they could not attain their end, since with them refractive and dispersive powers both varied directly with the density. They wanted also glasses in which one of those powers should rise with the density, or remain stationary, while the other fell. One long series of experiments with phosphate

---

1   See Stokes's communications to the British Association in 1871 and 1874.

glasses ended in failure because, though optically successful, they were affected by the damp of the atmosphere. In the end admixture of boric acid, and of baryta, solved the problem. Barium borosilicate glasses were the foundation of the Jena fortunes.

First warning of the German enterprise came to Spon Lane in March 1886 from a London customer, James Swift, who inquired whether the firm had heard of Abbe's new objectives, shown at the last meeting of the Microscopical Society. They greatly reduced, he said, the secondary spectrum, gave a perfectly colourless image in the telescope, and would "doubtless make a great change in optical matters." About the same time Stevens, of the firm's London office, called attention to an account of Abbe and Schott's work published in the *English Mechanic*. It was proposed, but only proposed, to send a man to Jena to investigate.

Two and a half years later came a letter from Voigtländer. Expressing great regret at the interference with his firm's long connexion with Spon Lane, he said that the essential advantages of the new baryta flint glasses, both light and dense, rendered their adoption obligatory, and further stated that the Government pressure limited his use even of Chance Brothers & Co.'s hard crown, "the superiority of which we have always highly acknowledged." And again Swift, as reported by Stevens, drew "a very gloomy picture as to the future of our optical glass," naming the Jena works as those of the future; their glasses thoroughly reliable for density, and samples always marked with their powers of refraction and dispersion. Chance Brothers & Co., he said, must take the matter into consideration, if they were to hold their position. Another firm of opticians, R. and J. Beck, wrote that they were making experiments which, if successful, might induce them to obtain most of their supplies from Jena. They wanted two glasses, differing considerably in refractive power, of which that with the higher index of refraction should have the lower dispersion. "Correction of the secondary spectrum is of little importance in photographic objectives."

In spite of all this, and of further letters from Voigtländer, it was not deemed advisable to take up a difficult manufacture of special glasses wanted but in small quantities. Not until 1895, as said, was experimental work on production of some of the Jena glasses undertaken. Next year was obtained a borosilicate crown glass, which found ready sale. In 1897 Arthur H. Lymn was engaged as scientific manager. He conducted experiments on the large scale during three years, giving particular attention to light flint and dense crown baryta glasses. But he failed to overcome the difficulties encountered, chief among them bad colour arising from impurity of the barium materials and the peculiar liability of these glasses to seed. In 1900 he was replaced by Walter Rosenhain, who had a brilliant scientific record, and is now a Fellow of the Royal Society. Yet neither was he successful in solving the problems presented. His research work on the large scale proved very costly; results for the year 1904 were pronounced disastrous; and he was desired to make no more experimental meltings but in the laboratory. Towards the end of 1905 he left to take up work at the National Physical Laboratory, and nothing more in the way of novelty was tried until the arrival in 1909 of the present manager, Samuel Lamb. In the meantime J. R. Wharton supervised the work more or less directly, charge of the laboratory being entrusted to W. J. Rees, previously Rosenhain's assistant.[1] Lamb, after he had made himself acquainted with general methods, and especially with the work on baryta glasses already done, was able to make progress. A gas furnace for two 38-inch pots was constructed for continuous working and the manufacture of large discs, up to 16 inches diameter, was resumed. It was decided to conduct the moulding and the annealing separately, to use an oil-fired kiln for the former process, and to reduce the glasses to shape by means of a pneumatic press, instead of rubbing them down, as hitherto practised, by hand.

Oil-fired kilns were tried also for the annealing, but in this case they were not successful and the coal-gas furnaces supplanted them. These, modified as found to be necessary, remain in use. For specially fine annealing electric apparatus enabled the temperature to be very exactly regulated. An acquisition of 1913 was a Belgian smoothing machine, which enabled grinding and polishing to be done with far greater rapidity. It proved of great value, when the pressure for war requirements set in. Also was adopted mechanical stirring of the molten glass, by hand a very laborious operation requiring the employment of eight to ten men. The first type of stirring machine came into general use in August 1914.

All this time, in face of the subsidized competition of the Jena glasses, the manufacture had frequently been conducted at a loss, and more than once it was proposed to abandon it. Only the sense of its importance to

---

1   Rees, after absence for a period from ill-health, left in October 1917 to take up work at the Institute [Department] of Glass Technology at Sheffield.

the nation induced the Company to persevere, in spite of the discouraging results and the apathy of Government. The nation reaped the benefit of this persistence at the outbreak of war.

That brought about an entire change in the position of affairs. At the same time that an enormous demand for field-glasses and cameras for military purposes set in, the supply from Germany was suddenly cut off. It came as agreeable news to the War Office that optical glass was made in England also. Soon the Company was urged to increase its output to a maximum extent. A second gas-furnace and appurtenances were got to work in August 1915, and a third in January 1916.

Fortunately, before the emergency arose, the difficulties of the new glasses had been to a great extent surmounted. A zinc crown and a dense barium crown were successfully placed on the market in September 1914, and a light barium flint in October. But there were new obstacles. That essential constituent of the glass, potash, had hitherto been obtained from the great natural deposit of its salts at Stassfurth, and no other supply was immediately available. Before long the British Cyanides Co. succeeded in producing a very pure carbonate of potash, but it cost £150 a ton as against £30 for the German article. It remains to be seen whether the said Company, by a new process, will be able to obtain a potash which may compete with the German product permanently.

Also from Germany had come the barium minerals. They could be obtained elsewhere, but not in quantities sufficient, and the prices asked in England for the purified salts were prohibitive. The Company, therefore, determined to purify the raw material themselves, calling in the expert advice of H. W. Crowther, formerly research chemist at the Oldbury works, to assist Rees. Plant was put down under their direction for producing pure carbonate and nitrate of baryta, both of which salts continue to give excellent results.

Shortage of labour in consequence of the War, a difficulty very soon acute, was met, so far as was practicable, by further substitution of mechanical for hand work. Electrically driven machines for setting pots in and removing them from the furnaces dispensed for this with the services of six men out of ten. The cost and labour of taking down the front of a furnace, every time that a pot had to be withdrawn, was obviated by fitting sliding doors. Other manual work was saved by making the stirring continuous, instead of conducting it at intermittent periods.

Another great difficulty was the supply of pots, in view of the number now required. The Company's own resources were insufficient, and the makers in the neighbourhood could not be relied upon to furnish enough of proper quality and fully matured. In the circumstances, it was determined to try open pots instead of covered, as more easily made and more quickly dried. They answered well enough for the crown glasses, but for the flint return had to be made to the covered.

With the various improvements introduced the length of the founds was materially shortened, and it became possible to produce from three furnaces the quantity of glass for which five had been considered necessary.

Negotiations with Government for an arrangement for the supply of optical glass, carried on in desultory fashion during the year 1915, were brought to conclusion after Mr. Lloyd George became Minister of Munitions, mainly through the instrumentality of the late Mr. S. A. Esslemont, C.B.E. It was agreed to enable the Company to raise its output to 14,200 lb. per month by the erection of new plant and buildings at Government expense, proper obligations being undertaken in return. Included was the establishment of a Research Laboratory, equipped with the most modern appliances and instruments[1] the whole cost of the installations amounting to some £40,000. A fourth furnace was got to work in August 1916, and a fifth in October. As the result, the output of optical glass rose from about 2,600 lb. for the first six months of 1914, before the war, to 92,000 for the first six months of 1918. Baryta glasses, light, medium and dense, and borosilicate crowns, had by this time become articles of ordinary manufacture.

The Research Laboratory came into full operation in June 1917. Two first-class men of science had been engaged, Mr. F. E. Lamplough, formerly a Fellow of Trinity College, Cambridge, and Dr. C. V. Burton. The services of the latter, however, were lost by his death from a laboratory accident at the Royal Aircraft Factory at Farnborough, a week after his appointment. Subsequently Mr Adams and Miss Mathews were engaged as assistants in the chemical and physical sides respectively.

The following notes on the work done in the Research Laboratory have kindly been supplied by Mr. Lamplough.

Attention was first directed to investigation of the causes of bad colour, opalescence, &c, in fluor crown glasses, resulting in a melting of one of these glasses being produced,

1   Among them is to be noted a special microscope presented by the firm's old customers, Messrs. Beck.

which was found to have a dispersion less than any listed by Schott u. Genossen.

Towards the end of 1917 we were suddenly confronted with the problem of producing glasses for a very large number of photographic lenses required by the Air Board. The different types of lenses necessitated the production of a light barium crown, two light barium flints, two dense barium crowns and an extra light flint, the last three glasses of the most extreme types made by Schott. The experimental work for these was carried out without assistance from analyses, as samples were not at once forthcoming, and the glasses were successfully made; two of them having properties rather more extreme than the corresponding Jena types, whilst the third glass (extra light flint) was decidedly more durable than the sample of Mantois glass submitted towards the end of the investigation. Some little difficulty subsequently arose out of changes of index which were found to occur, particularly with dense barium crowns, during moulding and annealing.

Problems relating to homogeneity and annealing have always been coming forward, and results of interest and importance have been obtained during their investigation. It has been found that the homogeneity of a glass in its behaviour to light is markedly affected by internal stress (a point upon which we had been considerably misled by a well-known optician), and at present the perfecting of the method of annealing is receiving serious attention.

Many improvements in testing glasses and in the rapid measurement of indices have been introduced, dispensing with the necessity of working the discs optically for the purpose of examination. The general reliability of these methods is shown by the fact that out of over 6,000 discs examined for veins in the Research Laboratory not one has been returned for this defect, and the tests have given confidence to opticians in the use of varieties of glass only recently obtainable in this country.

Research has also been carried out on didymium and other coloured glasses, and on the investigation of the subject of the relative dispersions of glasses. A paper on the latter subject by Miss Mathews and myself has recently been published by the Optical Society.

In conclusion may be cited some paragraphs from a paper, also from the pen of Mr. Lamplough, published in "Nature," March 27, 1919.

In considering the position of the industry after the war, it is obvious that there are resources in this country for the manufacture of all the optical glass which will be required by our opticians. Nor need there be any apprehension regarding the ranges of glass which will be available for the use of the lens designer. Without any notable exception, Messrs. Chance Brothers & Co. have been able, by their previous experience and by the work of their research laboratory, to produce glasses which, in their optical constants, cover the full range of glasses mentioned in the Jena list for 1913.

The further development of the optical glass industry would appear to be well provided for in view of the practical research work carried out by the manufacturer and of the more general work conducted by the British Scientific Instrument Research Association, recently formed under the direction of the British Optical Instrument Manufacturers' Association. To maintain the supremacy of the nation in regard to this manufacture, however, it is not only necessary to be able to produce the material of good quality, but it is further essential that it should be produced at prices which will compete with those of foreign firms. With the greater time which manufacturers will be able to devote to the subject with this end in view, there should be no difficulty in arriving at a satisfactory solution of this point.

However large the possible output and however perfect the quality of British optical glass, the future of the industry can be assured only if British opticians are able to achieve and maintain supremacy in home and foreign markets by excellence in the design and workmanship of their instruments of precision and by cheapness of manufacture of the more common optical products.

CHAPTER IX

COLOURED AND ORNAMENTAL

THE manufacture of coloured glass was practically confined, during the earlier years of the firm's existence, to green pot-metal, although in 1835 recipes were purchased for £5 5s. from a gentleman professing to be a Prussian manufacturer, and flint glass pots for the purpose were ordered to be placed at the ends of every furnace.[1] Stained glass, red, orange, yellow, and lemon, was produced, but only to the extent of some 20 to 40 sheets per week. In June 1838 a muffled kiln was ordered for these, and in 1839 the staining room to be enlarged.

By 1843 an Ornamental department had been established, and so much progress had been made that a stained glass window, partly designed by a Mr. Dobson, could be sent to the Houses of Parliament. We find also orders for a "geometrical delineating machine" and a pentagraph from Holtzapffel & Co., and inquiry made of that firm concerning a first-rate artist to supply designs.[2] In December 1844 the aid of Bontemps was invoked. The firm, James Chance wrote to him, must establish the manufacture of coloured glass on a proper footing, without delay, since one of their principal opponents was producing to such an extent as to interfere with their connexion. "We shall therefore require your instructions both as to mixtures and manipulations necessary for obtaining all the chief colours."[3]

In 1846 a patent was granted to James Chance and Henry Badger, manager of the department, for reheating sheets and articles of glass for staining, painting or enamelling,[4] and renewed attempts were made to obtain the services of a good figure painter and designer. Then, in May 1847, Edward Palgemayer was engaged "to carry out and superintend the manufacture of coloured glass" by processes of his own. It was discussed whether, in the interests of secrecy, the furnace to try his special mixtures

---

1   Board minutes of February 5 and November 6, 1835.
2   Letters of James Chance, August 15, 1843, and August 16, 1844.
3   James Chance to Bontemps, December 9, 1844.
4   English patent No. 11185. It included arrangements for flattening and bending sheets of glass.

ALEXANDER MACOMB CHANCE
(To face page 211)

should occupy an out-of-the-way corner at Spon Lane or be put up at Oldbury. Decision was in favour of the former course, on a site adjoining No. 2 house.[1]

First experiments, however, were made in the house itself. It was found

that prolonged founds and a general heat are quite essential for these materials, for otherwise the sulphuret is volatilized before the fluxing commences and the metal becomes hard to melt by an excessive loss of alkali; this would account for the apparent discrepancy that the hotter the furnace, the more difficult it is to melt this metal.

Then it was resolved "that this frit is uncertain in its nature and results, and is also very destructive to the pots and furnaces, so far as it has been hitherto tried." The experiments, therefore, were relegated to a small laboratory furnace, and Palgemayer was desired to write to Munich for a German workman to blow his flashed glass. In November he was given entire management of No 3, not only in respect of two pots for his flashed and pot-metal, but also of those for ordinary sheet and shades. At the end of the year additional room was provided for mixing the materials for "the Cathedral glass, the best green glass, and the light coloured glass," and it was resolved to have a third pot in No. 3 for experiment on the last. Attempts were made again to engage a competent artist in England or from Munich.[2]

Results were disappointing. All the coloured glass, James Chance advised, was too thin, much of it under 13 ounces. He would have its substance not less than 18 to 21 ounces, even at the cost of smaller sizes. He saw no reason for exceeding for ordinary stock, at the risk of bad workmanship, 38 inches by 28, and thought it absurd to flash the glass expensively. He found both flashed and pot-metal too light in colour. "It is not advisable," he wrote, "to be blowing all the colours in the wholesale way in which they have been made hitherto. We shall do ourselves much harm in sending out 2nd rate coloured glass "; procedure must be cautious and in the first place the right tints be obtained."Palgemayer ought to perfect what he professes before taking up a variety of experiments."[3]

After four months it was decided to suspend the manufacture of the light-coloured glass and to give Palgemayer three months' notice to

1   Lucas Chance to James Chance, May 19; Board minutes of May 5 and June 12, 1847.
2   Board minutes of 1847.
3   Memorandum of December 24, 1847.

terminate his agreement. Although, on his pledging himself to furnish good results, if not interfered with, he was maintained for a time in the exclusive and responsible management of No. 3 house,[1] on the coming of Bontemps he left.

Bontemps was the leading expert of the time in the manufacture of coloured and decorative glass. It was he who in 1836 had rediscovered the secret of producing the brilliant ruby glass of the ancients. Since 1829 he had applied himself specially to the painting of windows, and he had founded a school of the art at Choisy-le-Roi. He introduced at Spon Lane the processes of which he gives a full description in his book; enamelling on glass, engraving it by means of hydrofluoric acid, applying patterns by printing and lithography.[2] From June 1848 he undertook the superintendence of the Coloured and Ornamental departments, as said elsewhere, Badger continuing to manage them under him. As an early result, customers were requested to return all specimens in their possession belonging to the firm.

That the departments soon were busily employed appears from a memorandum by Bontemps at the end of the year, asking for more men. He reminded the Board that several leaded windows, each of which would bring in new orders, would shortly be finished; that he would soon have to put in hand the window for the Birmingham New Cemetery; and that he had only two men for the lead work, one of whom was the best figure painter, the other the machine etcher. He wanted to secure orders by sending out specimen work, stating that in six months he had only been able to make three examples for the purpose. Some of richer design, the cartoons for which were ready, he expected to be very useful, but as at present situated he could not undertake to complete them in less than six months. He asked for at least one more lead-worker.

The Board, on the other hand, had resolved on November 28 that Bontemps and Badger should submit a scheme for reducing wages. Opposing this, Bontemps pointed out that in starting a new manufacture a first difficulty always was to collect a proper set of men; that now "a crew of clever and quiet men" had been assembled; that more would be wanted, as orders increased; and that if wages were lowered, the best of them would seek other situations. His remedy was not to reduce wages, but to get more out of the men by piece-work. He himself, he said, had never

---

1   Board minute of May 10, 1848.
2   See the *Guide du Verrier*, pp. 335-6, 364-5, 710 *et seq.*

objected to a workman earning six or seven francs a day instead of five, if he did two days' work in one, and on this principle had obtained from twelve men, better satisfied, the same output as previously with twenty, and at less cost. "The men acquire more dexterity and quickness, and you become able to reduce the prices without reducing their reasonable living; you have more profit without lowering the comfort of your people." As to obtaining orders, Bontemps observed that very few glass dealers knew anything about ancient windows, or could supply information wanted, fix prices, or so on. The firm should correspond with customers direct. Circulars, suitably worded for the different cases, should be sent out not only to architects, but also to the bishops and higher clergy, both Anglican and Roman. Such a "capital work" as the New Cemetery window would help much, when it could be seen; it should be noticed in the leading newspapers and by the Archaeological Society. In a word, advertise.[1]

In another memorandum Bontemps dealt with the question of design, giving particulars of six different styles of work that he would like to execute as examples. In employing Mr. Wurm, he said, the firm had had in view placing their studio on the level of any existing on the Continent or in England.[2] But the art was not confined to Munich; it flourished besides at Metz, at Choisy, at Paris and elsewhere, and in England there were artists such as Pugin. Although the department was well equipped with men competent to carry out the work, there was no-one to supply proper designs but Wurm; it was better to give up figure painting altogether, than to turn out such a window as that of the Evangelists. Appreciating this, the firm had desired Wurm to obtain an artist from Munich. But there was only the royal studio there, employing two or three skilled artists at the king's expense; they would not leave assured positions, with the prospect of pensions, save for a very high salary and guarantee. And a man who had yet to learn his trade would be of no use. In France, on the other hand, there were several sources of supply, and in the present condition of affairs there men could be had at a cheap rate. For instance, it was fairly certain that a man who had executed Wurm's cartoons at Choisy would come for £10 to £12 a month, travelling expenses, and £15 a year for lodging and firing. Besides designing, he could turn his hand to other work, as he had done at Choisy, such as determining the proper price for each window, and so establishing production on a surer basis. There should also be a painter for the ornamentation, especially the architectural, who could help Badger too in laying out the work for the lead-men. Such a man was a German who had done such work at Choisy and would come for £8 a month.

In accordance with this advice of Bontemps, a skilled workman, Charles Snape, and an apprentice named Hill, were obtained from Messrs. Forrest and Bromley for the leadwork. Then Bontemps was given the superior charge of No. 2 furnace, and authorized to engage the French blower Eperthener specially for coloured glass, and it was resolved to advertise for painters again and to institute piece-work. In the autumn of this year, 1849, the firm was able to show results at the Exhibition held in Birmingham in connexion with the meeting of the British Association, on which occasion Bontemps read a paper on coloured glass. Their principal exhibit was the window mentioned, of three lights and in the style of the thirteenth century, destined for the chapel of the new Church of England cemetery at Hockley. It was not a little commended. One critic felt himself "bound to speak highly of the design of this window. The grouping is well managed, and the drawing, with one or two exceptions, is excellent." Although he thought the window "an illuminated picture rather than a means for the transmission of light'" yet he rated it "a fine work in conception, in feeling, in colour, and in execution," doing "as great credit to the designer and the workmen who have assisted in its execution," as to the "spirit and enterprise of the firm." It may be noted that the firm insisted, in opposition to the wishes of the Cemetery Company, on the window being designed by their own artist.

Like praise in regard to workmanship was accorded to two smaller windows in Renaissance style, though the designs did not please the critics. One of them wrote: "We greatly admired the one with medallions, but utterly eschew all admiration for, or approval of, the Renaissance style of glass painting, however meritorious the component parts of the design may, taken separately, be." And another, with nothing but praise for the large window: "the other two belong to a debased style of glass painting, when elaborate pictorial effects took the place of the simple mosaic." Apart from the style he judged them very excellent in execution, artistically painted, and the tints well chosen and harmonious." Besides was noticed "an exquisite panel flower-group, treated with great delicacy of handling and richness of effect." [1]

Another window, of which there is notice in this year in Robert Chance's diary, had been placed in a church at Eve Hill, Dudley.

At the Great Exhibition of 1851 the art products of the firm failed to interest the Jurors. There was one window with medallions, in thirteenth-century style, destined for a church at Leamington, another with armorial bearings of distinguished people, such as the Prince Consort, the Duke of Wellington and Lord John Russell, a third with figures of St. Peter and St. Paul, and a fourth, in the Renaissance style, with representations of St. George and Britannia. Also "A landscape and ornamental work, suitable for a dwelling-house," and "flowers painted and enamelled on a large sheet of glass, with borders ; the glass having been burnt in a kiln four times"[2]

In the previous year Bontemps, as has been noticed, had employed a portion of his time in Germany inquiring into methods of making coloured glass there. He had particulars to report about the manufacture especially of ruby and of yellow pot-metal. On his second tour in 1852 he saw Messrs. Gimiez, of Paris, about the preparation of a book of designs, and at Hamburg and in Holland he left specimens of the firm's work in landscapes.

At the Paris Exhibition of 1855 large sheets of coloured glass shown by the firm were commended for their uniformity of tint. A novelty was Bontemps' flashed glass etched with hydrofluoric acid so as to produce white designs on a coloured ground. A special process of printing, he says, simplified the work and made it better and cheaper. But another mode of ornamentation was criticized by George Wallis, of the Birmingham School of Art, as follows:

> The method, introduced by Messrs. Chance, of decorating window glass by the transmission of impressions from lithographic stones, either drawn in chalk or ink, was illustrated by a few contributions; but unfortunately the subjects and style of execution were not of such a character as to attract much attention. This was to be regretted, as with a judicious adaptation of a means at once simple and effective, when a proper mode of treatment is adopted, highly decorative effects may be produced in an economic manner.[3]

---

1   Aris's Birmingham Gazette, October 22; Midland Counties Herald, October 18, 1849.

2   Bontemps' *Examen historique et critique*, p. 54 ; Official Descriptive and Illustrated Catalogue, ii. 701.

3   *Rapport du Jury Mixte International*, pp. 931-2; *Reports on the Paris Universal Exhibition*, 1. 254 (Wallis), ii. 394 (Bontemps).

In 1857 was instituted an accurate register of the mixtures used for coloured glass, with the processes of manufacture and the qualities and quantities of the produce. Made at this time were yellow, blue, purple, green, signal-green and yellow-green pot metals; flashed ruby, blue and opal; green, yellow and rolled "cathedral" Masselin had been in charge of the department, in addition to his other duties, since 1851, and Lucas Chance had expected much from him.[1] In this year 1857 was introduced a new ruby glass, much harder than the old, as was found in the embossing. A furnace with four open 42-inch pots was built for it, but was very soon put out, the failure being attributed to want of time on Masselin's part to attend to it properly. But a similar furnace in No. 3 house failed also. Covered pots were then tried, and with success. With them, Henry Chance records, there was less loss in skimming, greater certainty of good results in colour, and increased production, since the furnace could be kept hotter during the time of working and so the pots be worked out lower. He writes :

> The new ruby, as it can be blown out, will yield as many cylinders per pot 46 x 32 as the old (lead) ruby yielded of 36 x 24, the brim ends cut off from the latter being almost enough, if blown out, to make a cylinder 36 x 24, and also a great deal of metal is saved by the system of blowing "poste" (described).

As concerns church windows, if we may judge from that erected in Oldbury parish church to the memory of William Chance—a window not long since destroyed by a gale and replaced by one in modern taste at the expense of his grandchildren—the productions of the firm at this time were among the worst of Early Victorian perpetrations.

It was now, or soon afterwards, that Sebastian Evans was placed in charge of the department. He also was an educated man, a Master of Arts, with a taste for journalism. A notable work of his design was the "Robin. Hood" window shown at the London Exhibition of 1862. It was a fine piece of work, but did not find a buyer; as stated on the painting of it preserved at the works, "owing to its unfortunate position in the building, it could not be properly appreciated." It lay by for some years in pieces, until

---

1   "With an educated man like Masselin, provided he be attentive and persevering, we shall raise up a manufacture of coloured glass such as never was seen before. ... If he proves inefficient, of course we shall put an end to the agreement. . . . Coloured glass requires a separate furnace and a scientific manager. Bontemps knows all that is necessary, and Masselin is the man to carry his knowledge out " (to James Chance, August 12, 1851).

Talbot received permission to collect them and reconstitute the window. Subsequently it found its way to the South Kensington Museum, where it remained until, a few years since, the authorities inquired whether the firm had any use for it. The reply being in the negative, it was understood that the window would be broken up.

Another large ornamental window of the time illustrated a passage in the "Idylls of the King." It is figured in a detailed description of the making of leaded windows at Spon Lane, published in the Illustrated Times for June 21, 1862. Generally, the firm's coloured glass shown at the Exhibition was reported to be "much better than the foreign and not much dearer. The foreign glass has a dead appearance, whereas ours is brilliant; their metal looks very stringy and the colour smeared."[1]

In the summer of 1862 Evans expressed desire to be relieved of his charge and to be employed only as an artist. George Muddock was appointed in his place, at a salary of £200 a year. Evans, meanwhile, had accepted an invitation to contribute an account of the glass shown at the Exhibition to the "Record" of it published by the "*Practical Mechanics Journal*," an elaborate work compiled by first-class men. There was nothing in what he wrote to offend, unless it were his statement that little progress had been made in the manufacture of glass since 1851 and that it still fell "far short of the standard of excellence demanded by the needs of science and advancing civilization" or the account that he gave of its various branches. The art products of the firm, for which he was himself responsible, he naturally did not criticise. However, the firm held his action to have been improper, and when he declined to admit the impropriety it was decided to terminate his engagement. He agreed to supply designs whenever required, the firm undertaking to be ready to execute any orders that he might put into their hands. From the beginning of 1863 Muddock's agreement was renewed for five years, at the same salary but with 7½ per cent, of the profits of the department in addition.

Among the minutes of 1865 occur acceptance of a proposal by him to enable the workmen to show examples of their industry at the forthcoming Birmingham Working Men's Industrial Exhibition, and a resolution to establish a special library for the department, and to supply it with art periodicals. But the manufacture of windows was nearing its end. In June 1864 it had been decided, says John Chance, to show no more "windows of

---

1   Private report by A. F. Jack, previously cited.

the modern kind, but to endeavour to produce Church work of a character to place us in the front rank," as in other branches of glass manufacture. On the other hand Alexander Chance, charged in October 1866 to examine and report upon the state of the Ornamental department, strongly advised the opposite. He had had, he said in a letter to James Chance, "special opportunities of judging of the actual state of things owing to Mr Muddock's absence during the past week." The reasons that he gave in his report for thinking that "Church work is not, has not been, and probably would never become, a profitable process of manufacture on these works" were the following:

1. Whatever excellency of design, colours and execution we may attain, people will always prefer the works of those who make this business a speciality.

2. Our prices must consequently be lower than the market value in order to obtain orders, and the profits are necessarily reduced in proportion.

3. Many of our competitors, like Clayton & Bell, Hardman, Wailes and others, are themselves the artists and designers and superintend personally the work. They thus save large sums that your firm must pay in wages. Hence another disadvantage and loss.

4. This personal supervision has great influence, also, on many who order Stained Windows. Mr. Gilbert Scott, alluding to this point, distinctly gave me to understand that as long as none of the Messrs. Chance turned their talents to his art, he should hesitate to send us orders, or recommend us as Stained Glass Artists.

On the other hand, Alexander Chance went on to say,

The ornamental work, called "Modern Work," succeeds very well with us and is decidedly profitable. For many years past it has received the attention of your firm, and is much more closely connected with your real business than Church Work. Nearly all our modern work is ordered by the dealers, whilst nine-tenths of our good Church Windows are supplied direct to private persons. The Church Glass Stainers do not, as a rule, do modern work, whilst Chance Bros. & Co. have acquired a reputation for this style of glass that allows them to obtain really good prices. It has resulted from the former fact that it is very much easier to keep men, whom we have taught to do our modern work, than it is to keep Church Glass painters. The latter are in great demand all over England, and dictate pretty much their own terms, whilst the former only know a peculiar kind of work, that few shops undertake, and in consequence are less sought after. We have always found this to be the case.

He recommended, therefore, that Church work should be given up entirely and for the future only "Modern Stained Glass" be made, embossed, enamelled, stained and stained enamelled; that T. W. Camm, H. Olden, Healey, all the lead-men and Clarke should be discharged, and such of the painters as might be deemed proper; and that Muddock should be offered entire charge and control at a reduced fixed salary of £150 a year, but with 10 instead of 7½ per cent, of the profits.[1] Were Clarke discharged, he said, Muddock would have ample employment and be obliged to superintend details personally, whence "more accuracy and promptitude in the despatch of orders might certainly be looked for." He thought that the terms proposed would not be refused since Muddock would with difficulty obtain elsewhere the £250 a year that he might expect.

In conclusion Alexander Chance remarked: "the time now employed by me at the Glass works, some three hours a day, in looking after the interests of Church Work, would be at the disposal of the firm, and any other work you would wish to give me I should do my best to perform."

The advice was taken, and the reduction did not stop at Church work. In June 1867 it was resolved to terminate Muddock's agreement, and a year later to give up the stained work also, confining attention to the " Embossed, Enamelled and Stained Enamelled Glass." There was no longer need of artists, and the principal men left, some of them, as Swaine Bourne, Camm, Holloway, and Samuel Evans, setting up in business for themselves. The work that was left was of small importance, and died in the course of years a natural death.

It is interesting to find that in the later sixties certain kinds of coloured glass, as opal, gold ruby and yellow-green, were supplied to the firm from outside the works—namely, by George Wood, of Birmingham, who had been for many years in the employ of Lloyd and Summerfield. The manufacture of this glass was supervised by William Edward Chance. In July 1870 Wood was brought to Spon Lane as coloured glass mixer. In spite of trouble with him at the end of the year he became manager of this department, but at the beginning of 1873 had notice to leave; and about the same time William Edward Chance took over the manufacture of this "antique" glass to his works at Oldbury. Wood was succeeded

---

1   "At the same time, I think it would be desirable that I (as being thoroughly acquainted with the "ins and outs" of the department) should now and then go thro' and see that no carelessness &c. is being allowed to take place."

as coloured mixer by Job Parish, and he in his turn by Forrester Lloyd.

As time went on, employment of old-type pot-metals and of flashed blue and opal for windows went gradually out of fashion, public taste turning to prefer the lightly tinted "cathedral," "figured" and "muffled" glasses, the manufacture of which, as has been said, rose rapidly to be of leading importance. Certain quantities of pot-blue in two shades and a pot-green continue to be made, but for the rest, production has long been practically confined to the signal-green and flashed ruby required specially for signal lamps. For the former the war brought a new and very large demand for the goggles required by the soldiers in campaigns in certain climates. Not that this glass was the best for the purpose, but it could readily be turned out in large quantities for the emergency, and for that sufficed. As soon as was possible new glasses, specially for eye-protection, were supplied, again on the large scale, in particular a greenish-yellow glass for use against the glare of snow.

About the year 1912 was taken up the subject of spectacle-glasses which should absorb the chemically powerful ultra-violet rays of the spectrum, supposedly injurious to the eye. A number of trial meltings were made, and some yellowish glasses produced effected the purpose. Soon afterwards the late Sir William Crookes, O.M., as a member of the Commission appointed to investigate the causes and prevention of cataract in certain industries, turned his attention upon glasses which should absorb both the ultra-violet rays and the hot infra-red, subjecting the eye to the visible portion of the spectrum only. In collaboration with him the Company worked out mixtures to produce two types of "Chances' Crookes glass," the one colourless, save for a faint bluish tinge, the other of a pleasant light neutral tint, which in addition gave some general absorption of the visible rays. Dr. Ettles was commissioned to observe the pathological effect of these glasses by tests upon patients, but no report from him has been received. A competitor has recently brought out a greenish Crookes glass, to which has been given the name of "anti-glare."

A cheaper glass of the same character, faintly tinged with yellow and known as "antifade," is of value for the lighting of buildings whose contents are liable to injury by the chemical action of the ultra-violet rays. Experiments in this direction were conducted about the year 1908 by Rees, in consultation with the Curators of the South Kensington and the Birmingham Natural History Museums. The latter, Mr. William H.

Edwards, made use of the glass in 1913 for protection of his private collection of moths and butterflies, and in 1915 the Birmingham Museum and some of the Water Colour rooms of the Picture Galleries were fitted with it. Unfortunately, it cannot be used for the pictures and show-cases themselves, on account of its appreciable colouring. Another application has been for aeroplane hangars, where it has proved efficacious in saving aircraft fabrics from the usual rapid deterioration.

Early in 1918 demand arose for dark glasses to protect the eyes of workmen, such as acetylene and electric welders, from light of extreme intensity. Investigation initiated in the Research Laboratory at Spon Lane resulted in the production of a glass of dark green colour and simple type, termed "Chances' Arc Screen." This gives a pleasing citron-colour to objects incandescent at a yellow or white heat, enhancing the effect of variation in radiating power by which alone objects in a furnace can be distinguished. The glass can be "plained" and worked with ease exceptional and unforeseen.

More recently a number of neutral or nearly neutral glasses have been made for sextant screens, varying from a very light tint to the intense dark colour required for screening the sun. A uranium glass of optical quality and fluorescent intensity greater than the similar Jena product has been made, also a glass which absorbs all but the last remnants of the visible violet but transmits a large amount of ultra-violet light. Experiments are in progress to satisfy a number of inquiries for special coloured glasses of high transparency and optical quality.

CHAPTER X

THE ALKALI WORKS

LACKING room for extension of their chemical plant at Spon Lane, Chances and Hartleys in April 1835 took on lease from the devisees of John Houghton, deceased, 8¾ acres of land at Oldbury lying between Park House Lane and the "Houghton" arm of the Birmingham Canal. The lease was for 99 years at £100 a year, with power of redemption at 20 years' purchase at any time within that period, an option exercised by Chance Brothers & Co. in 1841 for the sum of £2,050.

The immediate object was to erect plant to work Phillips's process for making sulphate of soda by furnacing salt with copperas, noticed in the first chapter. Operations on a large scale were in contemplation; Clay, the chemical manager at Spon Lane, submitted estimates for plant to cost £4,340 for a yield of 48 tons of saltcake and 34 of white ash per week, or £6,430 for double the quantities; chimney, engine-power and machinery being reckoned at the same amount in either case. For the lower sum there were to be eight saltcake and six black ash furnaces, a pyrites bed of an acre in extent, a wharf 50 yards long, and other plant to suit. But caution prevailed. The wharf was ordered, but for the rest only two salt-cake furnaces, a pyrites bed of a quarter the size projected, a cistern and a receiver, a leaden cistern or boiler for evaporating, a chimney costing £50 instead of £700, and a shed; the whole at an estimated cost of £490. Such was the humble germ from which, as the oak from the acorn, have grown the great works now of Chance & Hunt Limited. The work was put in hand at once. When it was near completion Clay advised erection of a cottage and counting-house, "as trespassers are already troublesome, and some person should be constantly in the immediate vicinity"; and this was allowed.

Failure of Phillips's process is indicated by a resolution of February

THE SPON LANE SCHOOLS 1845
(To face page 222)

1837 to erect a vitriol chamber, 55 feet long, 17 wide and 12 high, estimated to cost £560 and to suffice for a production of seven tons of saltcake per week. Also were ordered additional buildings, six low chimneys and flues, two "barilla" and three other furnaces, while Clay was instructed to look out for land in the vicinity suitable for houses for the workmen. Additions during the next three years included a crystallising house "at the north end of the present erection," a steam engine and boiler, two cranes, a lime kiln, a vitriol chamber about 12 feet square for experiments, two more saltcake furnaces, a white ash furnace and chimney, another barilla furnace, a furnace for rouge, and a cooperage. To be noted is an order for five tons of Glauber's salts. Lastly, in 1840, came a new and larger vitriol chamber, 120 feet by 19 by 18, notable because its burners, at as early a date, almost, as anywhere in England, were constructed to burn pyrites[1]—iron pyrites, that is, obtained chiefly from the "coal-brasses" of the Midlands, with finer qualities from Cornwall and Wicklow.[2] The greater portion of this early plant lay along Park House Lane southwards from the crystals house. Carpenter, after Clay's retirement, managed the works until 1841, when he was replaced by Peter Ward. As builder we find already the name of Thomas Jackson.

In 1844 two more vitriol chambers were ordered, of the same size as the last, partly to relieve Spon Lane and partly to increase production; also a saltcake furnace of John Lee & Co.'s new fashion.[3] When in the following year it was decided to transfer the whole of the firm's chemical manufacture to Oldbury a fifth chamber and more furnaces were put up, and the works became of real importance.

Manager Ward was active in pursuit of improvements and new

---

1   The innovation dated in France from 1833. In 1838 a monopoly of Sicilian sulphur, granted by the King of Naples to Messrs. Taix and Co. of Marseilles, raised its price from £5 to £14 a ton. Immediately a number of patents for burning pyrites instead were taken out in England, and Thomas Farmer began to use it in 1839. The monopoly was soon revoked, but the innovation by that time was established (*Reports of the Juries*, Exhibition of 1862, Class IIa, pp. 11, 12; William Gossage, *A History of the Soda Manufacture*, British Association, 1861. A later authority, Professor Roscoe, in his South Kensington Lecture of 1877 on "*Technical Chemistry*," stated that the price of sulphur was raised by the monopoly from £6 or £8 to £20 a ton).

2   James Chance in 1841 instructed Carpenter, manager at Oldbury, to send over twelve tons of the ordinary pyrites every week, as it was found to assist the burning of "our Wicklow and small Cornish."

3   "For the most complete apparatus for the production of saltcake manufacturers are indebted to Mr. John Lee, who proposed the substitution of a concave iron pan for the leaden one previously employed"; whereby much less corrosion and the possibility of using much stronger acid (Muspratt's *Chemistry*, 1860, ii. 908).

processes. Among failures may be noticed Adcock's plan of recovering the nitrous fumes from the chambers by drawing them through showers of water or other liquid by means of a spiral wheel; the schemes of Francois de Sussex for using nitric acid instead of sulphuric to decompose salt, and in other ways, and modifications of them devised by Ward himself and by Cowper, the firm's chemical adviser; and, perhaps, a patent washing powder made by mixing glue, gelatine or mucilage with soda or potash.[1] On the other hand, Sautter's process for absorbing the nitrous fumes in concentrated sulphuric acid was a success.[2] It was talked about in 1843, and two years later Robert Chance saw it at work at Prémontré. Other favourable reports having been received, it was introduced at Oldbury and remained in permanence; for long still known there by the name of Sautter, but generally by that of Gay-Lussac.

Another success, of importance probably unforeseen, for the principal object at the time was to utilise part of the waste and noxious fumes of muriatic acid, was the manufacture of bicarbonate of soda. Ward condensed the fumes in pipes of earthenware cooled with water and used the acid obtained to generate the necessary carbonic acid from limestone. To provide a site for the condensation the copperas bed near the vitriol chambers was removed. Very soon three bicarbonate caves were at work, and the manufacture remained a source of profit to the firm in permanence.

The muriatic acid nuisance was, indeed, a pressing question. As far back as 1836 complaint had been made of the damage done by the fumes from Spon Lane, allowed to escape into the air. Clay had then taken out a patent for condensing them by means of an underground flue, suitably packed,[3] but the difficulty was, how to dispose of the acid when condensed. The manufacture of bleaching powder was considered, but the project got no further than plans and estimates. At Oldbury the nuisance increased with the growth of the works, and complaints became frequent and insistent. In May 1847 the inhabitants, headed by the vicar, threatened organised action for damages. Their deputation was informed that any such threats would be set at defiance, but that endeavour should be made to do justice in each case of complaint, after due inquiry. Part of the acid was utilised, as said, and the manufacture of bleaching powder was again projected,

1   English patent No. 11189 (1844), in Ward's name.
2   English patent No. 9558 (1842), granted to Charles Maurice Elizee Sautter.
3   English patent No. 7196 (1836).

as also that of chlorate of potash. But nothing of that kind materialised. The only remedy seemed to be to build a chimney 200 feet high to carry off the fumes. But first had to be ascertained whether that was feasible on the site selected, undermined by coal workings, and for the present it was resolved to replace the existing chimney, 160 feet high and in a dangerous condition, by one of equal height but of much larger area at the base. Relief was afforded when opportunities began to offer for sale of the acid. There is record in 1849 of a proposal by Messrs. Evans & Askew to take it all at 70s. per ton, provided that it were free from iron and sulphuric acid, and of an agreement to supply a Mr. Bradley for three years. The condensing plant was now ordered to be trebled, and before long came other opportunities.

In 1846 Ward took out a patent for carbonating the soda ash liquors by means of carbonate of magnesia, and incidentally for manufacturing Epsom salts.[1] The process promising well on trial, a monopoly was secured of importation of "Tripoli stone" from the Levant.[2] In April 1848 it was resolved not to license other manufacturers, "because all the ash-makers would be induced to become soda-makers, and our competitors both in soda and soda-ash would produce articles equal to our own." In spite of these high hopes, the next notice of the process— August 1849—is of large quantities of the stone lying by unused. Carbonation and the manufacture of Epsom salts had alike been suspended, and attempts to supply sulphate or muriate of magnesia for manure had come to nothing. One of Lucas Chance's economical proposals of the year was to turn £5,000 worth of the stone to account in some way. Ward, acquainted, presumably, with the work of Dyer and Hemming, saw a possible solution in an ammonia process analogous to that for soda, which in modern times has proved so brilliant a success. The process was bound to fail from leakage of ammonia, if from no other cause, before the means were found to prevent that

---

1   English patent No. 11279 (1846). It contained, besides, provisions for making sulphides of magnesium and calcium and for treating for the manufacture of Epsom salts magnesium stone directly.
2   James Chance wrote on November 25, 1847, to Mr. J. W. Goodman, of the Minories, who had sent a sample of the stone from Turkey containing nearly 94 per cent, of carbonate of magnesia: "It doubtless comes from one of our own cargoes, as we are the sole importers of the article and have a patent for all known applications." To quote Muspratt's *Chemistry* (ii, 535): "Mr. Chance has lately imported a very pure natural carbonate of magnesia from the Mediterranean, to which he adds sulphuric acid, and proceeds in the usual way for the production of sulphate; and having, it is understood, secured the entire produce of the mine, he is likely to become a formidable competitor to the other manufacturers."

leakage. The only alternative was to resume the manufacture of magnesia and of Epsom salts, under Ward's patent.

Another idea of Ward's, interesting to find at this early date, was to dispense with chambers for the manufacture of sulphuric acid. Robert Chance records the fact in his diary, but does not say what were the means proposed. A novelty patented, from which much was evidently hoped, was a reverberatory furnace of which the arch was heated by causing the waste heat to pass over it.[1] The firm undertook to bear all expenses in connexion with it, and Edward Chance, now directing the works and one of the patentees, was to keep for himself all royalties on licenses granted and all profits made elsewhere than at Oldbury. He was specially authorised to offer licences to Messrs. Tennant and John Lee & Co., and it was ordered that the invention should be applied to all the saltcake furnaces immediately, and to the black-ash furnaces as they required renewal. Next year the manufacture of soap was contemplated, but not undertaken. But that of rectified sulphuric acid was begun, on a scale of ten to fifteen tons a week, and five pyrites burners were ordered to be substituted for a like number burning sulphur. There was a proposal, too, to absorb the nitrous fumes in lime, with subsequent recovery of nitrate of soda. For other matters, a laboratory was established in the office buildings, and there is first mention of cabin boats to carry ash and soda to Liverpool and bring back "stone" from Nantwich.

For all this enterprise, the works were a source of loss instead of profit to the firm. In January 1847 William Chance termed the management "infamous" ascribing the blame to his brother undertaking more than he could properly attend to.[2] It had already been decided to put Edward Chance in control. "Let him have it complete, ledger and everything" his uncle had written, expressing desire for relief from the responsibility. He undertook to buy the nitre and brimstone, to see to economies, to go with his son Robert periodically to inspect, and to supply the necessary assistance with the books at balancing time. He claimed, after a thorough investigation of ways and means of reducing costs and increasing produce, to have "put fresh life into the concern" and expected a production of 80 tons of soda per week.[3]

---

1   English patent No. 12067 (1848), taken out in the joint names of James and Edward Chance. It included an improvement in the glass manufacture, drawing sheets into the annealing kiln by means of tongs, instead of pushing them in, as heretofore.

2   To James Chance, January 18, 1847.

3   Lucas Chance to the same, November ri and 15, 1846.

Accordingly it was resolved, "That Mr. Edward Chance should devote all his time to the Oldbury Works, with a view to his becoming the responsible Manager of that Concern, being subject to the control of Mr. R. L. Chance in the mercantile part, and to that of the Board only in the manufacturing Department."Ward remained as manager under him, directed to "consider himself on all occasions as responsible for any thing connected with the Oldbury manufactory."[1]

The works had now grown so large that more land was required. In 1848 Ward was instructed to negotiate with Dugdale Houghton for the purchase of about four acres beyond the canal arm, and when Edward Chance submitted a statement showing that the deposit of vat waste and rubbish on other land of Houghton's, on the other side of Park House Lane, would effect an annual saving of £293, it was decided to purchase all of it that could be got by annual payments of £60 per acre, £75 being deducted from the first instalment on account of the brook. Houghton agreed to sell 4½ acres, so that "Blue Billy" mount may be dated from this time. The first-named purchase was agreed just two years later. Lastly, in 1851, Houghton offered "the remaining piece of land on the side of the arm of the canal opposite to the Chemical Works," and the offer was accepted. There were the usual legal delays, but formal conveyance of all these lands for the sum of £8,211 was at last effected in June 1853. The part beyond the canal arm extended to Trinity Street and measured 8¼ acres, but two acres at the south end were re-sold for £1,320 9s. 3d. to Messrs. J. & E. Sturge, who now removed thither a portion of their works from Birmingham, and who before very long handed over the undertaking to Messrs. Albright & Wilson.

Besides these purchases, it appears from a Board minute of 1858 that the firm held from Dugdale Houghton, on a lease terminated at Michaelmas 1857, the whole of what remained of his "Park House Farm," some 46 acres, and that they occupied about the half thereof. Park House itself and sundry fields were situated at some distance from the works to the south-west, but the bulk of the property lay eastwards of the "Houghton" canal arm, and what Chance Brothers & Co. occupied must be taken to have been part of this. There was a new lease of some part in 1859, but soon after that date Houghton must have disposed of almost all the property, for what the firm acquired of it subsequently was bought from other owners.

---

1   Board minutes of November 26 and December 16, 1846.

Ward retired from his post towards the end of 1852. The last thing done under his management was of the first importance, introduction of the manufacture of salts of ammonia. It was agreed with Mr. Bradley, before mentioned, in the autumn of 1851 to take over his sal ammoniac business. Next year Ward took out a patent for improvements in sublimation, for recovering the ammonia left in the gas liquor after distillation, and for obtaining it directly from coal gas by treatment thereof with acids.[1] Just before he left estimates were ordered to be prepared for production of a ton a week of carbonate of ammonia. Plant to obtain the ammonia from gas liquor was put up in the first instance on land at Swan Village, leased from the Birmingham and Staffordshire Gas Company.

The new manager was Henry Samuel Rayner, and at the same time Henry Chance joined in the administration. In April 1853 he was given charge of the office work, and in September entire control of the alkali processes. Objects of expenditure in this year were new vitriol chambers, more muriatic condensers, a hoist, cooperage buildings, large extension of the crystallising plant, another bicarbonate cave, a supply of pure water from the mines, additions at Swan Village, and two more furnaces for saltcake for the glass works, raising the production of that article to 50 tons per week. The hoist must have been the simple and efficient water-balance lift, designed by James Chance, so long in use for transfer of the vat waste to the "Blue Billy mound, while the cooperage extension was intended to work James Robertson's patent for making casks by machinery.[2] It was agreed to take a licence for this, exclusive within twenty miles of Birmingham, and Edward Chance was authorised to arrange about the machinery and for a cooper to go with Robertson to Glasgow to learn all about varieties of casks. A good demand from Burton was expected, but the venture failed. In October manufacture for sale was suspended, and in May 1854 the use of Robertson's machinery abandoned altogether. In December it was offered for sale to a Liverpool gentleman for £1,800.

A youth now distinguishing himself in knowledge of chemistry was Thomas Cotterill, the same who in after years occupied so prominent a position in the offices. While yet in his apprenticeship a patent for improvement in the manufacture of soda was taken out in his name, and it was proposed to give him and Rayner separate departments.

---

1   English patent No. 295 of 1852.
2   English patent No. 12225 (1848).

The patent, No. 74 of 1853, was for getting rid of the sulphide in the vat liquors by means of native protocarbonate of iron, finely powdered, half a ton of this being required to a ton of soda ash. The process was not successful; at the end of the year it was suspended on the ground that there was no apparent saving.[1] It was arranged to employ Cotterill in making analyses, with permission to do the work at home. When free of his apprenticeship, in June 1857, he was engaged for five years at a salary beginning at £105, so that clearly he had shown himself a youth of value.

Still the works were unprofitable. At the end of 1853 William Chance stigmatised the last half-yearly balance sheet as "very painful." He thought that his son Henry might be more advantageously employed elsewhere, and blamed Edward for not spending more time at Oldbury. He hoped that Lucas and James Chance would watch vigilantly what went on, "for otherwise I fear it will be an out of pocket affair, as it has been for many years."[2]

It was no want of willingness to work that prevented Edward Chance from attending at the works so regularly as was desirable, but ill health. Indeed, for a man with weak lungs their atmosphere must have been trying in the extreme. Of the family ability and business capacity he had his full share, and he actually worked much harder than his health should have permitted. He and his brother Henry were continued in their charges, and before long they and Rayner put a very different complexion on affairs. By 1857 good profits were being made, and this was maintained for several years. There was great expansion of the ammonia and vitriol plants; bicarbonate production was raised by some ten tons per week; and the manufacture of superphosphate for manure was undertaken. A minor but important success was the application of the waste heat from the blackash furnaces for salting down the vat liquors.

There were also failures, as with an early form of rotatory black-ash furnace by Elliot and Russell,[3] a new arrangement of fire bars devised by John Bird, employed at Lucas Chance's fireclay works at Himley,[4] and Shanks's process for carbonating the red liquors, which proved too costly when the carbonic acid required was generated by the now valuable mu-

---

1   "It being borne in mind that the difficulty in regard to the mother ash from the old process has to be overcome " (Board minute of December 8, 1853).
2   To James Chance, November 11 and December 17, 1853.
3   English patent No. 887 of 1853.
4   English patent No. 2358 of 1854.

riatic. Interesting matters of 1857 are a contract with John Corbett, of Bromsgrove, for a supply of salt for seven years at 11*s.* a ton delivered, and an arrangement with Hunt of Wednesbury for him to take all the vat waste for six years, subject to six months' notice to discontinue. This, it may be supposed, had to do with the sulphur recovery processes now being tried by Gossage and others.

The firm was put in the way of manufacturing superphosphate by Dr. Thomas Richardson, of Newcastle, who had several patents on the subject.[1] He had been called in to assist in 1853, when it had seemed possible to save the magnesia manufacture by employment of his process of subjecting the impure carbonate or hydrate, suspended in water, to the action of carbonic acid, the pure salt being obtained from the resulting soluble bicarbonate. The process, his patent stated, was specially adapted for "the impure hydrate of magnesia, which is a waste product in what is known as Ward's process for carbonating soda-ash."[2] The end in view was not attained, and the manufacture was finally abandoned, but Dr. Richardson was retained as consulting chemist at a salary of £50 a year, "with a view to obtaining information respecting improved methods of making alkali upon the Tyne."

Edward Chance reported the demand for superphosphate for manure to be already very large and fast increasing. He saw every prospect of an extensive and profitable manufacture, offering great local advantages and increased consumption of sulphuric acid. Dr. Richardson undertook, for a fee of £100, to put his process into working order at Oldbury, sending a working foreman to superintend. It was determined in December 1856 to make two kinds, one for cereals, containing a percentage of ammonia, "the other of a simple character, similar to that made and sold by Dr. Richardson." In July 1857 a contract was made with James J. Headley, of Cambridge, for a supply of coprolites, 30 tons a week at 41*s.* a ton delivered, option being given to double the whole quantity of 500 tons contracted for before the end of the year. The option was exercised, and the manufacture was carried on successfully for a number of years.

Increased supplies of gas liquor for the ammonia process were being sought already in 1853, when to cope with the surplus winter supply it

---

1   Presumably this was the Dr. Richardson, F.C.S., F.R.C.S., who in 1862 was Professor of Chemistry at the University of Durham.

2   English patent No. 1441 of 1853. It included an improvement in the manufacture of "Venetian Red."

was resolved to construct a reservoir at the works to contain 8,000 cubic feet. In April 1854 negotiations were opened with the Wolverhampton Gas Company for 91,000 cubic feet a year at 13s. per 1,000 gallons, and in September it was resolved to erect plant beyond the canal arm,[1] such as would enable to be worked up into saleable sulphate or muriate all the liquor which could not be dealt with at Swan Village. In March 1855 arrangements were made to cease manufacture there, the Gas Company agreeing to deliver their liquor at Oldbury at the same price. The land, however, was retained on a lease of fourteen years, terminable after seven years by either party by one year's notice or by payment of a sum equal to the value of a year's supply of liquor. During the next four years we hear of more stills, subliming pots and boiling down plant to meet the increasing demand, and of more liquor sought from Wolverhampton, Walsall and Kidderminster. A curious resolution, not likely to have been carried out, was to use the coal pits for storing the liquor in preference to keeping the water in them for a reserve supply.

In 1861 the attention of Henry Chance, now in sole charge at Oldbury, and of Rayner was directed to the necessity of collecting the vapours escaping from the subliming pans. After two years it was resolved to remove this operation to the new side of the works, to enable a scheme for the purpose to be carried out. Yet two more years elapsed before Rayner received £50 for having mastered the problem, with another £50 for his services in connexion with the lawsuit of Powell versus Chance, of which below. In 1867 Henry Chance was authorised to seek provisional protection for the manufacture of muriate of ammonia from the waste calcium chloride of the bicarbonate process. But nothing appears to have come either of this or of another project of the same year, to save the cost of bringing gas liquor from Birmingham by erecting a small ammonia plant on land to be acquired as near as possible to the Pagoda Works of the Birmingham and Staffordshire Gas Company.

The vitriol chambers ordered for erection in 1853 were six in number; four of them, calculated to burn 50 tons of pyrites per week, with a consequent increase in saltcake production of 10 tons weekly, to replace existing ones, the other two, on the new side of the works, to burn on economical principles either brimstone or pyrites. Edward Kendling was engaged to take charge; we find the two new chambers known as his.

---

1   The existing plant was situated at the back of the Schools, behind the crystals house.

They were provided with larger Sautter's columns of the newest design. Generally, brimstone was burnt for acid for sale, pyrites for that used for decomposing salt. All the pyrites that the home counties could supply was bought up; finer qualities, as before, being obtained from Cornwall and Wicklow. Opportunity for sale of the acid was importantly increased by the advent of the Messrs. Sturge. A first contract with them, of February 1853, was for 8 to 10 tons weekly of brown oil of vitriol at £3 15s. per ton, another, of October 1854, for all that they required at £3 10s. A third, for the former quantity at the latter price, was made in September 1855 with Arthur Albright. Moreover, use was found for the burnt stone in its application as material for the beds of puddling furnaces. This was patented, and supplies of the burnt stone were purchased from other manufacturers, as Adkins & Co. and William Hunt.[1]

In December 1854 it was desired to cancel Kendling's agreement, but on investigation it was found that neither had the firm the power, nor was the dissatisfaction with his management well founded. Instead, an agreement for one year was made to pay him for making rectified acid up to 15 tons weekly 7s. 6d. per ton, and for any further quantity 5s.; for brown oil of vitriol of 1·750 strength 1s. a ton; and for packing and filling the acid into carboys and delivering them into boat or waggon 2d. per carboy. He remained in charge till June 1857.

In that year and in 1858, to meet the increased demand for sales and for the superphosphate, five new chambers were built, one old one being dispensed with. There was a new Sautter's column, calculated to save its cost of erection, £425, in six months, and the leaden concentrating pans were ordered to be altered to Brooman's system, recommended by other alkali makers as saving in outlay and repairs and as getting rid of a great nuisance. Now also copper pyrites from Spain was used for the first time. A contract was concluded in March 1858 for an eight months' supply, 500 tons per month, at 40s. per ton delivered at Gloucester, the vendors allowing 5s. per ton for carriage of the burnt ore to Swansea or St. Helen's.

Not that the use of sulphur was given up for many years yet. On the contrary, it increased, and the firm chartered vessels of its own to bring the brimstone from Sicily. In 1858 the weekly consumption amounted to 15

---

1   But not from the smaller makers, lest they should be encouraged to increase their consumption of pyrites (Board minute of July 27, 1854). The patent, apparently, was that granted to James Boydell, of the Anchor Iron Works at Smethwick, No. 279 of 1854.

tons, and in the latter half of 1863 to 24. The corresponding figures for pyrites were 142 and 135 tons, equivalent to about 57 and 54 of brimstone in saltcake production. During 1864 the supply of the latter amounted to 1,137 tons, at a cost of £7 1s. 7d. per ton delivered.

Two years later Henry Chance was authorised to take up on behalf of the firm 200 £10 shares of the Tharsis Sulphur and Copper Company.[1] Then, in 1868, was considered extraction of the copper from the burnt ore on the works. Mr. Thomas Johnson, of Runcorn, proposed to form a limited liability company with the firm for the purpose. This was declined, but it was resolved (August 1869) to set up plant at Oldbury and to pay Mr. Johnson £100 when it was completed in accordance with his plans and to the firm's satisfaction. The matter does not appear to have gone further.

Two more chambers had been erected in 1865, raising the number to some fifteen. There may be noted a proposal by Peter Ward, now of Bristol, to divide them up by glass partitions, to promote formation of the acid,[2] but this, it is remarked in a valuable treatise on the state of the chemical manufacture embodied in the Juries' Reports of the London Exhibition of 1862, was neither novel nor of use.

At that Exhibition the firm gained a medal for "fine samples of soda products, and particularly fine specimens of sal ammoniac." Their exhibit included samples of sulphate, carbonate and bicarbonate of soda, caustic soda, superphosphate of lime, ammonia salts, and sulphate of iron.[3]

Lastly may be noticed on the subject of sulphuric acid the continuous method of concentrating it in a series of glass vessels, introduced at Oldbury in 1871. A patent for it, No. 1243 of that year, was taken out in the name of Henry Chance, as a communication from Junius Gridley, of New York.

Messrs. Sturge and their successors, Messrs. Albright & Wilson, rendered other important assistance to Chance Brothers & Co. in requiring a large supply of muriatic acid. The first contract, of December 1852, was for 25 tons weekly, of 33° strength, at 20s. per ton. Another of January 1855 was for 30 tons a week at 10s. a ton, the acid being now delivered through a pipe and so the expense of carboys and cartage saved. The abomination

---

1   Board minute of September 17, 1866.
2   English patent No. 1006 of 1861.
3   *Reports of the Juries*, pp. 39, 178.

of former times was now an article of value, and measures were taken to increase production. The old stone condensing jars were replaced by the columns packed with coke, which Gossage had brought into use nearly twenty years before. By the end of 1853 120 tons per week were being condensed for sale or for consumption in the different processes. In 1859 it was resolved

> that a close furnace for roasting S Cake be erected at once with a view to pass ultimately the whole of the gas from our roasting furnaces through condensing towers unconnected with the main chimney, the present system of working with open furnaces and coke rendering the complete condensation of the gas impossible.

In spite of all this, claims for alleged damage to crops, trees and fences in the neighbourhood continued to be pressed. Yearly payments in compensation were considerable. In 1856, for instance, £120 was paid to Mr. Bowlby, the vicar, for damage to his house and garden. Besides which, various plots of land affected were taken on lease, to avoid compensation. As an example of this, in January 1854, it was resolved to continue to rent seven acres of Willett's land at £4 an acre, and to re-let on the best terms obtainable. And there were other nuisances; the sulphuretted hydrogen from "Blue Billy" on account of which cottages near the mound, rendered untenable, had to be bought or rented,[1] refuse from the pyrites beds running into the brook, and calcium chloride discharged from the bicarbonate process into the canal. On this last account had to be satisfied in 1858 a claim by Messrs. Chapman & Grainger for £327 10s. for damage by the contaminated canal water to their engines and boilers, a sum of which Albright & Wilson paid one third.

One case resulted in a lawsuit, interesting chiefly from its consequences. Plaintiff was John Powell, an Oldbury draper, who on a previous occasion, in 1855, had claimed and had received £8 in compensation for alleged damage to his goods. At the first attempt he lost his suit. Said the Alkali Inspector, Dr. Angus Smith, in his Report for the year 1865: "It was unfortunate that Mr. Powell chose to attack a point so carefully defended, and to prove an escape where the greatest care had been taken to prevent one" a fact established by Dr. Smith's own minute examina-

---

1   On this account in the years 1854 to 1858 two cottages close to the "spoil land" were bought from Mr. Clarke for £125, with £25 allowed in compensation for past damage, and it was arranged to rent the following: seven from Mr. Dufheid, four from Joseph Holloway, eight from James Sadler, three from Richard Williams, and three from John Dunn.

tion of all possible points at which muriatic gas could escape. Although, as he says, "it was evident, from the appearance of gardens around, that an occasional escape of deleterious gas did take place from some work in or near Oldbury," yet frequent tests, the most searching that could be devised, declared the innocence of Chance Brothers & Co. Proof of this he found in the fact that in certain "small chambers through which all the air passes after the gas has been condensed from it . . . one may remain unhurt and even at ease."

After eighteen months Powell returned to the charge, having in the meantime formed a "Ratepayers' Protection Society," five members of which were on the local Board of Health. This time he won, although, in Henry Chance's opinion, his evidence was weaker than on the former occasion, and that for the defence stronger.[1] The suit was only for £4, but the issue was deemed of such importance as to require application for a new trial.

This being refused, a veritable panic ensued in the town of Oldbury on a rumour, which had foundation, that Chance Brothers & Co. were thinking of removing their works elsewhere. Powell and his friends were stigmatised as "Artful Dodgers," and a "Tradesmen's Protection Society" was speedily formed in opposition to his, with a committee of nearly a hundred tradesmen and ratepayers. Their notice set forth that they were "obtaining signatures to a Memorial to the Messrs. Chance, expressing sympathy with them in the action against their works, now being taken by certain individuals actuated by motives of a selfish and sordid character," and "regret to learn that such annoyances may eventuate in the Messrs. Chance removing to another locality."[2]

After a crowded meeting in the "People's Hall" the Memorial was presented at Spon Lane. It was replied:

> The Memorial just read is extremely gratifying to us, and will not fail to make us pause before deciding as to the removal of our Works, especially as the statements expressed in it are those of the influential and larger portion of the inhabitants and manufactur-

---

1   Letter to James Chance, January 10, 1867.

2   The poster ended:

Property owners—Would you have your houses tenantless? Then follow "The Dodgers."

Tradesmen—Would you like to see your shops like some in the town always empty ? Then follow "The Dodgers."

Ratepayers—Would you have additional burthens heaped on your shoulders? Then follow "The Dodgers."

Look at this! ! The Alkali and Phosphorus Works have in their employment Men, who, with their Families, number nearly 3000! These occupy 509 Houses, and deal with the Tradesmen of Oldbury!

ers of Oldbury. It is quite true that we have at times entertained the idea of removing our Chemical Works to a more suitable locality, partly, we admit, from the annoyance caused by unjust attacks upon us, but mainly because an inland district like this is unfavourable to the development of a trade which depends in a great measure upon foreign demand. We have, as is adverted to in the Memorial, used every known means, and spared no expense, to render the processes at our Works free from any just cause of complaint; and the Government Inspectors have borne public testimony to the fact that no Alkali Works in the United Kingdom excel our own in the successful application of contrivances and apparatus for preventing the escape of noxious gas.

After which reference was made to the circumstances in which the adverse judgment had been given.

The sequel was seen in an election to the Oldbury Board of Health in March. The five candidates of the "Tradesmen's Protection Society" headed the poll, the five of the opposing body stood at the bottom.

In later years escape of all noxious gases from the works was effectively controlled by the system of tests first instituted by Alexander Chance in 1870, afterwards made continuous, and finally perfected by the use of Mactear's self-registering apparatus. Escapes were then reduced to the absolute minimum.[1]

Alexander Chance was appointed in June 1868 to be "Managing Director of the Alkali Works under the supervision of Mr. Henry Chance." The latter of late years, since his brother James had devoted himself almost exclusively to his lighthouse work, had had more to do at Spon Lane, besides which, the Oldbury Works had grown too large for divided responsibility. Profits there since 1864 had been falling steadily, and it was not all due to depression of trade. Alexander Chance was 24 years of age, and save for present want of technical knowledge of the chemical processes well fitted by overflowing energy and vitality and by brilliant business capacity for the charge entrusted to him. In the commercial conditions of the alkali trade he was well posted as the result of a business tour in the United States undertaken on behalf of the firm in the summer of 1866.[2]

1   See Alexander Chance's paper on "The Utilization of Waste Products in the Soda Industry," read before the Birmingham Philosophical Society on March 8, 1883.

2   Points of his report were that low test ash and bicarbonate were in great demand in the United States, and that the products of Chance Brothers & Co. were preferred for their greater purity, their bicarbonate selling for £2 to £3 more per ton than any other brand. He advised that far too little of it was sent over. With sal ammoniac, on the other hand, he found the market to be overstocked. An intimate account was given of the character and status of American firms engaged in the trade.

FREDERICK TALBOT
(To face page 237)

Arrived at Oldbury, he soon found that all was not right, that in the vitriol manufacture, in particular, there was heavy loss unexplained. Relations between him and Rayner became so strained, that in March 1869 the latter sought an interview with the Board on the subject. The partners naturally were inclined to support him, after his long and for the most part satisfactory service, but figures submitted by Alexander Chance left them no alternative but to desire his resignation. He expressed willingness to confine himself to superintendence of the chemical processes, but in the end it was agreed that he should go.

It proved difficult to replace him, and the sole conduct both of manufacturing and selling was too much for one man, however able and hardworking. Taking stock of the vitriol returns in November 1869, Alexander Chance was aghast to find a loss in that department of over £2,000 in four months. He saw no alternative but to offer his resignation.

> The revelations [he wrote to the Board] which I have placed before you in the last few days disclose such a startling state of affairs that I feel it my duty to disregard all self-interest in the matter, and candidly to tell you that I am prepared to place my resignation in your hands if, after giving this matter your serious consideration, you may desire to cancel my present agreement. ... I find myself compelled to confess my inability to explain how such a loss could have occurred, unless some one has designedly contrived it from motives of personal jealousy or revenge.

Then, after showing how he had been pressed, the economies that he had been able to effect in departments that he understood, and the absolute necessity of having a competent technical manager:

> If you require me to master the various processes, and perfect them step by step, I must be relieved of current duties, but I frankly admit that my predisposition is decidedly in favour of commercial negotiations, &c, rather than of the technicalities of manufacture. ... If I had the leisure at present, I should at once organize a system of tests and checks upon various processes now entirely neglected.

And finally he put the question :

> Is there scope for my energy and ability, such as they are, to be employed to your advantage and to my advancement at Oldbury ? On the one side appears, in startling relief, past inability to conduct processes successfully, without proper assistance and overburdened with work, and an alleged distaste for manufacture, as I fancy was inferred from the recent discussion on the copper extracting question. On the other, a taste and

an ability for effecting combinations, such as those by which the prices of ammonia and of rectified acid have been greatly enhanced, and a determination to cut down expenses, as shown in the office and in the various departments in the works, in the buying of materials and in the reduction of stocks.

We must, he wrote to James Chance, ascribing the principal loss undoubtedly to bad working of the chambers, engage a thoroughly practical foreman without a moment's delay." This necessity he emphasised by pointing out that the ammonia process, in the capable hands of Daniel Deeley, was doing very well.

Fortunately for the firm, the proffered resignation was not accepted. Alexander Chance was desired to go on and see what he could do with the help of a practical manager. Such a one was found in December in the person of Henry C. D. France, destined to remain at Oldbury till near his death. Results for the year 1870 proved very different. Comparing them with those of 1864, taken as a standard year, Alexander Chance was able to show loss of alkali, at the "very satisfactory" figure of 10½ per cent., no greater; 8,700 tons of vitriol and 7,450 of saltcake made at decreased costs of 6s. 4d. and 2s. 8d. per ton respectively; and like reduction in the costs of soda ash and crystals. He could point in the vitriol manufacture to better chamber production, saving of nitre, an exceptionally favourable purchase of foreign pyrites, an association of Birmingham makers to increase prices of the acid and reduce that of home pyrites, a large reduction in consumption of the expensive brimstone by confining its use strictly to the production of rectified acid (the figures showed 427 tons burnt in 1870 as against 1,102 in 1864), other great reductions in cost of repairs, and a profit of £2,300 gained without outlay by arrangements for sale of the burnt copper ore. And this, though during the year four old chambers had been palled down and only one large new one erected in their place. Other departments also showed up well. Fine profits from the ammonia process Alexander Chance ascribed to "increased vigilance and economy in manufacture on the one hand—for which great credit is due to its energetic and clever manager, Mr. Daniel Deeley—and greatly enhanced prices on the other." The superphosphate, he said, a thriving process up to 1864 but since then gradually dwindling to be almost profitless, for the last two years by vigorous effort had brought in £1,200 a year. While ordinary bicarbonate was quoted at £10, the firm's special quality, although in limited demand, commanded £17 per ton. He could not recommend reduction of quality, to increase sales, on account of the heavy cost of carriage to a sea-port. Of two products of small importance: "Except that the copperas is a necessary result of the use of home pyrites, and that the washing-powder furnishes light employment to women in distress, widows and orphans or children of our workpeople, these two processes might have been passed unnoticed." Generally, the report went on:

In a scientific and systematic point of view these works may be said to have made a rapid stride in 1870.

All vitriol is now delivered by measure from the chambers to every process, and the cisterns at each saltcake furnace are provided with measuring taps, of which the foreman keeps the key, and so now he is responsible for any acid wasted in the process.

A system has for some months past been in operation for testing daily the quantity of muriatic acid gas escaping in the salt cake flues. Thus escapes of gas may be discovered and arrested without delay. The Government Inspector is much pleased with this system.

Our new manager, Mr. France, possesses many qualities which render him fitted for the post he occupies. He is thoroughly straightforward, and having been born and reared in the midst of the best chemical works in England, he has acquired much practical knowledge, which has been turned to good account.

Then about the muriatic acid, no longer "a source of anxiety as refuse" but one of profit:

As regards litigation for alleged nuisance, our opponents have been rapidly pacified in 1870. John Powell has become as docile as a lamb under the soothing influence of a friendly order now and then, whilst William Underhill now rents his acres from those whom he regarded not long ago as his natural enemies.[1]

And in conclusion:

The results recorded in this report have only been obtained by sheer hard work, and by hard work alone can they be maintained and still further increased. So long as I remain with you, and I trust it may be for life, you may continue to count upon my whole energies being devoted to your service.

And so they were. For year after year under Alexander Chance's di-

---

1   It may be noted that in June Thomas Cotterill, or in his absence Keene, were authorised "to accompany Mr. Powell outside the Alkali works, whenever the latter may wish to point out what may appear to him to be an escape of Muriatic Gas from the Works" (Board minute of June 20, 1870).

rection the works continued to be markedly prosperous and a model of good management. He gathered round him a staff of exceptional capacity and trustworthiness. In the offices such men as Thomas and Charles Cotterill, John Duncan, James Robinson, W. R. Smith, or Henry Goodyere, the cashier, were not readily to be matched. Horace William Crowther rendered exceptional service in the laboratory. Leading men in the works, besides France, were Charlie Thomlinson, the engineer, his brothers Robert, who succeeded Deeley at the ammonia process, and Joseph, foreman of the fitters, Edward and William Jackson, William Hopkins, foreman of the carpenters, Thomas Orange, in charge of the vitriol chambers, Levi Evans, John and Nehemiah Bradley, James Beach, George Birch, Henry and Richard White. Nor must that remarkable man William Wall, the storekeeper, be forgotten. Alexander Chance himself received recognition of his services from 1874 in the form of a proper salary, was admitted a partner in the new firm of Chance Brothers in 1879, and finally succeeded his cousin Henry Chance as Chairman of Chance & Hunt Limited.

It would be tedious to enter into details of the progress made under his direction. Sufficient to say that all that was possible was done to keep the works abreast of the times. Particular notice may be taken of a proposal in December 1875 to substitute for the old Leblanc alkali process the newly developed ammonia-soda method, which, said Alexander Chance at the time, "if it could be generally adopted, would effect a complete revolution in the alkali trade." He advised that 100 tons of ash per week had been made by the process for two or three years past at Solvay's works in Belgium, that another factory was in operation in France, that Brunner & Mond had been turning out 60 tons per week for many months at Nantwich, and that Richards, Kearne & Gasquoine were about to start a works, with like production, at Sandbach. "Saltcake, black ash, wet vats, brown ash, together with their various pans, vats and furnaces, as well as the cumbrous and pestiferous "Blue Billy" would all be swept away, and in their stead gas liquor and salt brine become the principal reagents to be treated."

Objections to the adoption of the new process were the price of salt at Oldbury, 13s. 3d. per ton as against 6d. for the equivalent amount of brine in Cheshire; a probable dearth of ammonia in consequence of the enormously increased demand, were the process adopted generally; and especially the total loss of the chlorine. It was deemed inadvisable to make the venture, at all events at Oldbury. As years went on, and Brunner, Mond

& Co. developed their enormous industry, the salvation of the Leblanc makers was the very substance that once had been their bane, muriatic acid. And that had not availed them, but for the successful endeavours directed by Alexander Chance at Oldbury for recovery of the sulphur in the vat waste.

It had long been known that the calcium sulphide of the waste could be decomposed by carbonic acid, but the difficulty was that, unless the sulphuretted hydrogen produced were of approximately uniform strength, it could not be utilised. For the amount of air required to burn it could not, under conditions constantly varying, be regulated.

In 1883, after two years of hard work and the expenditure of £10,000, Chance Brothers succeeded in carrying out the beautiful process devised by Schaffner and Helbig. By this, by boiling the vat waste with magnesium chloride, the whole of the sulphur was obtained in the form of pure sulphuretted hydrogen, which could readily be burnt to produce sulphurous acid for the vitriol chambers. There remained in the vats calcium chloride in solution with magnesia in suspension. Through this mixture was blown carbonic acid from a lime-kiln, with the result that the lime was precipitated as finely divided carbonate, suitable for the manufacture of cement, and the magnesia reconverted to chloride.

Chemically the process was a complete success, but not commercially. It reduced the cost of sulphur by a half, but the pyrites owners lowered their price to suit. Consequently it would not have paid alkali makers to put up plant to work the new process. They benefited, and Chance Brothers among them, but the latter had borne the cost. Left with the plant on their hands, including a powerful carbonic acid pump and three large boilers set on end, used for the carbonating, they determined, before dismantling these, to try the direct effect of carbonic acid on the vat waste on the large scale. The result was unexpected. For an hour or two the gases issuing from the carbonators were free of sulphuretted hydrogen; when that appeared, it increased rapidly to double the amount equivalent to the carbonic acid that was driving it out; then it decreased again until the calcium sulphide of the waste was completely converted to carbonate.

The reason was soon perceived. Sulphuretted hydrogen is absorbed by calcium sulphide in the presence of water to form the soluble bisulphide, so that none could issue until the sulphide in the last vessel was saturated with it. But in decomposing the bisulphide, the carbonic acid drove out double its equivalent of sulphuretted hydrogen. By the time that that began to appear at the vent, there was an accumulation of it in the vessels, and gas drawn from the second of them contained a fairly regular percentage of it.

To prove the process, a circuit of seven vessels was arranged, of which five were worked at a time. As soon as sulphuretted hydrogen began to appear at the exit of the fifth, that was closed, and the third vessel was drawn upon. There the gas was found strong enough to burn in air. So long as it remained of that strength it was drawn off and collected in a gasholder; when it would no longer burn, the exit was closed. Meanwhile the sixth and seventh vessels had been filled with waste, and another cycle of operations began, starting from the third. The first and second were emptied of their contents, completely carbonated, and made ready to be the last vessels of a third cycle. In this way was obtained continuously a gas containing always 30 per cent, or more of sulphuretted hydrogen, and this could be burnt for sulphurous acid for the vitriol chambers, or half-burnt to produce sulphur itself in the Claus kiln. The process was in use before long in practically ail the Leblanc alkali works of Europe.[1] At the Paris Exhibition of 1889 its value was recognised by the award of one of the three "Grands Prix" given in the class, the others being allotted to Solvay for his ammonia-soda process and to Pechiney of Salindres.

The continual growth of the works had necessarily involved from time to time the acquisition of more land. In May 1860 was bought from the Rev. John Evans the square piece of about acres lying between "Blue Billy" and the "Valentia" canal arm and known as the "cinder meadow" Then, in August 1862, were added two small plots, with cottages, in Park House Lane, near the canal bridge. In 1866 were purchased the "Oxford Works," previously held for some seven years on lease, and the "Britannia Iron Works." These additions carried the firm's property between the "Houghton" arm and Trinity Street northwards to abut on the main Birmingham Canal. Next year was acquired the greater part of the land between Park House Lane and the "Valentia" arm, to the north of what was held on that side already. In 1871 purchase from Ralph Gough of an outlying piece of land of about 7½ acres at Tat Bank carried the property

---

1   For a full description see the paper read by Alexander Chance before the Society of Chemical Industry on March 5, 1888. The patent for the process was No. 8666 of 1887, taken out in the names of Alexander Macomb and James Frederick Chance.

beyond Trinity Street, where all then was open country. It was at once let to William Underhill, whence the reference to him cited. About the same time and in 1872 was bought most of the land on that side of Trinity Street now held, besides a piece extending to the Titford branch of the Birmingham Canal exchanged, when the Oldbury Railway was made, for that other piece to the south of it subsequently used for the site of the cyanide works. In 1874 was acquired the length of Trinity Street itself that now ran through the premises, and in 1875 the 11 Old Malt Shovel Inn" and thirteen houses on Tat Bank Road, this piece of land adjoining that let to Underhill. Next, in September 1877, came purchase from the Willett trustees of land east of Park House Lane and north of the Canal. When a few years later the Oldbury Railway was brought through this land, a piece cut off at the north end was laid out, on the initiative of Alexander Chance, as a recreation ground for the people of Oldbury. The opening ceremony took place on June 27, 1885, and the ground was maintained at the firm's expense until, in 1897, it was given to the town.

In the years 1882 to 1884 the firm's holding of the land between Park House Lane and the "Valentia" Canal arm was completed, save for a small piece at its south-eastern corner, firstly by the purchase of 3½ acres at the Park Street end, and secondly by that of the "Seven Stars" public house and other property along the Birmingham Canal. Another public house bought and closed in 1887 was the "Locomotive Inn," at the north end of the firm's premises beyond Trinity Street. In the same year the land enclosed by the Birmingham Canal and the Oldbury Railway, fronting Park House Lane, was sold to the Aluminium Company for their new works. Lastly, so far as we need concern ourselves, came in 1889 the purchase of some 7½ acres of building land on the far side of Tat Bank Road, in 1892 of about 6½ acres beyond and adjoining the property at Tat Bank previously held, in the following years of certain freeholds in Park Street and Langley, and in 1896 of other land at Tat Bank, occupied after the outbreak of the War by part of His Majesty's Munitions Factory.

To turn now to the social work of Alexander Chance at Oldbury, work to which his deep religious conviction, his generous nature, and his constant anxiety to meet the wants of others attracted him unceasingly. Institutions established before his time, Schools, Dispensary Fund, Mission Relief, he fostered, and he added to them others ; in the first place, in 1870, a Burial Fund, maintained, like the Dispensary Fund, by the contributions of the workmen themselves, and then their Convalescent Home at Quinton, in an airy situation but four miles from the works. Of this we have the following account in a pamphlet issued by the firm in 1881:

> Sick and Burial Clubs are generally to be met with in connection with most large works, but we believe that the maintenance of a Convalescent Home, for the exclusive benefit of the Work-people of one single Firm is, if not altogether exceptional, of very rare occurrence.
>
> The idea originated in 1871 with a member of our Firm, by whom a small cottage in the country was rented, and maintained for a time by private friends, as an experimental Convalescent Home. After our Work-people had seen the benefit which some of their fellow-workmen had derived, when recovering from sickness, by a short residence there, the scheme for establishing the Quinton Convalescent Cottage Home, on a permanent basis, was, in the autumn of 1872, submitted to our men at a mass-meeting in the Works. To the credit of our Work-people be it said, they immediately recognised the merit of the scheme and warmly supported the undertaking. They readily assented to the proposal that each should thenceforth pay 1*d*. per head per week towards the maintenance, and that they should contribute towards the cost of building a Cottage Convalescent Home. On the 15th April, 1874, the new Home was opened. As shown by the following statement, the cost of the building and furniture was defrayed half by the Work-people and half by the Firm, the former handing over a balance which had accrued during a term of years from the Dispensary Fund, and making further grants, as required, during the progress of the building, which were met by grants to the same amount from the Firm, who also paid for the freehold site.

The site and boundary wall cost the firm close upon £200, while for the building expenses the men found £572 and the firm the same amount. Cost of maintenance during the first six years was met by the men to the extent of £808, the firm subscribing £21 a year and keeping the cottage in repair. The account proceeds:

> The Home contains eight beds, and is open all the year, but is used generally only from March to November.

None but our Work-people and their families can be admitted, for whose exclusive use it is intended. The management is entrusted to a Committee of seven members, of which Mr. A. M. Chance (who originated and carried out the scheme) is chairman, the six remaining members being appointed as follows:— three from amongst the foremen, selected annually by the foremen, and three from amongst all the other men, appointed by ballot, at the time of electing the doctors each year.

In addition to the value of this Institution as a Convalescent Home, it is much prized by the Work-people as a kind of country house, which they may visit when they feel disposed to do so, in their leisure hours. Several members of the Firm have made special gifts, in order to add to the pleasure of all who visit the Home: and frequently on summer evenings, and generally on fine Saturday afternoons, parties of Work-people stroll up to spend a few hours there. A summer arbour, a covered bowling alley, and a picturesque shrubbery, are some of the gifts in question; and these have added much to the pleasure of the inmates. . . .

As has been stated, our Work-people, during the past six years, have contributed towards the current maintenance of the Home directly 77 per cent., indirectly 10 per cent., leaving only 13 per cent, to be supplemented by the firm. The knowledge that they themselves maintain their Home almost entirely exercises the best possible influence upon them, they feel a natural pride in it, as was aptly expressed on one occasion in the reply of a workman, who was asked by a gentleman driving by to whom that pretty cottage belonged, when, drawing himself up to his full height, and letting his pride get the better of his grammar, he said, " It belongs to we, sir."

We regard this scheme as capable of adoption by large employers of labour throughout manufacturing districts generally, and as tending alike to the advantage of employer and employed.

The friendly feeling thus fostered by the masters leads to a better understanding, and tends to lessen the differences which too often exist between capital and labour; whilst the spirit of self-reliance on the part of the men, created by the knowledge that their own money mainly supports the Home, is obviously of the greatest benefit. The "penny per head per week" is not missed by any, but the value of its silent teaching is great. It illustrates forcibly the benefit of co-operation, and is a direct incentive to thrift.

After the passing of the Employers' Liability Act in 1880, a larger scheme was launched, establishment of a Provident Accident Fund in lieu of its provisions. Every man employed was to be obliged, as a condition of his employment, to contribute a penny a week, and the firm an amount equal to the total of their contributions. The fund was to be managed by a committee of thirteen, six foremen and six ordinary workmen, with a representative of the firm as chairman. All accidents of whatever kind, and however caused, were to receive compensation adjudged by the committee immediately, without any of the delays and uncertainties attending the operation of the Act. That a large number of accidents would fail to benefit by its provisions was prominently in Alexander Chance's mind in proposing the scheme. He foresaw also the mischief which litigation would inflict on the good relations between the firm and its work-people.

Firstly he set the scheme before the foremen, who thoroughly approved of it, and then explained it to a mass meeting of all the workmen. They accepted it unanimously, and it was started at once. After two years the fund had so prospered that workmen, in whose houses there was infectious disease, could be compensated out of it for not coming to spread the infection in the works. After five years the accumulated balance amounted nearly to £800.

This pecuniary success was largely due to a consequence not, perhaps, foreseen, a remarkable reduction in the number of accidents. During the first five years there were but 232, almost all of a trivial nature. For nine years, at least, there was not one fatal accident, whereas in the like period, before 1881, there had been seven. Undoubtedly a sense of personal responsibility for mutual interests made the men more careful.

In mission and religious work, and in the cause of temperance, Alexander Chance was zealous. To promote the latter by example he donned the blue ribbon himself, until detriment to his health obliged him to relinquish total abstinence. His leading helpers, France in the van, joined loyally in the crusade, and before long the drunkenness, that had been rampant, was of small account. The furnace-men found the oatmeal water, supplied to them free of charge, far more thirst-quenching than the beer of which they had formerly consumed such quantities. Of great service to this and other good work was the Langley Temperance Club and Institute, again a work of Alexander Chance. It comprised reading and recreation rooms, and was in all respects a model Working Men's Club. There was a Literary and Debating Society, and frequent concerts, lectures and entertainments. In an early report France, chairman of the committee, was able to write: "The Committee feel certain that judging by our success in the past our prospects for the future are very bright."

One thing that Alexander Chance had much at heart was the provision of proper housing for the work-people. Already in 1872 he obtained authority to inquire into the cost of eligible sites. In March 1876 he submitted a scheme for building twelve cottages as an experiment, concluding his memorandum on the subject: "The present high price of materials and labour renders the erection of cottages for work-people almost impossible on commercial principles, but the need of cottages in Oldbury was never greater. Is it not then a case where philanthropy may step in and help? "The purchases of land at Tat Bank were made largely with this object, and a scheme was promulgated to enable the workmen to become owners of their cottages. In 1886 the number owned by the firm amounted to forty-three. A new church provided for the spiritual needs of the Tat Bank colony.[1]

In 1890 the firm was converted into a private Limited Liability Company under the name of the Oldbury Alkali Company Limited. Henry Chance was Chairman, Alexander Chance Deputy Chairman, Managing Director and Secretary. About the same time was undertaken a new manufacture, that of cyanides. Messrs. Albright & Wilson and others joining in the undertaking, a new company was formed to conduct it, the British Cyanides Company Limited, now prosperous under the able management of Kenneth Macomb Chance, second son of Alexander. Lastly, in 1898, the company was amalgamated with another firm of importance and long standing, Messrs. Hunt & Sons of Wednesbury, and since that time the joint concern has retained prominence under the style of Chance & Hunt Limited. Alexander Chance was Chairman of the new Company from 1901 to 1912, when he retired to spend his last years in quiet at Torquay. There he died, worn out with work, on November 22, 1917.

---

1   See Chapter XII.

CHAPTER XI

THE SCHOOLS.[1]

Chance Brothers & Co. have their place among the pioneers of national education in England. Already in 1834, when the necessity of it had but just forced itself upon the attention of Government, the partners resolved that the establishment of schools for the workmen's children should have earliest consideration. In 1837 a plan for buildings was approved; schoolrooms to accommodate 250 boys, girls and infants, a lecture room, and residences for a married couple and a single woman as master and mistresses. But differences of opinion on the subject of religious teaching delayed execution. At last, in December 1842, Lucas and William Chance signed the following memorandum of agreement:

1. A building to be erected, to include Infant, Day and Sunday Schools, Committee or retiring room, library and reading rooms, with a house for the Master and Mistress of the Schools.

2. The building to be erected, fitted up, and maintained or kept in repair at the expense of the concern, and all taxes and levies to be paid by the concern, gas also to be supplied gratuitously.

3. R. L. Chance to have the entire control and management of the library and reading rooms, and to defray all charges attendant thereon.

4. W. Chance and J. T. Chance to have the entire control and management of the Infant, Day and Sunday Schools, and the rooms appropriated for these Schools to be under their exclusive control. W. C. and J. T. C. to pay all the charges of management, including the salary of master and mistress, but to be at liberty to make such a charge as they may see fit for education.

5. As the partners differ in religious opinion, it is distinctly understood and agreed that they are pledged to avoid on public occasions reference to each others' sentiments as far as practicable.

---

1   Quotations in this chapter are from Talbot's Sketch of a History of the Schools, from 1845 to 1887, privately printed in the form of a letter to the firm when they were handed over to the School Board in the latter year.

6. No lectures on politics or religion to be delivered without authority in writing of two of the present partners, but this regulation is not intended to exclude R. L. Chance from holding the religious meetings of his body on a Sunday, in either of the rooms under his control.

In determining the site, reference should be had to the rooms being lighted by means of gas.

Yet it was a year and a half before a final decision to build schools to accommodate 200 boys, 150 girls and 100 infants was taken.[1] Immediately thereon Messrs. Bateman & Drury, architects, were instructed to prepare plans for buildings to include three schoolrooms on the "British and Foreign" system, 18 feet high and with capacity for the numbers named, two classrooms, about 25 to 27 feet by 18, attached one to the boys', the other to the girls' schoolroom, but so contrived as not to interfere with extension of the building, an enclosed yard with separate accommodation for each schoolroom, houses for the surgeon and for the master and mistresses, and a library and reading room. Special directions were given that the ceilings should be considered with reference to sound and that the windows should be at least five feet from the ground.[2] It turned out, however, when plans and estimates were received, that the cost of all this, with eight cottages to be built besides, would be excessive; £4,000 for the schools and a lodge, £900 for each house, and £1,400 for the cottages; a total, with ten per cent, for extras, of some £8,000. James Chance pressed upon his father reconsideration of the plans; postponement of the building of the infants' school and houses much less pretentious. Davies (the works' surgeon), he wrote, said that it would suit him better to live in lodgings at West Bromwich than to have the trouble and expense of a house to himself, and he could not be expected to inhabit one worth £900. If a surgeon's house were to be built at all, it ought to be moderate and unpretending, "otherwise the people will begin to complain of subscribing to the luxuries of the Doctor." And the master's and mistresses' house was "too grand in extent for the class of persons whom the schools could afford to pay." In fact, two of the cottages, made somewhat better than the others, would suffice for the teachers, while the surgeon's house should not cost more than £350. Too pretentious a scheme, he thought, might "defeat one ob-

---

1   Board minute of May 15, 1844.
2   Instructions to Mr. Bateman, in James Chance's hand, May 18, 1844.

ject which we have in view; that is, setting an example to our neighbours of providing means for educating the lower classes." He had explained to Bateman

> how anxious we were to prove that good schools might be erected for a moderate outlay, and particularly that we should be very desirous of avoiding any appearance of show . . . I should be decidedly opposed to that degree of economy which would give to our schools a workhouse, or workshop exterior, but I feel certain that a pleasing and appropriate elevation may be obtained for £2,500 as an ultimate outlay, including an infant school and all the conveniences now proposed.

Whether a third schoolroom were added in the future or not, "the buildings would still be well proportioned, convenient and complete"[1]

These proposals were acted upon in the main, with provision for the three schools. The final outlay worked out at £5,300, the cost of the land being reckoned in addition at £1,030. The boys' school was ready for opening in October 1845 and those for the girls and infants in April 1846. At the former date the following notice was issued:

> Notice is hereby given, that the Boys' School in Spon Lane will be opened for the reception of Boys of seven years old and upwards on Monday the 20th instant.
>
> A preference will be given to Children of the Workpeople connected with our two establishments, but if the School is not filled by them, it will be opened to the Public.
>
> The Instruction given will include the following, viz:—Reading, Writing, Arithmetic, Grammar, Geography, the elements of Mechanics, and other branches of knowledge as soon as the children are sufficiently advanced to receive them; for which the charge will be to the children of our workpeople 3*d.* per week no extra charge will be made for books, slates, &c, used in the School.[2]

---

1   James Chance to William Chance, August 1844.
2   On the principle of exacting fees Talbot wrote: "When the Schools were opened, it was a subject of discussion whether they should be free, or a weekly fee be charged. The conclusion arrived at was, that it was desirable in the interest of education itself, and in justice to other schools, that a fee should be charged. That this decision was wise is proved, I think, by the uniform experience of the country; for without doubt, nothing has more deepened the appreciation of people in the work of education than the payment of a weekly fee for their children. In this neighbourhood, in 1845, there were two free schools. They both failed in attracting children of the best type, but after a while they were rebuilt and re-organised, and now, as fee-paying schools, both are successful. It is important that the reason for this payment of fees should be understood, for undoubtedly there are many erroneous notions afloat on the subject." The fees appointed did not, of course, meet the expenses of the Schools. The deficit was made up in the earlier years by William and James Chance, afterwards by the firm.

Payment to be made in advance regularly every Monday morning, or the children will be dismissed.

The hours of attendance will be from 9 to 12 in the morning, and from 2 to 5 in the afternoon in the Summer months, and till dark in the winter months; and punctuality of attendance will be rigidly enforced. The Boys must come to the School neat and clean, and be subject to such regulations as may be found desirable for promoting the welfare of the Schools.

The Master (Mr. Talbot) will attend at the Boys' School on Thursday, Friday and Saturday next, from 9 to 2 o'clock, to receive applications from our workpeople only, in the first instance, for the admission of their Boys. A Girls' School and an Infant School will be opened as early as possible.

Our anxious desire is to give the Children of our Workpeople the advantages of an education suited to their position in society, and especially to connect with it the inculcation of sound religious and moral principles.

The Schools have been erected at the joint expense of all the Partners, but the management of them rests exclusively, by mutual consent, with the undersigned.

William Chance

James T. Chance.

Smethwick, October 10th, 1845.

P.S.—A parent must in all cases attend when application is made for admission.

To promote the objects in view it was resolved in January 1846 "That this Board, taking into consideration the great importance of educating the boys engaged in this manufactory, are decidedly of opinion that no boys should be employed until 12 years of age." This enactment would not count for much in modern days, but it was a great step forward at a time when much younger children were still forced to work in mines and factories.

As stated in the notice, the schools were open to others besides the children of the firm's workpeople. In later years the strangers formed the large majority.

Sound moral and religious training was the basis of the education .given from the first. Time after time in his letters William Chance insisted on the necessity of this. He and his son were fortunate in at once securing the services of Frederick Talbot as head master, a man of rare qualifications, thoroughly in sympathy with their views and zealous to give effect to them. One pupil of his, Mr. Samuel Nicholls, now the head of a prominent firm of glass merchants, who entered the infants' school in 1848, and came under his rule when "passed into what was then called the big boys" at seven years of age, wrote of him recently as "a veritable martinet, but withal a most excellent schoolmaster." Although, this gentleman said, he had had six boys of his own at leading public schools, he had found "nothing better than the character Mr. Talbot endeavoured to impart to the boys. Fear God, tell the truth, and don't be afraid of work, was the ruling feature of his curriculum." Duty, honesty, cleanliness and good manners were the essentials of Talbot's teaching, and scores of men in good positions throughout the world confess obligation for their success in life to his strict discipline. In his retrospect of 1887 he wrote:

> Regularity and punctuality in attendance on the part of the scholars have from the first been insisted on, as a condition of being allowed to continue in the School. Absentees have always been promptly enquired after, and the practice has been attended with the best results. Truancy has never been known to succeed, ... I believe it would be impossible to obtain a higher average of attendance with punctuality, than is obtained at the present time in the Boys' School. . . .

The leading principles on which your Boys' School has been conducted have had as much relation to the formation of character and habits as to the acquisition of knowledge. Regularity and punctuality in attendance, neatness and cleanliness of person, prompt obedience to law, energy and accuracy in work, truthfulness in word and deed, and respect for others, have always been inculcated. To give due effect to these principles the discipline has been undeviatingly regular and strict, and it is a source of satisfaction to know that they have not only been supported in their integrity by the managers, but that in the main they have received the approval of parents. At the present time there are no more regular and well-behaved children in the School than the sons of those, who were themselves the former subjects of this discipline and training.

The instruction of your School in the principles of religion has always been regarded as an essential portion of its course of education. Formerly there was much more time given to the teaching of Biblical knowledge than there has been of late years . . . Lessons from the Bible are, however, still a part of the School course, and are both explained and enforced. The Schools have been, from the first, perfectly unsectarian.[1]

---

1   "It would, I believe, be an immense gain to education, if Inspectors were again authorised to examine specifically into the moral instruction given in Schools. I know of nothing myself at present

The first official inspection of the Schools, made by desire of the teachers and on the invitation of the firm, took place in December 1847. The inspector reported them to be conducted with great vigour and success, and "quite a pattern to the neighbourhood." In March 1850 the Rev. J. P. Norris, the newly appointed School Inspector for Staffordshire, Cheshire and Shropshire, made his first visit. He wrote in his report: "The most complete set of Schools in my district. The group of buildings externally resembles a College, and has, I believe, cost the Messrs. Chance £5,000 altogether. The Boys' School is a spacious Gothic hall, admirably organized." So much was he impressed, that he appended to his report a good portion of a pamphlet of Talbot's, explaining his methods,[1] and a plan of the boys' schoolroom, 60 feet by 33, showing how they were placed in galleries and controlled by the master from the floor. In his next report he stated books and apparatus to be liberally supplied, and marked instruction and discipline "very good," the standard of instruction " fair," methods 11 excellent," and master and mistress " of very fair attainments and great ability in administration." He added: "The peculiar excellence of this school is the organization, method and discipline by which the master is enabled to maintain an easy ascendancy over nearly 200 boys in one room. The superior value of class galleries over any other arrangement, where the numbers are large, is here strikingly illustrated" On which Mr. Talbot:

It is interesting to know that the class galleries, with seats only, thus reported on, are now used in all large schools, and their value everywhere acknowledged. A model of the schoolroom, with its internal arrangements, was contributed by request to the South Kensington Museum, where it was exhibited for many years amongst the educational appliances.

Perhaps this model was one which Talbot was desired by the Board, in May 1854, to send, with specimens of the apparatus used, to an Educational Exhibition of the Society of Arts at St. Martin's Hall.

Headmistress of the girls' school in 1851 appears to have been Miss Thorp, but in 1853 Mrs. Duff had entered on her long and admirable charge. The infants, until October 1853, were under the care of Miss Twist,

authorized by the Code, that can at all be compared, in respect of intellectual training, to a well-devised scheme of moral instruction; and there is nothing that I have missed more in the annual inspection, than the catechising in religion and morals, which was a marked feature in the visits of some of the older Inspectors."

1   Printed, he says, "for private circulation by the master of the school at his own printing-press."

of whom said Mr. Nicholls: "a more loveable, sympathetic teacher I think it would have been impossible to find." Her successor then or later, Miss Annita Teague, who died but recently (July 1918), yielded to her in those qualities no whit.

Of pupil teachers Talbot says:

> In 1846-7 Minutes were published by the Committee of Council on Education, by which valuable aid was offered to Schools that were in a condition to take advantage of it. The principal Minute was that which provided for the employment of Pupil Teachers at the public expense. Your Schools were some of the first to take advantage of this Minute, and from that time to the present about sixty young persons have been trained as teachers in the three Schools. Probably as many more Scholars had either been sent, or have found their way from these Schools into others in the neighbourhood, to become Pupil Teachers, and the three Schools cannot have contributed to the teaching power of the country less than a hundred teachers. In looking over the entire list, it is most satisfactory to find that a good proportion of the number are still engaged in the profession, and that two, at least, have taken the degree of B.A. at the London University.

It was not long before it was found necessary to erect schools at Oldbury also. An estimate for buildings to accommodate 150 boys, 100 girls and 150 infants was ordered to be obtained in October 1850. Education was to be given on the "Irish system," and application to be made to Government for a grant in aid. Mr. Robertson, head master designate, was desired to obtain complete statistics about the education of the workpeople. In March 1851 the estimate of John William Smith for £720 was accepted, William Chance undertaking to make definite arrangements with the architect.

Robertson retained his post for ten years. William Chance wrote of him (March 6, 1851) as "energetic and clever," but wanting Talbot's ' order and regularity." In 1858 he was presented with £30 for his services in connexion with the Provident Society. But his application to have his house rent-free was refused, on the ground that his salary of £120 "was more than sufficient to secure the services of a first-rate master."[1]

The Day Schools were supplemented by Evening Classes for older boys,

---

1   Talbot and he each received, in addition to the above named salary, £23 on government certificates; but while the former had some £45 besides on account of capitation fees and pupil teachers, Robertson enjoyed but about £13 from this source, while Talbot also had his house rent-free.

at Spon Lane from 1846 and at Oldbury from 1853, with the help of assistant masters. Want of success at first was owing to the very need of such opportunities of learning. In a report of October 1848 Talbot attributed failure in the previous summer "as much to the overwhelming amount of work as to the inefficiency of the master engaged for its performance." In point of fact, he said, nearly 200 boys and youths, aged from 12 to 20, had presented themselves, half of them "destitute of the meanest elements of an ordinary National School education," and the number had proved unmanageable. It was true that as many or more boys were commonly in attendance at the ordinary schools, under the management of one teacher, but the circumstances were different.

> In the one case the School is opened with a few boys, who are speedily reduced to order and from whom the best qualified are selected to take charge of and to instruct others subsequently admitted. These latter gradually fall into the confirmed habits of the mass, and thus it is that our large Schools present an appearance of order and industry, which we should in vain expect to find, for a long time at least, if those Schools had started with their hundreds instead of tens.

In the Evening Schools, Talbot went on, the system of putting certain scholars in charge of the others had failed, for the latter would not acknowledge the authority of those who, they said, "only work in our shop." The evil, he thought, would be remedied in course of time, especially if those only were admitted who really desired to learn. In the meantime it was important to devise some method by which one master might accomplish the firm's purpose more effectually. He proposed to divide the scholars into three classes, those employed in the glass houses, those in the plate works, and those in other departments; each class being again divided into boys and adults. The three classes would attend at different times, the first on unoccupied mornings between 12.0 and 2.0, the second, who left off work early, for an hour and a half from 5.0 or 5.30, and the third, not free till 6.0, in the evenings.[1]

At a subsequent date notice issued that for the future the Evening Classes would be open on Mondays, Wednesdays and Fridays for instruction in Reading, Writing, Spelling, Arithmetic, and Geography. There were also to be Advanced Classes for Reading, Grammar and Composition, for Part Singing, for Mensuration, and for Algebra, conducted respectively by

---

1   Report dated October 20, 1848.

Messrs. Talbot, Hanson, Marshallsay, and Dugmore, and Special Classes for Chemistry under Mr. Jones, for Book-keeping under Mr. Simpson,[1] and for Drawing, including Freehand, Mechanical, Perspective, Model, Water Colour, &c, under Mr. Sounes. Fees were to be twopence per week for the Ordinary and Advanced Classes, and twopence per lesson for the Special.

The Evening School, Talbot tells us, "was held with fair but varying results from 1846 to 1859, but at times, and especially from 1860 to 1874, it was in a very flourishing condition." During the summer of 1863 he conducted an examination of the boys under 17 years of age employed at the Glass Works, the results of which, he says, gave "a great impetus" to the attendance. The general subjects were Reading, Writing, and simple Arithmetic, the boys in the Ornamental department being further examined in Drawing, and those in the offices in Decimals, Grammar, Spelling and Correspondence. Out of 382 boys employed 314 presented themselves, of whom 108 had received education at Spon Lane, 169 at other schools, and 37 nowhere.[2] The percentage marked "good" was 25·4, "fair" 28·3, "bad" 46·2. These were not great results, especially as the standard, says Talbot, was purposely made low, the mark "good" being but the equivalent of the third class of an elementary school. Some, he goes on, who had obtained it could have passed higher, and all boys not so marked were required to attend the Evening School. "One of the most pleasing features" of that work he found in the presence "of a considerable number of young men, who sit side by side in the various classes with the youngest boys, and give for the most part the most regular attendance and careful attention to the lessons." He emphasised the importance of the works managers pressing the boys to attend regularly, proposed to provide tea and coffee for those living at a distance, and noted that a small library was at the service of all, and tables covered with literature for those who awaited, after work, the opening of the School.[3]

A notice issued in October 1863 shows that the Evening School was now conducted in two divisions, each meeting from 7.0 to 9.0 on three

---

1   Of the Ledger Office. Fancy, says Mr. Nicholls, above quoted, who attended the Night Schools after entering the works as a boy in 1854, "a lad from the seven-storey being taught bookkeeping!" Another who taught in the Schools was Mr. Genner.

2   In illustration of the unsectarian character of the education given, 85 of the boys were of the Church of England, 14 Roman Catholics, 90 Primitive Methodists or Wesleyans, and 76 of other denominations; while 49 attended no place of public worship.

3   Report dated September 23, 1863.

LONG SERVICE MEDALLISTS 1916
(To face page 256)

evenings in the week and subdivided into four classes. To quote its further contents:

All boys employed in the Glass Works under seventeen years of age, who cannot pass a good examination in Reading, Writing and Arithmetic, are expected by Messrs. Chance to attend the Evening School: and all others are strongly encouraged to avail themselves of the opportunity afforded by the School for improving and extending their education.

The Instruction comprises Reading, Writing, Spelling and Arithmetic in all the Classes, and the First Class is further specially instructed in the following subjects, which are prescribed for the next examination of the South Staffordshire Association; the Gospel History, English History, the Geography of the British Isles, and in the Rule of Three and Vulgar Fractions.

An Examination of all the Scholars will be held at Christmas and Midsummer, and a Prize of a Book will be given to the two Scholars in each Class who reach the first and second place respectively in such examination. Certificates will also be issued to ail Scholars who pass a good examination in any three subjects, and they may be revised every half-year, after further examination.

An Examination of the most advanced Scholars of the Evening School of the district will be held by the South Staffordshire Association in March next. The names of all successful candidates will be afterwards published by the Association and Messrs. Chance will give two Prizes, of One Sovereign and Half a Sovereign respectively, to the two Boys from this School who shall attain the first and second highest positions in the First Class List (Order of Merit).

A Book or other suitable Prize will also be given to all Scholars who shall have attended the Evening School upwards of 100 times from September 1863 to June 1864.

Occasional Lectures will be delivered during the winter on Monday Evenings, at eight o'clock.

A School Library of nearly 400 Books is opened free to all regular Scholars. Books must be returned weekly, on Monday Evenings.

The School opens at six o'clock every evening, and Boys may quietly assemble and read, or amuse themselves with various books and periodicals, or with games of chess, draughts, &c, provided for them, till seven o'clock. School work commences punctually at that hour, and no Scholar can be reckoned present who does not reach School on any evening before half-past seven o'clock.

Boys may wash themselves in the Lavatory attached to the School, but it is strongly recommended that, where at all practicable, they should all go to their homes for refreshment, before coming to School. Clean hands and faces are essential to comfort in School, and a neat and clean suit of clothes is becoming more and more the ordinary dress of the evening Scholar. It may be hoped that every Boy, whose working dress is necessarily too dirty for School, may in time be provided with a proper suit of clothes. There can be no doubt whatever that a boy's progress in learning is very much helped forward by his feeling clean and comfortable.

In 1872 it was resolved to make attendance at the night schools, at both works, as far as possible compulsory, and to that end to deduct the fees from the wages of all boys and girls between the ages of 13 and 18 required to attend.

In competition for the certificates and prizes given by the Society of Arts for proficiency in various subjects the pupils of the Evening School were very successful. Says Talbot on this head :

> An interesting feature in the history of your Evening Schools was the formation, in 1859, of classes for the preparation of some scholars in certain subjects for the annual examination provided by the London Society of Arts. That Society, in the first place under the direction of its President, the Prince Consort, had provided for several years on a liberal scale a scheme, whereby persons over sixteen years of age could be examined in a certain number of no less than twenty-five subjects, and could gain certificates of three grades of proficiency in each. To distinguish the best scholars in these subjects, there were attached to each of them money prizes of the value of £5 and £3 respectively.
>
> The examination, which is now well-known, had not been introduced into either South Staffordshire or Birmingham in 1859; the first local meeting in connection with it was held in your Schools early in 1860, and at the examination in the Spring eleven candidates from your Evening School were examined and gained certificates, one of them taking the first prize of £5 for excellence in Logic. In 1861 seventeen certificates were gained by twelve scholars, and the one who gained the first prize in 1860 was again successful in gaming the first prize for Latin and Roman History, whilst two others gained in the same 3/ear the second prize of £3 for Bookkeeping and Arithmetic respectively. In 1862 ten pupils gained certificates, two of them taking the first and second prizes for Arithmetic, and in 1864 another gained the first prize for Drawing. It is worthy to note that nearly ail the young men who thus took certificates and prizes, and whose career is known to me, have done well in life; several are still in your employment, in responsible appointments; two others are the Principals of large establishments and one who gained the highest distinctions is now the respected Manager of a large industrial establishment in Birmingham.
>
> In connection with the above scheme it may be stated that the necessity of establishing a Local Board for the holding of the Society of Arts' examination in Spon

Lane originated the formation of a Local Board, which was large enough to hold examinations under the same scheme in other places, should its services be required. The formation of that enlarged Local Board was the commencement of the South Staffordshire Association for the Promotion of Adult Education and Evening Schools, and for twenty years, from 1860 to 1879, that Association, under the Presidentship of the late Lord Lyttelton, took a most active and useful part in helping forward, in various ways, the interests of educational institutions of all kinds, but particularly of Evening Schools. Whilst forwarding, as it did, the particular objects aimed at by the Society of Arts, it supported a valuable scheme of its own; this scheme provided for the annual examination of the best evening scholars of the entire district, and offered in connection with it a large number of valuable prizes and other encouragements to regular attendance and steady progress; for many years the certificates and prizes provided by the Association were the objects of keen competition, and there was no School that was more successful in the race than your own.

And the same, Talbot goes on, with liberal prizes given by an Association founded by Mr. Seymour Tremenheere

to encourage parents to keep their children regularly at school, and for longer periods . . . The most regular and best instructed scholars of the principal Schools of the district competed for these prizes, and your Schools sent in candidates for several years. In 1856 the Boys' School succeeded in winning forty pounds in money and book prizes, the largest amount obtained by any one school during the existence of the fund. I refer to it for the purpose of noticing the career of as many of these boys as occurs to me. From a list of the nineteen prize boys for that year I find that one is the sub-editor of an old newspaper in the South of England, one the headmaster of a large Board School in London, three are managers of manufactories, three are clerks, one is a manufacturer, one a commercial traveller, one a successful accountant, two have died: of the remaining six I have no certain knowledge.

Talbot laid great stress on the value of examinations for keeping the boys up to the mark. "Probably," he says,

the greatest advantage of education that has resulted from the Minutes of the Education Department, next to the provision for the training of Teachers, has been the system whereby individual Scholars are examined as to their progress from year to year; and the arrangement for this being done effectually constitutes the great distinction between the Codes since 1860, and those of an earlier date. It is a source of satisfaction to me to know, that in a paper which I contributed to the Educational Section of the Social

Science Association, at its first meeting in Birmingham in 1857,[1] I urged very strongly the importance of this point.

Of another examination in 1866, attended by 310 out of 398 boys and youths employed in the works, the results, says Talbot, are interesting enough to be recorded, because the examination was undertaken at the same time as that of a large number of other young people, who were employed in several of the largest Ironworks of the district. The results of this collective examination, suggested to others by the plan adopted by yourselves, were tabulated in a special report, which was circulated by the South Staffordshire Adult Education Association; its purpose was to call the attention of the employers of juvenile labour, and also of the Education Department, to the importance of stimulating the evening instruction of adults. The circulation of the report had a very stimulating effect on the attendance at Schools throughout this district, and after some time the Education Department modified several of the clauses of the Code which related to Evening Schools, on a representation made by a deputation from the above Association to the Lord President of the Council. . . .

There was revealed in this examination the fact, that though your Firm had made great efforts to extend and improve the education of the neighbourhood, yet that a large proportion of the children, who had now grown up and commenced work, had not availed themselves of the opportunity afforded. This led you to increase the efficiency of the Evening School by the engagement of additional Assistants, and, after a time, to open in a separate room a Special Day School for youths employed in the Glassmaking Department. For several years this last experiment was successful, and the youths were regularly examined, together with those of the Evening School, at the annual visit of the Inspector of Schools. In 1868 he examined altogether 164, in 1869 the same number; in the two succeeding years the numbers had risen to 243 and 248 respectively. Soon after this time there arose a great demand for the labour of youths, owing to an improvement in trade, and a considerable number of the elder scholars were attracted into other works; many of those who remained were disturbed by the same cause and lost interest in their educational progress; and a change of Assistant Masters, which occurred at the same time had a further effect in diminishing their interest in the School. It was

---

1   Published in the *Transactions* of the Association for the year named. Talbot cites from it in his Sketch, pp. 8, 9.

therefore finally, but reluctantly, relinquished.

This experiment of a Special Day School for youths employed in the Glass-making Departments was particularly interesting. The Rev. H. R. Sandford, the Government Inspector of Schools at that time, took a great interest in it, and endeavoured to obtain for it the recognition of the Education Department, but only to find that there was no legal enactment, and no section of the Code, that dealt with it. Mr. Robert Baker, too, the Chief Inspector of Factories for this part of England, took a note of the experiment, and in his report for October, 1868, he commends very strongly the plan of organizing a periodical examination of working lads by their employers, in order to demonstrate to them their need of continuing their education. Mr. Baker had in his mind, as I have reason to know, the plan adopted by your Firm, and he copied in full, in the same report, a printed card, which had been affixed in the various rooms of the glassworks, prescribing such an examination and encouraging young people of both sexes to take advantage of it, and of the Schools opened for their improvement.

Of the good effect of the experiment, moral or otherwise, on the conduct of the lads who attended the School regularly, those who watched it most closely can speak with the utmost satisfaction. They improved quickly in manners, dress and general deportment, many learned the rudiments of their education for the first time, and several of them attained to very considerable proficiency in the ordinary subjects of their instruction. A goodly number were well-instructed youths when they came to the School, and it was observed that many of these were the most regular in their attendance and the steadiest in their application to their lessons. From a printed list, now before me, I find that in 1871 sixty boys of this special School were examined by the Inspector in the fourth, fifth and sixth Standards, whilst forty-three others were examined in the lower Standards. Of the sixty, thirty-one had attended the School over 100 times, whilst of the remaining twenty-nine only three had made fewer than 60 attendances.

Freehand drawing was made a definite branch of instruction in the day schools from 1848. In 1851 Mr. Norris reported: "Half the school learn to draw, and some have arrived at great proficiency in drawing scrolls and other forms used in ornamental art." In January 1855, on the suggestion of Mr. Wallis, of the Government School of Design in Birmingham, a drawing master was engaged to attend on one evening in each week, and in 1858 all the boys and girls employed in the Ornamental department of the Works were ordered to attend this drawing class, to be held now on two evenings in the week. A year later Talbot was presented with the sum of £20 "to mark our approval of his zealous efforts in connection with the drawing classes, and our sense of the distinguished success which has been the result, as shown by the late examination of the School of Art at Birmingham"[1] Before 1857, says he, a number of youths had passed the second grade examination of the Government Science and Art Department and had received certificates.

At this time a further step was taken by your Firm; a special class for drawing was organized and connected directly with the Birmingham School of Art, as a Branch School, and taught by one of its Assistant Masters. The result was very successful; pupils from this class gained certificates, and finally medals from the Department. In 1859 three were gained, in 1860 six, and in 1861 eight, and in the latter year one set of drawings was selected for competition for a national medallion. The Drawing Class thus inaugurated, and recognized by the Department of Science and Art as a Branch School of Art, was, I believe, the first of the kind in the provinces. The class continued with success and attracted many scholars of both sexes for several years, and a considerable number of Pupil Teachers from your own and neighbouring schools obtained, by its means, full drawing certificates. Of the young men who were pupils of the above class, and who earned its distinctions, it is interesting to note that three, at least, are the Principals in as many glass-painting establishments; one is the Head Master of one of the most important Schools of Art in Ireland: another is the Head Master of one of the first Technical Schools in England; and another is Science Master in one of our great Public Schools.

In 1863 and 1864 the drawing classes were conducted on Thursday evenings by Mr. D. W. Raimbach, head master of the Government School of Art in Birmingham. Notice was given in October 1863 of an examination to be held in December, when the prize of a book would be awarded "to the most proficient student in each Class, who shall execute the best Drawing within a given time," and of another in February 1864 by an Inspector from the Government Department of Science and Art. In connexion with this the firm offered (1) a prize of three guineas to any student who should succeed in obtaining a National Medallion for a work

---

1   Board minutes of March 25 and April 29, 1858, and March 3, 1859.

JAMES TIMMINS CHANCE
at the age of eighty
(To face page 262)

executed during the year ending March 10, 1864;[1] (2) for youths engaged in the Ornamental department only a prize of two guineas for the best water-colour design for a panel in modern stained glass, the form to be a diamond, of length 13½ inches in each side, and this prize to be awarded by Mr. Raimbach and Mr. Muddock (manager of the department) jointly; (3) a case of instruments or box of colours, of the value of half a guinea, to every student who should succeed in gaining a Government Medal at the examination; (4) three cases of instruments, or boxes of colours, of the value of ten, seven and five shillings respectively, for boys and girls under 16 years of age; and (5) two books of the value of five shillings each for girls of any age employed in the works, who should "produce in the School the best copy, in outline, of an elementary example, No. 35." Award of these prizes was limited to such as had attended the School for at least three months previously. The distribution of medals and prizes was to take place at a public meeting in the Schools in June, when there would be an Exhibition of all works of merit done during the year.

A similar notice was issued in August 1864 in regard to examinations to be conducted in February 1865 by a Government Inspector, and in April by the Society of Arts. It was stated that that Society's first prize of £5 for freehand drawing for the year 1864 had been awarded to H. Hill, of the Ornamental department. For a private examination to be held in May the firm again offered prizes similar to those of the previous year, with an additional "Workman's Prize of five guineas for the best design for a Window in Modem Stained Glass." And so in other years.

Principally responsible for all this progress and success was Talbot himself, of whose fertility and activity of enterprise Mr. Norris, by this time an archdeacon, wrote in jest: "This amusingly characteristic letter from your many-sided School Master reminds me delightfully of old days, when he would have volunteered to command the Channel Fleet, or preach in the Nave of St. Paul's, as lief as give his gallery lesson in Spon Lane." Much he originated, and what he did not originate, that was good, he ardently seconded. The need of physical exercise and recreation was not forgotten. In November 1856, for instance, he submitted a large scheme, advocating it in the first place, to suit Lucas Chance's particular predilection, as calculated

---

1   It was noted to this: "The largest number of Medallions given in any one year by the Department of Science and Art must not exceed one hundred throughout the whole Country. The works that obtain them are therefore of the highest merit."

to remedy the deficient use of the library and reading-room. He proposed that a scheme on foot for providing dining-rooms for the employees should embrace a central hall, fitted up as a gymnasium, with four rooms opening from it; one for a lavatory and baths, another for refreshments, a third for a reading-room and library, and the fourth for conversation and smoking (were that allowed), provided with games and open only to men of mature age. And further a series of good and cheap concerts, lectures by first-class men, and prizes for special attainments in such subjects as drawing, music, mechanical skill acquired in leisure hours, essay-writing, &c. He proposed besides the hire of a field for summer amusements, such as cricket or quoits, and permanent acquisition of a piece of land to be let out in garden plots, for the produce of which exhibitions would be held and prizes given annually. All this to attract the younger men, who had few opportunities of recreation, from seeking the excitement, which they must have, in vicious places of entertainment.

Fifteen months later the Board resolved, on Talbot's initiative, to provide the boys with a playground and a gymnasium. Another idea, successfully carried out, was the institution of a brass band. "This Band," he says, "was commenced in the Schools in 1856, and the members, who all belonged to the Glass Works, contributed for their instruments over £150. The Band finally became attached to the Smethwick Rifle Corps." Another was established at Oldbury, at Robertson's instance, the firm contributing £40 towards purchase of the instruments.[1]

There also the work was conducted on like principles and lines, both at the boys' and girls' schools at the works and at the infants' school at Langley. As head master of the first, Robertson was succeeded in 1861 by Shaw Hanson, a man who could compare in earnestness and liberality of mind with Talbot. Their efforts were sedulously encouraged by the partners, and the ladies of the family took special interest in the work. There were school treats for the children, various entertainments for older people, and annual gatherings of former scholars. Harmony and good feeling were the consistent note in everything.

On receipt of the Government Inspector's Report for 1872 it was resolved to record in the minutes of the Board their appreciation of the

---

1   The cultivation of music, then as now, was by no means confined to the members of the brass bands. There was a string band, in which Genner was a performer on the violin, and a glee club, with Solomon Cutler as a tenor soloist.

efforts of Talbot and the mistresses at Spon Lane, "which have maintained the Schools in their present satisfactory condition for so many years." But now School Boards had been established, and the firm was conceiving reluctance to bear the expense of maintenance and to pay the education rate as well. Talbot's application for a new classroom to accommodate the increased number of scholars was refused on this ground, and it was ordered that only so many should attend as could be taken in. However, next year (1874) it was arranged to convert the boys' cloak room into an additional classroom.

Nothing more requires comment until it was finally decided, in March 1886, either to close the schools at the end of the year or to transfer them to the local School Board on such terms as could be arranged. Negotiations ended in December with an agreement to lease them to the Board for five years at £75 a year. On this severance of Talbot's connexion with the firm he was accorded a pension of £100 a year, to include any Government pension and the value of his house. Miss Teague, mistress of the infants' school, was similarly provided for to the extent of £50. Talbot, expressing his sincere thanks, intimated his intention to retain his post under the Board, as he thought he might do service to the School in its transition stage; he would feel it proper to relinquish the work, when he felt unequal to it.[1]

It was now that Talbot composed his "Sketch of the History of the Schools," from which so much on special subjects has been quoted. What may be termed his general paragraphs run as follows:

> The decision of your Firm to effect a change in the management of your Spon Lane Schools suggests to me that the time is suitable for the publication of a brief record of the work which has been done in them from their opening, and of the connection of that work with the history of education in this district during the period of forty-one years.
>
> In 1845, when your Schools were opened, National Education was attracting more attention than had before been accorded to it; following the example of the National Society, and the British and Foreign School Society, the Government since 1832 had assisted by a grant in the erection of Schools and School Houses; in 1839 it had established the Committee of Privy Council on Education, whose office it was to administer this grant, and also to ascertain the state of education in the country: this latter duty was undertaken by a few Inspectors, who visited certain schools and districts and reported upon their condition and prospects to the Council, and thus to the country.

---

1   Letter to James Chance, February n, 1887.

About this time (1839) The Factory Acts had effected an improvement in the education of children employed in the cotton manufacture, and in many cases the great employers of labour in cotton mills had gone beyond the requirements of the Acts, and had made some efforts, by the building of Schools, to extend education amongst their work people.

In this neighbourhood too, without the pressure of legislative enactments, several employers had made some provision in the same direction; but it was your Firm that first set the example of erecting a complete set of Schools, with residences for the Teachers ; a set of buildings which even now, in the presence of magnificent schoolrooms erected by the Boards, still vie with them in general adaptation to their purpose, and especially in respect of spaciousness, light, and ventilation.

The Schools . . . have with the orginal outlay, subsequent extensions, and expense of annual maintenance, cost your Firm not less than £20,000; and now that national effort, by means of the Education Act of 1870, has placed upon the rates the burden of supplementing the cost of elementary education, it is reasonable that your Firm should seek to be relieved from the responsibility which they undertook under circumstances so different.

But whilst thus stating a very sufficient reason for the course you have seen it proper to take, no one can be more sensible than I am of the loss which the mention of this course suggests. The loss, indeed, can only be duly estimated by one who has known personally, through all these years, the sacrifices, the unfailing interest, and the devotion to the highest claims of education which have characterized the management of your Schools; and I cannot do better than record here the language of one who in past years knew your Schools well, and who has ever taken an interest in their welfare. In a private letter received lately occurs this passage: "I am indeed sorry to learn that the immediate cause is the transference of the School to a School Board—that it should be closed (the alternative you mention) I will not believe possible. No School Board will maintain it with the same munificent and watchful personal care, which it ever enjoyed under the Messrs. Chance. Schools connected with great firms of industry were ever, and will long, I trust, continue to be, characteristic of England. France used to point proudly to Messrs. Schneider's Schools at the Creuset Iron works, and Messrs. Chance's Schools might well rival them." Whilst I am fully sensible of the truth of these kindly remarks, and conscious that they reflect, to a great extent, the feelings of many parents who themselves or whose children have been educated in the Schools, the hope may be nevertheless expressed that the contemplated change will in time conduce to a wider interest being taken in the education of the neighbourhood; and that the work which has been thus far undertaken by private individuals from high and disinterested motives will, in the future, be none the less well performed by a Board, elected by an intelligent constituency.

From 1845 to the present time the number of boys admitted to your School has been over 5,000, and taking the number of girls and infants to be at least as many, the total number is thus raised to 10,000 children. A large number of these, especially in the earlier years of the life of the School, were mere birds of passage from one school to another, but as a rule your Schools, especially of late years, have been singularly free from this objectionable class of scholars, and the above number, therefore, represents a very substantial gain to the ranks of educated people; whilst a large proportion, probably as many as a third of the whole number, have left the Schools with a thorough acquaintance with all the essentials of a sound elementary education. The above numbers are, of course, exclusive of admissions to evening schools and classes.

I have thus sketched the history of your Boys' School, and of the other educational institutions which have been more or less connected with it, from their opening until the present time. The period through which your Schools have passed has been a remarkable one, and in the future it will be remembered as that in which Elementary Education received, for the first time in this country, national recognition and national support. Simple and partial as were the first attempts of Parliament to affect the condition of education, they have become at last, together with the operations of School Boards and the enlarged efforts of the denominational bodies, adequate and effective for the purpose. But whilst the machinery thus exists, the great work itself is far from complete, and the latest report of the Education Department shows that greater sacrifices must yet be made before National Education can be pronounced either universal or thoroughly efficient. It is an honour of no small degree to have been connected, in the best of all ways, with the great educational movement throughout its experimental phases; to have assisted, however feebly, in promoting sound conclusions as to what the country really requires, and to have shown to some extent that schemes of National Education should be co-extensive with the entire period of youth; further, that the success of such schemes must depend more upon the Teacher, his love for his work, his competence, his conscientiousness, and his freedom from needless restrictions and supervision, than upon any other single agency. And, surely, in these respects your Firm has set a noble example; your plan has been invariably to place entire confidence in your Teachers; to afford them all reasonable aid and encouragement in the prosecution of their work; and from first to last the relations, which have subsisted between yourselves and them, have been in the highest sense cordial and complete. To this must be attributed the success which has attended your Schools, and which, with God's blessing, has conduced, and will conduce to the religious, intellectual and physical benefit of thousands who have been educated in them. Men and women now labouring in many ranks of life, in their various callings, and in every quarter of the world, will remember with gratitude the name of your honoured Firm, and the share which the education you have so liberally provided has had in fitting them for their various duties.

In closing this brief record of your Schools, I feel bound to refer to some of those who have assisted in conducting them, and to whose labours their success is largely due. Of Mrs. Duff, now deceased, hundreds of women can testify to the excellent training which they obtained in the Girls' School from her conscientious and laborious discharge of duty, and from her constant regard for whatever tends to the formation of good habits and character ; of Miss Teague, still engaged in the good work, hundreds more will remember with gratitude the admirable way in which, as infants, they were trained in habits of reverence, order, punctuality, neatness, attention and obedience. Of the large number of boys who have passed through the School there have been none better behaved, and none more successful, than those who have had the advantage of the early training of the Infant School. To Miss Worthington, Mrs. Duff's successor, Spon Lane people are indebted for one of the best Girls' Schools I have known; for quiet order, for an evident love of work, and cheerful obedience to duty, I do not think the School can be excelled. Of many others, who have rendered good service in the Boys' School, I may mention Mr. Hanson, the highly respected Master of your Oldbury School; his energetic services in connection with the Evening School, twenty-five years ago, and in particular, in carrying on Young Men's special evening classes for singing and other subjects, are still gratefully remembered by many; Mr. Hale, who was most serviceable during the experiment of a special school for glasshouse youths, to which reference has already been made; Mr. Marshallsay, who will be long remembered for his cheerful and arduous services in the crowded Evening Schools of twenty years ago, as well as for the interest he took in the athletic exercises of that time, and who is now prospering in Canada, and is also an esteemed member of the local legislature; and lastly Mr. A. Wheeler, who to many gifts as a teacher united a singular love for his work and a skill in teaching which I have seldom seen. To all these, and to many others with whom it has been my good fortune to be associated during all these years, and whose names and special services will occur to many, I beg leave to present my heartiest thanks for the cordial way in which their help in the Schools was rendered, and for their love for the great work in which we have one and all been so long and so well engaged.

Talbot continued his mastership under the School Board for five years more, until 1892, and soon after his retirement was very properly elected a member of the Smethwick Board.

When the five years' lease to the School Board was drawing to an end, the firm offered to sell them the land and buildings for £4,000. There was no sale then, but in 1898 a renewed offer at £3,500, after first refusals, was accepted, the firm undertaking, if the Board would take over the boys' and girls' schools and build a new infants' school, to contribute £500 towards the expense. The property, excepting the master's house and the lodge, was conveyed to the Board at the end of the year, and the playground opposite the schools let to them at £1 a quarter.

Talbot wrote to James Chance about the negotiation (May 18, 1898):

> You will be interested in knowing that our School Board is summoned for a special meeting on the 12th for passing resolutions to purchase the Schools for £3,500; to make application to the Education Board for their consent to the purchase; and to make application to the same Department for permission to borrow the said sum. The Chairman will move the resolutions, and the V. Chairman, I suppose, will second their acceptance.

> Curiously enough, the V.C. is the member who was originally most opposed to the scheme of purchase, and under whose tuition, it is supposed, the statement that Chance's Schools were the most expensive under the Board was made at the meeting to approve my candidature for membership. But the V.C. has become a convert, as well as all the other malcontents, and I have no doubt of the resolutions being carried unanimously. I enclose herewith a statement of sums borrowed for Schools built by the Board, and have jotted the interest at 3 p.c. opposite each. Of course this sum ought to be added each year to the annual cost of the school, but that has never been done, and hence the fallacious conclusion as to Chance's Schools, with rent £75 added yearly, to which they have come in the relative cost per child. That is to be corrected in the future, and when the Infant School has been enlarged with accommodation for 300, the relative position of the school to others will be altered. The percentage of attendance at Chance's Boys' School is highest under the Board, and the Head Master tells me that he believes it to be the highest in England. He does not know that it was the highest long before he was born. People have been trained to it, "one soweth and another reapeth." And so it must ever be. Well, thank God, dear Mr. Chance, for any good we have ever done in this and higher and better ways: I feel that a grateful heart is the backbone of religion.

The statement showed the cost of building various schools under the Board to have ranged from £4,300 to over £15,000, entailing interest from £129 to £462, as against the £75 rent charged for Chance's Schools. In another letter, announcing the unanimous adoption of the resolutions,

Talbot cited the Smethwick Weekly News as of opinion, "and in this we all concur," that the Board had made a very good bargain.

In the new building accommodation for the girls was increased by the addition of a storey to the infants' school, the infants being transferred to the old girls' schoolroom. The opening ceremony took place on October 16, 1900, when opportunity was taken to unveil a portrait of Talbot by a Smethwick artist, Joseph Gibbs, a gift from George Chance to the School Board. Making the presentation, George Chance observed that the history of the Schools was the history of a large part of Talbot's life. He remembered his father saying, on the occasion of the opening of the Park, that if there was one man he envied it was Talbot, through whose hands had passed so many hundreds, perhaps thousands of lads, whose success in life was built on his foundation. Expressing the wish that the portrait might remain for many years in the School, he intimated that if the Board thought proper to remove it, the matter lay entirely in their hands. Accepting the gift, the Rev. G. Astbury, Chairman of the Board, paid eloquent tribute to the value of Talbot's life-work. The portrait now hangs in the Smethwick Public Library.

Satisfied with the alterations, the firm duly paid over the £500 promised towards their expense in February 1901.

A year later, on February 24, 1902, Talbot's useful life ended, seven weeks after the death of James Chance, his patron, collaborator and friend.

After School Boards were abolished, the Schools were carried on under the direction of the Education Committee of the Borough Council until 1914. Then, says the *Smethwick Telephone* in an account of the closing of them and of the Rabone Lane Schools, " Messrs. Chance again came forward with a liberal offer, and the Council decided to allow them to re-purchase, the amount being £2,000." The actual amount paid was £2,250. For a time the buildings were turned to war purposes, but then was carried out a scheme considered in 1887, and again in 1898, to utilise them for dining and recreation rooms, a scheme formerly negatived because it was felt that their use for schools should be continued so long as they were required for that purpose. The requirement having lapsed, in 1917 was opened in them the Canteen, of which account is given in a later chapter.

APPENDIX.

INTERESTING tributes to the value of the Schools, and particularly of Mr. Talbot's teaching, were paid after his retirement at a Dinner attended by more than forty of his old scholars, on February 10, 1893. Mr. John Egginton, presiding, said :

> There are few in this room who can go back in their personal recollections to the early forties, . . . and only those who are able to contrast this part of Staffordshire now with what it was in those early years can fully measure the debt the district owes to these schools and to similar organisations which followed them. . . . Messrs. Chance have always been most fortunate in the selection of their teaching staff, and we "boys" especially owe our thanks to the unerring instinct with which they, happily for us, discovered Mr. Talbot. From the first the standard of instruction was very high, and the education given, using that term in its widest and truest sense, was amongst the best that money could have bought anywhere. The religious training was constant, earnest, and practical, and bore pointed reference to the lad's present needs, to the future wants of his manhood, and to his working, commercial, and home life. Its formularies were the Lord's Prayer, the Ten Commandments, and the Apostles' Creed, of which said good Richard Baxter, of Kidderminster, "The older I grew the more they furnished me with a most plentiful and acceptable matter for all my meditations." I have always considered the religious instruction given in these schools to be the true type of that religious education best suited to all our national public schools. But behind and beyond all this, and enforcing it in the best of ways, by example, we felt as lads, and we know now as greyheaded men, that our master was himself deeply imbued with this truly religious spirit, and it was on this account that the lessons he taught have been clinging to us, and influencing our lives for good, ever since. It was the case of that fourteenth-century teacher so beloved of Chaucer:
>
>> " Of Christes law and His Apostles twelve
>> He taughte, but first he followed it himselfe."
>
> You all remember those wise mottos, painted up in quaint old English text above the frieze rail of the school, showing us so well for what purpose we went to school and so admirably illustrating the essential spirit of what was taught. These are some of them: " Lying lips are an abomination to the Lord"; "A false balance is an abomination to the Lord"; "Better it is to get wisdom than gold"; "He that heareth reproof getteth understanding." Well, that all the generations of boys from 1845 till now got the reproof we are quite sure. We trust we also got, as we might have done, the understanding. . . .

Then we had excellent Sunday Schools, and for many years Sunday services in the Boys' School, which led ultimately to St. Paul's Church being founded. All this beneficent work was founded and continued by the members of the Chance family, and throughout the personal interest taken by them in the various organisations was most marked and helpful. . . . My own teacher in the Sunday School was Mr. Edward Chance, who walked often over to Spon Lane on Sunday afternoons, for there was then no railway, and one of my now most cherished volumes is the Sunday School prize he gave me in 1849. . . . We all know and feel, but cannot adequately express, the thanks we owe to three generations of the Chance family; the district round, those who are assembled here, the thousands of men and women, pupils of these and of the Oldbury Schools, who cannot be here, scattered, as we know them to be, in all countries and dimes. Successful beyond measure has the work in these schools been. The old boys have made their mark in every profession and calling in this country, in our colonies and dependencies, and in foreign lands. Many hold conspicuous positions of trust, honoured far and wide in their work and their lives. Many more in the "sequestered vale of life," though not so widely known, are equally honoured and worthy. Nor must we forget the great number of women and girls who have gone through the schools, multitudes of whom grace and adorn, as only women can, every relationship of life. What we all might have been, but for these schools and the institutions attached to them, we do not know. But this we do believe, that for many of us they were the only opportunity we would have had, under God, of beginning an upward, a better, and in every sense a more successful life, than what seemed to be then round us and before us. The most and the best of what we are and hope to be, whether in respect of spiritual, intellectual, or social attainment, we lay at the feet of the work done for us in early life in these schools.

Similar sentiments were expressed by Mr. Samuel Nicholls, in the vice-chair, Mr. Genner, and others. Mr. Nicholls took it that "Mr. Talbot held a high and proud position in South Staffordshire and the Midlands. He held the premier position of any schoolmaster of an elementary school."

Mr. Smallman, an ex-member of the Manchester City Council, re-marked that

as one who had received the rudiments of his education at Chance's Schools, he should always hold those who were associated with them in the highest esteem and regard, particularly their old friend, Mr. Talbot. He frequently looked back with the greatest delight to the time when he was a scholar at Messrs. Chance's Schools. Whatever success he had attained in life, he might safely say that it was due to the instruction which he received in the day and evening schools. It was a source of great delight to go to the school evening after evening, and he was assured that the instruction which was there imparted had been of the greatest value to him.

Mr. S. C. Rock, proposing Mr. Talbot's health, said that

he had been since in the same profession, and he only wished to follow in Mr. Talbot's steps. . . . He had always found him a man of discipline, and he thought that they could feel thankful for it, now that it was over. It was not very sweet, but it had brought forth good fruit. . . . The noble aspirations which Mr. Talbot always set before them he should never forget, and he hoped, when he had accomplished the years of instruction Mr. Talbot had, that he should have the privilege of seeing such a noble band, reared by himself, as Mr. Talbot had that night.

Supporting the toast, Mr. B. Robinson referred to "three principles which Mr. Talbot specially inculcated: punctuality, truthfulness, and cheerfulness."

Mr. James Robinson, of the Oldbury Works, proposing "Our Guests," recalled the valuable work done by Mr. Shaw Hanson at Chances' Oldbury Schools. Mr. Hanson, in reply, hoped that he

should live long enough to see a number of his old pupils who would have the same feeling for him as those present had for Mr. Talbot. His brother, John Hanson, who now held an important post in Yorkshire, was always pleased to attribute much of his success in life to the training which he received at the Spon Lane Schools.[1]

---

1   From the *Smethwick Telephone* and *Oldbury Weekly News* for February 18, 1893.

CHAPTER XII

THE PROVIDENT SOCIETY AND OTHER SOCIAL WORK
—OLD AGE PENSIONS —
WEST SMETHWICK PARK

THE Schools, if the principal, were by no means the only achievement of
Chance Brothers & Co. in the domain of social progress. At a yet earlier
time they took measures to limit drinking, to restrict, so far as was possible, Sunday labour, and to establish dispensaries; later they built, or
helped to build, churches in the district and set on foot regular systems of
pensions and of poor relief. The special social work of Alexander Chance
at Oldbury has been recounted. Not least of benefits was James Chance's
gift and endowment of West Smethwick Park.

William Chance took the initiative in the establishment of a Provident
Fund at the Works in the year 1841.

A memorandum preserved among his papers tells us that the workmen
had set on foot a Sick Club amongst themselves as far back as 1820, but
that after five years it had split into separate clubs for the glass-makers and
cutters, the former seceding "on the ground of the disproportionate number
of cutters' claims upon the sick list," and that membership of the cutters'
club, which made no provision for old age, was by 1841 reduced from 60
or 70 to 45, while that of the glass-makers remained at about 100. It had
been desired, the memorandum goes on, that Sick Clubs (Glassmakers)
should be established throughout the Kingdom, making Smethwick's the
Parent Institution. A large number of the men having joined other Societies, which provide for the wants of old age, and many more having gone
into the Odd Fellows' Society, it remains to be ascertained how many are
inclined either to support the St. Thomas' Club as a Branch thereof, or
in forming one upon nearly the same principles at the Works. Many of
those who belong to the two before-named would be precluded from the
St. Thomas' on the score of age, being past 50.

The writer favoured preservation of the two clubs so long established.

"If any good could be done by adding to their benefits, it were good. I think all who do not receive the benefit of medical attendance would most readily join in paying an extra penny or even 2*d*. per week for such an advantage."

Having informed himself of the working of the "St. Thomas' Provident Institution" through its chief surgeon, Mr. W. Sanders, William Chance set down his first ideas in the following rough notes:

A principal consideration, where the medical men reside in reference to the residences of our own men.

To circulate among the men an extract from St. Thomas' Report.

Then to call them together and ascertain what numbers will enter, &c.

Perhaps we had better support one at Smethwick, one at Oldbury, and one at W.B.,[1] rather than undertake it ourselves.

Medical attendance club would, I think, be general.

Sick Pay club to those not already in clubs.

Savings Bank, if objectionable in the Works, might be carried on in connexion with Mr. Simcox.

It is not intended that our club should be a branch of the St. Thomas'.

The first thing to ascertain will be the parties who are disposed to join either or all of the several societies, in order that we may determine whether it will be desirable to have a separate society, or form one in connexion with Mr. Simcox.

When these ideas had taken shape in his mind, William Chance sent out the following printed letter to the workmen, dated August 26, 1841:

Having received an application from a Gentleman in the neighbourhood to unite with him in the formation of a Provident Society, and believing that such an institution will be highly conducive to your comfort and happiness, I am desirous of calling your attention to the subject.

Provident Institutions, or, as they are now commonly called, Sick Clubs, are well known throughout the country, and it would be difficult to find a district without one. The object for which they are established is, "To enable the labouring classes to provide for Sickness, Old age, and Death," and it is one which must recommend itself to the attention of every prudent and industrious mechanic, who is dependent on his own exertions for support.

Before, however, joining an institution of any kind, a careful man will necessarily ascertain how far the principles on which it is conducted will enable it to carry out the benefits, which it professes to have in view. Many clubs, after a few years' exist-

---

1   West Bromwich.

ence, have, from various causes, been broken up, and the members deprived of all the advantages to which their subscriptions entitled them. In some cases this has arisen from the public houses, where the meetings were held, having changed hands, in others from want of an efficient superintendence. But the main cause of the failure of so many clubs will be found in the imperfect system on which they were conducted. For instance, persons of all ages and trades have been permitted to derive the same benefits by paying the same subscriptions, and thus no calculation made of the greater liability to sickness in a man of fifty than in one of twenty. The experience of those who have devoted their time and consideration to the subject clearly proves that no club established on such principles can stand.

The institution, which it is now purposed to establish among you, will be conducted on a system by which the evils alluded to will be avoided, while all the advantages of other clubs, with many additional privileges, will be secured. This system is not a novelty yet to be tried ; it has been most satisfactorily tested in various establishments in Birmingham, Manchester and other places.

It was explained, then, that the Institution would consist of three divisions, a Medical Attendance Club, a Sick Pay Club, and a Saving Club, to any or all of which anyone would be allowed to subscribe. By a subscription to the first of a penny a week its members would be "entitled at all times to receive Medical Advice and Medicines (without further expense) from one of the Surgeons attached to the Institution." For the second weekly subscriptions to ensure 10s. weekly in sickness and £5 at death were graduated from 4d. at the age of 20 to 7½d. at the age of 39. By the Saving Club it was intended to provide means to the careful and prudent, of putting by small weekly savings against the time of need. How frequently is money squandered in health, which, if husbanded, would prevent the necessity, in times of suffering, of applying to parochial or gratuitous assistance? Any sum from one penny, but not exceeding five shillings, will be received each week. When these weekly payments have accumulated to an amount exceeding five shillings, interest will be allowed, and the members may withdraw their money at any time by giving a week's notice.

After which William Chance pointed out that the tables were calculated on scientific principles and thoroughly reliable, "being sanctioned by the first actuaries in the Kingdom" that the meetings would not be held in a public house, encouraging wasteful and noxious expenditure on ardent spirits, and that the payments will be included in one sum, and no extra charges made for secretaries, funerals, &c. Time will thus be saved in keeping the accounts, and expenses materially diminished.

In conclusion, let me urge you to consider the advantages proposed by this Society. Medical Advice and Medicines, Weekly Pay during the time of sickness, a sum of money at death, and a constant inducement to economy and prudence. Sickness will come; prepare for it now. Hard times will press upon you ; put by from your earnings in prosperous times, and be ready for them. Insure for your wives and families support, when you shall be no longer able to work for them. The weekly payments necessary to insure these advantages are so small, that you will not feel them inconveniently.

I shall take an early opportunity of calling a meeting, to explain to you more fully the details of this Institution.

The meeting was held, but only the first division of the scheme, a Dispensary Fund, came into being. Subscriptions to it were fixed at a penny a week for all unmarried men under 21 and all females receiving more than 5s. a week; twopence for older men and for all that were married. A personal ticket entitled the holder to attendances and medicines; a personal and family ticket to the same for his wife and such of his children as were not employed elsewhere.[1]

The first surgeon engaged was William Hammond, in February 1843. He was required to spend daily, excepting Sundays, an hour at Oldbury and an hour and a half at Smethwick for the purpose of prescribing to the workmen; to attend their wives and families at their houses; to find all medicines and appliances, excepting leeches; and to report regularly to the Board on the state of the workmen's houses, "with a view to having such sanitary regulations adopted as may tend to promote the health of the workmen and their families." All this for a salary of £100 a year and 10s. 6d. for each midwifery case. With the sanitary condition of the workmen's cottages he very soon had grave fault to find.

His appointment aroused considerable jealousy at Oldbury, where Thomas Richard Cooper had hitherto been attending to the workmen's medical needs. He represented that he had been doing this gratuitously and with assiduous care from the time that the works were opened, looking forward to the establishment of a club and his own appointment as

---

1   Directions on the workmen's cards ran:

a. If at work, never leave your work before the Surgeon sends for you.

b. If not at work, and able to attend at the Surgery, be there at a quarter before four o'clock.

c. Return as soon as possible to the Surgery any bottles which you receive thence.

d. In midwifery cases inform the Surgeon at least one month before confinement.

its surgeon. His plaint, however, was for the present disregarded.

Hammond soon found the work thrown upon him, and the expenses to which he was put, too great. He reported after eight months more than 2,000 cases treated at an average cost to himself for medicines of about 9*d*. each, and set down that of the horses, which he was obliged to keep and of which he had worn out two, at £70. He had no time, he said, to build up a private practice, and to continue to perform his work would require an assistant. With such help at £100 a year he brought his annual expenses for (say) 3,000 patients to £282 15*s*.

Agreement failing, he was replaced, when his twelve months' engagement was out, by William Davies, of Bristol, on a three years' agreement, subject to a trial period of three months, and at a salary of £300 a year. He undertook the entire medical and surgical charge of all the workpeople and clerks at Smethwick and Oldbury and their families, on conditions arranged similar to those with Hammond, binding himself to devote his entire time to the work and to undertake no private practice whatever. The firm agreed to fix a boundary within which its employees must reside or forfeit the benefits of the fund, and further to erect a house for Davies on its own land fronting the Oldbury Road.[1] The three months' trial does not appear to have resulted to his satisfaction, for in July the medical charge at Oldbury was intrusted to T. R. Cooper, above mentioned, who was bound to attend at the dispensary twice a day or, if otherwise particularly engaged, to be represented there once in the day by a skilful assistant, " being held responsible for prompt and judicious treatment." He was to receive £50 for one year to attend the workpeople, at present about a hundred, with a proportional increase for any greater number.[2] Davies retained the work at Spon Lane. In 1846, desiring to undertake private practice, he was permitted to employ a qualified medical practitioner as an assistant, on the undertaking that the workpeople should continue to receive "the same careful and anxious attention as heretofore." He was required to continue residence in the works, and was allowed £30 a year towards the expense of keeping a horse.[3]

The magnitude of the work is shown by figures for the years 1845 and 1846. In these years Davies treated respectively 3,223 and 3,182 cases, 101

---

1   Board minutes of February, 1844 with a copy of the signed agreement, of date February 13.
2   Copy of his agreement, July 23, 1844, ibid
3   Board minute of July 10, 1846.

and 102 of them midwifery, while during 1845 11,600 persons received medicines at his surgery. At Oldbury in the same year Cooper treated 1,620 cases, 45 of them midwifery. Visits and attendances for nine months only he reckoned at 6,379. In later years, with the growing number of workmen, these figures were largely increased, and more than one surgeon was required at each works.

Some dissatisfaction among the men, meanwhile, had made it necessary to issue the following notice, which contains several points of interest.

DISPENSARY

Owing to the long absence of our Mr. W. Chance, the collection for the above Institution, attached to our Works, has been delayed, it being necessary for the partners to consult together, on account of a most unjust complaint having been made by some of our workmen of the deduction from their wages towards defraying the charges incurred in carrying on the Dispensary.

This Institution was orginally established for the benefit of our workpeople, in consequence of our having information that many of them had incurred heavy bills for medical attendance, which they were unable to pay, and on account of which some of them had been thrown into jail. We were anxious that our workpeople, with their wives and families, might have the services of a surgeon of first-rate talent and education at the lowest possible rate of charge, and that the whole of his time might be devoted to our manufactory, having found that in cases, where the surgeon is engaged in private practice, the working classes are generally more or less neglected.

It is a great satisfaction to us to believe that the establishment of this Institution has been attended with the greatest advantage to multitudes of our workpeople and their families: no less than 3,004 cases have been attended by Mr. Davies from February 19, 1844 to December 31 of the same year, a period less than eleven months; 785 of which, *besides* 80 midwifery cases, have been visited by the surgeon at the houses of the patients, and no family has been charged more than one shilling per month for attendance (including midwifery cases), medicines, leeches, bandages, and everything else required.

So far from the establishment being advantageous to ourselves, we have sustained considerable loss by it, having already paid £196 more than we have received ; and if we had merely a regard for our own interest, we should be glad to give it up ; but being anxious for the welfare of our workpeople, and being fully persuaded that the Institution has been invaluable in numerous cases, we are desirous of continuing it,provided it has the support of our entire establishment.

On Friday next we shall collect the monthly contribution, and everyone will have to sign an agreement to support the Institution : if any persons refuse, their names will be handed to Mr. Davies, with directions to decline any further attendance upon them and their families ; and in case of any further difficulties being thrown in our way we shall put an end to the Institution altogether, when our workpeople will have to pay about £800 or £1,000 per annum, for attendance and medicines, to private practitioners, instead of about £200, which they have hitherto paid towards our Dispensary.

Chance Brothers & Co.
Smethwick Glass Works, June 1845.

Subscriptions having thus been made practically compulsory, the Fund prospered uneventfully and paid its way well. In May 1853 there is notice of a new dispensary to be erected near the reading room. Next year Davies' salary was raised to £500, as from the beginning of 1852, and in 1856 he was allowed further to have any surplus that might accrue over a reserve fund of £200. In 1857 he had as assistant a Mr. Whitehouse, with whom, however, there was trouble, when he developed an ambition to make glass himself. There was also a Mr. Guttridge, given in December 1865 the sum of £20 in recognition of gratuitous service in cases of lithotomy. In 1869 the balance to the credit of the Fund enabled payments for the year to be remitted. As examples of the benefits derived from it may be cited the sum of £14 given to George Dunn, to enable him to send two children to London for operations, and £2 to William Tooth, a mixer, to take him to a Convalescent Home at Rhyl.

In January 1868 was established a subsidiary institution, "Messrs. Chances' Glass Works' Provident Club for young persons of both sexes from 12 to 18 years of age, and women."Members of 12 to 15 years of age paid an entrance fee of is. and id. a week afterwards, insuring thereby in sickness 4s. per week for four months and 25. per week for four further months, should the sickness continue, and £2 in case of death. For members 15 years of age and upwards the entrance fee was 2s. and the weekly payment 2*d*., the corresponding benefits being 6*s*., 3*s*., and £3. It was provided that one-third of their total payments should be refunded to members on their leaving the Club, less amounts received for sick pay, and that all males ceasing to be members on attaining the age of 18 should in addition be entitled to a sum of money equal to the admission fee and one month's subscriptions, provided that they joined any well-established club in the neighbourhood and had been members of the Club for at least two years.

Other doctors employed at Spon Lane from January 1879 were John Manley and M. and A. Jackson, while a year later Davies handed over his practice at the works to his partner, Isaac Pitt. An election of 1881 showed the last-named to be a strong favourite ; his firm polled 1,224 votes, the Jacksons 134, and Manley 105.[1] In the new rules of 1893 the doctors named were William Frederick Marsh Jackson and Arthur Atkinson Jackson, John and John Herbert Hawkins Manley (father and son), Isaac Pitt, and Cyril Herbert Sharpe. The last-named had begun work in January 1891 and continued it till the end of 1907, when his place was taken by J. A. L. Pierce. Since the changes of 1912 Dr. Pitt continues to attend the pensioned members of the Society.

At Oldbury in 1881, the year that the Provident Accident Fund was established there, management of the Dispensary Fund was vested in a committee of the foremen, with a member of the firm as chairman. Each year the firm nominated two or more qualified doctors for the ensuing year, the Committee appointing. Of those chosen, each workman named for himself the one that he preferred, and each doctor was paid a fixed annual sum per head at a rate determined by the Committee.

There was also an Accident Fund at Spon Lane, from which regular payments were made until in July 1898 the Workmen's Compensation Act came into force. Its income was derived from workmen's fines. Shortly before the date mentioned the Board were informed that a large majority of the workpeople were for contracting out of the Act, on terms similar to those in force at Oldbury. Accordingly a scheme was prepared on the basis of the Company paying three-fifths of any compensation allowed and its employees two-fifths. When, however, it came to balloting on the proposal votes were decisively for the provisions of the Act. Figures taken at the end of the year showed £18 10*s*. paid during its first half

1   John Chance's diary.

from the Accident Fund, and £12 7s. 6d. during its last six months under the Act. And further, that whereas in the latter period £13 15s. would have been paid to twenty-eight persons under the "old system"' and £32 12s. under the arrangement proposed, actually only ten out of the twenty-eight had received benefit at all. On the closing of the Accident Fund the workmen's fines were handed over to the Provident Society. In January 1893 that society was registered under the Friendly Societies Act of 1875 as "The Smethwick Glass Works Provident Society, composed of persons over 16 years of age."Its objects stated in the new rules drawn up were to provide by voluntary contributions of the members, with the aid of donations, for the relief of members and their families during sickness, in the following manner:

> By providing the members and their families with medical attendance, medicine, drugs, and surgical appliances.
>
> By providing for the admission of members or their families to Hospitals, Convalescent Homes, or other like institutions, by the purchase of tickets, or by annual subscriptions, or by donations, or otherwise.
>
> By affording relief to the family of a member during the existence of any contagious or infectious disease in the family of such member.

The Directors of the Company, and such other persons employed at or in connexion with the Glass Works as should be nominated by the Board, were to be honorary members, each paying an annual subscription of a guinea. Payments by ordinary members, not to exceed fourteen in any one year, were fixed at 4d. for each female, the same for each male receiving less than 12s. a week in wages, and 8d. and is. respectively for male members receiving from 12s. to 16s. and over 16s. weekly wage. "Provided that the contribution of each pipe-warmer, pusher, and rolled plate table boy, whose contribution, according to the foregoing table, would be more than 8d., shall be 8d. only."

The Society was to be managed by a committee consisting of fourteen elected members, seven of them managers or foremen and seven ordinary workmen, and of others ex officio—namely, one of the Managing Directors of the Company, or his Deputy, as Chairman, and any Treasurer, Secretary or Trustee appointed. For each year two or more duly qualified medical practitioners were to be appointed by the Committee, to be paid, for each member of the Society on their lists, at a rate per annum to be settled by the Committee. Henry Talbot, the Company's principal manager, was the first Secretary, and the first Trustees George Chance and Macaulay.

In 1896 the Society was registered under the new Friendly Societies Act of that year, and its scope was extended to provide medical attendance for such of its members as were incapacitated for work by old age or permanent disablement; the Committee to allot a sum for this purpose up to £10 a year, and the Company to contribute the like amount or half the actual cost.

In July 1903 there was amendment of the rules. The members were now placed in three classes, Ordinary, Honorary, and Pensioned, the first class consisting "solely of persons over 13 years of age, for the time being employed at or in connection with the Glass Works." The Company undertook to continue to pay half the cost of allowances for infectious diseases (this had been done from October 1883) and, as before, of medical attendance for the pensioners, and further a subscription of not less than £50 a year to hospitals and the like for the benefit of members of the Society; this in addition to £8 8s. a year paid by the Company for the Directors and the principal manager as honorary members.

After the passing of the National Insurance Act of 1911 had to be set before the members the question of making the Society an Approved Society under its provisions. This was not done, partly because the Board declined to give the guarantee required by the Insurance Commissioners, partly because many of the employees preferred to receive the benefits of the Act through other Friendly Societies to which they belonged. The Society continued as of old, though with benefits reduced, the Company contributing as before. Further resolutions of the Board of this year (1912) were to pay the entrance fees for foremen wishing to join the "Foremen's Mutual Benefit Society," and two-thirds of the cost of the bandages and other first-aid appliances used in the Works. The average annual amount since paid under this head has been £18 17s. 8d.

In January 1913 the rules were amended to suit the new conditions. The principal change was in adaptation of the scale of contributions to the reduced benefits. The number of them now was not to exceed twelve in the year, and the amounts were reduced to 1d., 2d., and 3d. respectively, instead of 4d., 8d. and 1s. as before. Under these rules the Society continues to prosper.

Another institution of early date was that of a Reading Room and Library, of whose origin and career Mr. Talbot has left the following account:[1]

In January 1852 a meeting of managers, clerks and others was held to consider the best means of providing such an Institution. A committee was appointed, and an appeal to your Firm for assistance was most promptly and generously met by your erection of a commodious Reading Room, with Rooms for a Library, and a contribution of several hundreds of valuable books. These formed a nucleus, which in time was extended to a Library of nearly 3,000 volumes, in all the leading departments of general literature. The subscription was one shilling per quarter, the rooms were kept open from ten to ten, and a librarian was constantly present to perform the needful duties of the institution. At the best period of its history it numbered about 300 members, and for several years it was the centre of much activity. Lectures were delivered, concerts were given, soirees were held, and a workmen's annual excursion, which I believe was the first in this neighbourhood,[2] was organized in 1856 and proved very successful, not only in giving to a large number of workmen with their wives and families a most enjoyable recreation, but also in increasing the Library. In the course of thirteen years the commission derived from this annual excursion realized nearly £100, and this sum was devoted to the enlargement of the Library. It was also in connection with this institution that your Firm paid the expenses of the visit of about 1,200 workpeople in your employment to the International Exhibition of 1862. The establishment of the penny daily paper, and the formation of Free Libraries in the adjoining towns, coupled with the fact that a large number of your workpeople resided at some distance from the Works, led gradually to a diminished attendance at the Reading Room; and in 1870 the number of subscribers had so much decreased that your Firm, who had all along borne the expenses of the Librarian's salary, fire and gas, and repairs, concluded that the need for the institution no longer existed, and, with much reluctance, concurred in its being closed. The Library, which had unfortunately suffered considerably in loss of books, was then removed to the Schools, where it was preserved for some years; and finally, by a resolution of the Committee of the Reading Room and Library, which was called together for the purpose, the books were sold by auction in 1881, and the proceeds of the sale, which amounted, less expenses, to £23 11s. 7d., were presented to the Committee of the Free Library at Smethwick as a contribution to the Library Fund, and duly acknowledged in the report of the institution for that year. The

1   In his *Sketch of the History of the Schools*, pp. 15, 16.
2   In Mr. Genner's conviction it was the first.

Reading Room and Library, like the Schools, were open to the neighbourhood.

At the time of the sale of the books in 1881, and therefore also previously, the library was known as "Mrs. Robert Chance's." When, in 1916, the School buildings were converted for use as a Canteen, as hereafter related, opportunities for study were offered again.

Among other schemes proposed for the welfare of the workpeople were a Convalescent Home (1874) on the example of that lately established for the Alkali Works, and an Institute to be erected on the firm's vacant land at the corner of Spon Lane and George Street (1877). The former proposal failed to secure a majority of the workmen's votes. The latter was of an ambitious nature, the plans covering more than half an acre and comprising large men's and boys' dining and recreation rooms, a playground and a gymnasium. But they were never put into execution.

The firm's custom of awarding pensions to old or disabled workmen of long service and their widows appears to date its establishment on a settled basis from the year 1866, when regular entries in the Board minute books begin with the names of John Grigg, employed for about 40 years and incapacitated, and Benjamin Rock, employed for about 50, from the time, therefore, that the first glasshouse was built. There are isolated earlier entries, but before the year named such benefactions were mostly by the partners privately, and especially by those two most generous of men, Robert and John Chance, who for many years already had systematically distributed relief in the neighbourhood of the works and in Birmingham. One of those, whom they employed for this work, was Thomas Manwaring, who, when the firm took over the responsibility, became its official almoner.

From 1866 entries in the minute books of pensions granted are regular and frequent. As years went on, regulations for guidance in the matter were adopted. Thus in 1900 were made eligible for pensions all men of 25 years' service and upwards, no longer able to work, and their widows, subject to consideration of age and merit. In 1914 the pensioners numbered 105.

Robert and John Chance continued their benefactions, separately from the firm, during their lives. As a crowning act, John Chance placed with the Company in 1898 the sum of £4,000 in trust to continue his private pensions and allowances after his death; as they lapsed by the deaths of the recipients, the surplus income and ultimately the capital fund to be handed over to the West Bromwich Hospital.[1]

Manwaring continued his devoted labours until his own retirement with a pension in 1893. Soon afterwards, in May of that year, he died. His work has been carried on with equal devotion to the present time by his daughter, Miss Rebecca Manwaring, who had previously assisted him.

The standard of 25 years' continuous service is a low one in the firm's experience. Most of the pensioners could show records far longer than that minimum. As examples may be cited, previously to 1900 and besides the two mentioned, William Sparkes, lighthouse engine driver, with more than 50 years; Henry Howell, bricklayer, 45; Samuel Lashley, about 40 ; Benjamin Moore, about 38; Joseph Stokes, mixer, about 35; and among the foremen Emmanuel Sorrill; Daniel Smith, with 46; James Cotterill, with 50; David Bates, with 54; and Enoch Hall, with no less than 60. Indeed, in the matter of long service of their men, Chance Brothers & Co. can certainly vie with, and probably outdo, any other firm in the kingdom. When, in 1913, the Weekly Dispatch offered medals to men who could show continuous employment with one firm for a period of more than 50 years, a count was made at Spon Lane, and those who might claim them were found to number nearly seventy, thirty-five of them still at work. Charles Harvey could show 65 years' service as stoker and engine-driver, and the record of never having been late; Benjamin Perkins, crown glass maker, 64; Levi Blackham, sheet glass cutter, 63; Samuel Lunn, Charles Evitts, William Woodward, each 60 or more. Mary Stanley, just deceased, had completed 63 years. But all were outdone by the veteran rolled plate manager, James Simcox, with his 70 years' continuous service. This was a surprise for the proprietors of the journal, but they offered six medals, which were accepted and duly allotted. Afterwards the Company decided to distribute silver medals themselves on like conditions. Presentation of seventy-three of these was made on January 4, 1916, on the occasion of the tea at which the old people are accustomed to entertain annu-ally the Directors of the Company, a festivity inaugurated in 1905 by

---

1   It is proper to take note of John Chance's long continued interest in and services to this institu-tion. He joined the Committee in 1866, when the hospital was but a temporary outpatients', infir-mary. Principally by his efforts it was rebuilt in 1870 as a District Hospital, and in 1882 the circular wards were added. Later at his own expense he erected a new outpatients' department, enlarged the dispensary, and presented a much needed library. His last gift was a new casualty room, fitted with all the latest improvements. In this, when it was opened after his death, a tablet was placed to his memory.

George Lewis, its cashier.[1] Since then several more have been presented. In the first half of the last century excessive drinking was still the national habit, looked on as neither disgraceful nor injurious. In fact, a liberal consumption of alcoholic liquor was held to be advisable for health, and doctrines of abstinence were yet in embryo. Manufacturers then took but little heed of their men being drunk, provided that it was not at their work, and measures taken were but to that end. Such was a resolution at Spon Lane in 1841 to limit the consumption of "mixed beer" on the works to 36 barrels per quarter. Five years later, when it was found that a whole barrel of beer had been consumed one Sunday and that in one case a man had procured double orders, it was decreed that Withers alone should authorise the import. Next year the allowance was fixed at three 36-gallon casks per week, and then it was ordered that no drink should be brought into the works excepting for the blowers and gatherers and founders' crews, when at work, but not for such extra work as pot-setting. In 1848 it was resolved that no drink at all should be introduced on Sundays, and ordered that no person should remain at the head of a department who kept a public house.

Such orders became unnecessary, as time went on. Attention was rather given to the reform of habit; one important step in this direction was the provision of dining rooms and temperance canteens within the works. In modern times the efforts of temperance reformers, and public regulations, have gone far to extinguish the vice of drunkenness in England.

Restriction of Sunday labour to such only as was absolutely necessary was particularly the aim of William and James Chance. They brought the

---

1 On the occasion of the presentation Joseph Armstrong said: "I rise on behalf of the recipients of these beautiful and artistic medals to propose that our very best thanks be tendered to the firm for their expression of kindness shown to us in the gift. . . . Just about sixty-two years ago I began to work for the firm, and during that long period I have never had occasion to engage in a strike ... I believe that they are always ready to recognise regularity, punctuality and devotion to duty, which ought to characterise every honest workman. Memory is a wonderful thing, and as i stand here to-night there pass before my mind's eye like a panorama view the faces and forms of those members of the firm that have passed away into the unseen world. There was old Mr. Robert Lucas Chance, little in stature but big in heart and soul. Then comes Mr. Edward Chance with his hale, thin figure and bright and smiling face. Then Mr. James Timmins Chance with his commanding voice and figure, whose work and influence have been felt the wide world over . . . Last, but not least, there comes before me that great, good man, Mr. John Homer Chance, great because he was good and good because he was great. He has left behind him an influence that is lasting. . . . He was, in my boyhood days, my Sunday School teacher, and in the schools near the works yonder, which exerted a wonderful influence on my life, he lived to bless the world in which he lived, and blessing others, he was himself blessed. These men were pioneers in this and the surrounding district in secular and religious education, and those schools, empty though they are to-day, stand out as a monument to their kindness and generosity" (The *Weekly News*, January 8, 1916).

question to an issue one Sunday in 1841, by visiting the works to see what was actually being done. Lucas Chance was away in London, but his wife, attending with her family their usual religious meeting, rendered him the next day the following account:

On arriving at the Works yesterday we were hailed by the intelligence that William and James were there ascertaining the number of men employed that day, and preventing others, such as clerks, &c., from being engaged. After the meeting Robert brought a message from them to me to request I would go to the Hall to see them, and as I conceived this measure a very politic one on their part I went cheerfully. My feeling, previously, had been that of vexation in finding they were so engaged during your absence, and I was glad to have that removed by a frank avowal, on their part, of what they had done and intended to do. The object of my interview was to satisfy me, and through me you, that in order to keep the manufactory on, and the fires in, it was not necessary to found on a Sunday, and consequently the employment of so many men on that day was superfluous, and a mere matter of gain. William admitted that in founding three times a week instead of four they should of course diminish their profits, but he thought, if the matter were fairly sifted and every point weighed, that it would eventually prove to be rather an advantageous measure, than the reverse. The Sunday foundings were frequently very indifferent, in consequence of less attention being paid to them, and James asserted that since he had the oversight of the concern he could not conscientiously complain of bad work on that day, because he felt that it was a hardship to the man to be so employed and wrong in itself. Both William and he agreed in opinion, that by means of less hurry and driving at the Works, that is, taking things more calmly and investigating them more thoroughly, a larger proportion of good glass would be obtained than is the case at present, and consequently the concern would really nourish as much with three foundings a week as with four. But if such should not be the case, and the result was a diminution of profit, still, as it was a question of mere money, it ought not to be put in competition with the comfort and general good of the workmen employed; and so strongly were they both of opinion that the reform ought to be carried into effect, that they said, decidedly, it must be so, but that it was their wish to go hand in hand with you and do the thing amicably. I told them candidly, though I differed from them as to the motives, I could not but agree to the eligibility of what they proposed, and I thought you would cordially unite with me in wishing the workmen to have the same privileges as other men and to enjoy the leisure of one day in the week; and consequently I should use my influence with you to that effect. I added, moreover, that I should be glad of any measure that gave you less to do; and James said he would very gladly relieve you, if more were put into his hands.

The paper, which accompanies this, contains the particulars of James's observations, whilst at the Works yesterday. ... I must mention also that the men employed yesterday one and all stated they considered it a grievance to be deprived of their leisure, and that they considered it wrong to be engaged as they were on the Sunday. James also says that with the assistance of two men, who may be exchanged, the tires can be kept up well that day.

In the paper mentioned, James Chance supplemented a list of the men employed, forty-five in number besides "Frenchmen preparing their blocks, boys, and glass blowers about for Sunday purposes" with the following remarks:

In case of any contingency in the crown houses the glass makers are obliged to be fetched, even if in church, or at meeting. Jno. Gittins asserts that he has not had more than 3 Sundays to himself for 18 years. Of course the same applies in different degrees to the founders and their crews. For altho' the teasers change at 2 o'clock every 12 hours, still they can find no time to attend to religious duties, since they have to clean themselves, rest and eat before they are fit for entering a place of worship; to say nothing of the necessary absence of all disposition to improve the few hours which they might snatch for themselves, arising from their working without intermission.

The constant absence of so many fathers from their wives and families on the only day on which an union is generally possible must tend to great demoralization; and so long as such barbarizing causes are allowed to continue it is ridiculous to institute schools or other means of instruction. For what the children are taught there they see openly violated in practice by their parents.

Again, consider the influence of the bad example of so eminent a manufactory on the principles and practice of all around it, and how the religion professed by its proprietors must become hypocrisy, thereby, in the eyes of the world.

While every one of our chimneys, nearly, are sending out continuous volumes of smoke, scarcely any smoke is to be seen for miles around, although the iron forges, &c. are very numerous.

Even allowing the Sunday for founding, still there is so little time for contingencies with 4 founds a week, that generally Mr. Withers, his sons, and the glass blowers are not relieved till 12 on Saturday night, and since the glass makers have to begin early on Monday morning and to work for 12 hours a great portion of Sunday must be taken up in sleeping and preparing for work. Several men were noticed to-day to be coming in and out in order to enquire what time the founding would stop.

Having received his wife's letter with a copy of the above, Lucas Chance pointed out objections to the change proposed and referred discussion to the Board room. That he offered no determined opposition appears from the resolution adopted (February 23):

> The Board taking into consideration the religious opinions of Mr. W. Chance and Mr. J. T. Chance, which lead them to the belief that no persons should be employed on a Sunday in any way whatever excepting in cases where the existence of the establishment requires attendance, such as keeping up the fire, hereby determine to adopt plans with as little delay as possible for carrying out the views entertained by Messrs. W. and J. T. Chance throughout the Manufactory both at Smethwick and Oldbury.

It does not appear that the resolution was well observed. In 1848 Withers and his sons had to be given strict injunction that all men obliged to work must have alternate Sundays off, and that on no pretence whatever must any be brought in merely to expedite work, or in any case but of sudden emergency.[1] In February 1852, in consequence of earnest representations by the Rev. A. Williams, of Smethwick, it was resolved that no founding should take place on Sundays in any of the sheet houses, excepting No. 2, unless by special sanction of the Board.[2] This, however, proved to be impracticable. The heat of a glass furnace, once it is lighted, must be kept up continuously; in any case the "teazers" must be there to stoke it, and the founding employs but few more men. With the pot-furnaces the cost of an idle day could not be faced, nor would it so have been possible to get in the regular number of journeys in the week. While all other work on Sundays was as far as possible avoided, it had to be the practice to use that day for founding, in order that the glassmakers might get to work

---

1   Board minute of May 23, 1848.
2   Board minute of February 5, 1852.

again early on the Monday.

The Chances have always been active in promotion of religious obser-
vance in the neighbourhood of their works. Lucas Chance from a very early
time, and his son Robert after him, conducted regular Sunday meetings
of their Separatist brotherhood, firstly at the Glass Works, and afterwards
at the Chapel which is now St. Andrew's Mission Room. William and
James Chance and others of the family furthered zealously the work of
the Church of England. As a principal result of this, three churches have
owed their existence wholly or in great measure to the liberality shown in
providing the necessary funds or land; those of St. Paul at West Smethwick,
of St. John the Evangelist at Tat Bank, and of the Good Shepherd in West
Bromwich.

Up to the year 1858 there was no church in the immediate vicinity of
either Works. In 1857, in view of the needs of the fast-growing population of West Smethwick, the firm decided to build that of St. Paul. James
Chance laid the foundation stone on June 15, and the building was completed in the following year. In January 1860 St. Paul's was constituted a
new parish. The first vicar was the Rev. James Sheppard,[1] the next, from
1864, the Rev. Henry Stowe, the third, after Mr. Stowe's retirement in
1890 with a pension, the Rev. Thomas William Wilkes. He, on his appointment in 1901 to be Vicar of Wednesbury, was succeeded by the Rev.
George Arthur Anning.

Mr. Wilkes and Mr. Anning both worked with great energy and
success. A first thing carried out by the former was to raise a fund for repairs to the church, long neglected and very necessary. In February 1891
Chance Brothers & Co. resolved to find £450 for this purpose in three
years, provided,, that the sum of £1,150 were raised within that term, and
further to contribute £20 a year towards a curate's stipend. The external
repairs required were completed in 1894 at a cost of £900. Internal improvements included a fine organ, which, Mr. Talbot wrote, "adds much
to the efficiency of our services."[2] Further internal restoration was carried
out in 1899 and 1904.

First endowment of the living, to the extent of £29 a year, resulted
from a grant of £450 by the Ecclesiastical Commissioners in 1860 to meet

---

1   Subsequently Vicar of St. Thomas', Preston, and of Rookhope, near Darlington, where he died
towards the end of 1891.
2   To James Chance, December 27, 1891.

a benefaction of the like amount. In 1885 the Commissioners made a further grant of £160 a year on the ground of the increase of population in the neighbourhood. In 1918, again, in consideration of the abolition of pew rents, they granted £30 a year in augmentation of the Vicar's stipend. Lastly, in 1919, a sum of £250 having been raised on the promise of £100 by George Chance, if the parish would subscribe the balance, and this sum being doubled from the Diocesan Funds, and the £500 again by the Ecclesiastical Commissioners, income Was further increased by £40 a year.

In 1897 a Parish Room was built on a site adjoining the Church, affording accommodation (in the words of an appeal of January 1898 for subscriptions to make up the sum of £1,511 16s. 8d. expended) "for Sunday School purposes, Bible Class, Mission Services, Ambulance Classes, Young Men's Club, Gymnasium, Mothers', Sewing, Girls' Friendly Society and other Meetings." The firm and members of the family contributed £510 towards this object. Next year (1898) George Chance enabled the St. Andrew's Mission Room, above mentioned, to be opened, by purchasing the building for £1,200 from the trustees of Robert Chance, deceased, and letting it to Mr. Wilkes.[1] After the coming of Mr. Anning he transferred the ownership to him and others for the like sum of £1,200, of which £700 remained on mortgage. Of the balance nine-tenths were subscribed by members and relatives of the Chance family. Finally, in 1908, George Chance released the mortgage of £700 as a free gift, and the property was vested in the Birmingham Diocesan Trustees.

In 1908 also was built a new Parochial Hall, at a cost of about £1,100. It was opened by George Chance on November 14. Of this money, £250 was provided from the Bishop of Birmingham's Fund, £300 by the trustees of Alfred Roberts, and £100 by Chance Brothers & Co.

There remains for mention the new Vicarage, in Park Road, completed in 1905 at a cost of £2,860. To this sum George Chance contributed £760 and his two brothers £100 each, the balance being obtained by grants and by sale of the old Vicarage.

Patronage of the living was vested in 1858 in the Rev. T. G. Simcox, James Chance and others as Trustees. In 1885 were appointed as new

---

1 "I bought the Chapel to give Mr. Wilkes time to raise the money. ... I have no intention of giving the Chapel, because people take so much more interest in a thing which they have made some sacrifice to get" (George Chance to James Chance, February 9, 1899).

Trustees James, John, Henry and George Chance, and the Ven. J. H. lies, Archdeacon of Stafford. The last, on his death in 1890, was replaced by Frederick Chance, and the first three in 1902 by Arthur Chance, K. A. Macaulay and Edward Chance. In 1917, Macaulay retiring, the Ven. C. E. Hopton, Archdeacon of Birmingham, was appointed.

The present Vicar, who succeeded Mr. Anning in 1918, is the Rev. Albert Fordham Giddings.

At Oldbury Alexander Chance, throughout the time of his direction there, was earnest in promotion of Church work in the neighbourhood. On his initiative great help was afforded to it by the firm from time to time, both in Oldbury and at Langley. But he felt that opportunity for divine worship should be provided nearer to the Works, and particularly for the large number of the workpeople living at Tat Bank. Services were held for many years in the Infant School there, and at length in 1897 foundation stones for the Church of St. John were laid by Mrs. Perowne, wife of the Bishop of Worcester, and by Mrs. Alexander Chance. The sum of £1,400, which the building cost, was raised by public subscription, but the site was given by the firm. The Church appertained to the Vicarage of Oldbury.

In 1915 the land at Tat Bank was taken over for the Government Munition Works, and the Church, sold for the sum of £1,200, was converted for use as a laboratory. In place of it was built in 1916 in Darby Road, Richmond Hill, from the proceeds of the sale augmented by public subscription, an edifice intended to serve as a Mission Church until a permanent Church and Vicarage could be erected, and afterwards as a Parish Room in connexion. The land, affording space also for the future Church and Vicarage, was the free gift of George Chance.

Achievement of the Church of the Good Shepherd was largely due to the indefatigable efforts of Mr. W. A. Roberts, of West Bromwich, in raising funds. Of the £3,500 collected, Chance Brothers & Co. contributed £300 and George Chance £900. The foundation stone was laid by the Earl of Dartmouth on October 28, 1908. Previously, from 1905, had been used what is now the Parish Room. Curate of the new district from 1908, and then Vicar of the parish formed in 1910, was the Rev. William Edward Wibby, now Vicar of Ogley Hay, with Brownhills, Walsall. He was succeeded in 1917 by the Rev. Andrew Kirk.

James Chance in 1893, after the death of his daughter Katharine and partly thereby actuated, took it in mind to institute some large scheme for the benefit of the people among whom he had spent the best years of his life. There had recently been a movement, in which Councillor Samuel Smith was a leading spirit, to provide a recreation ground for the children of West Smethwick, and local manufacturers had been approached on the subject. When, therefore, George Chance was consulted by his father in the matter, he advised the gift of a Public Park, and on this James Chance determined. Land to the southward of St. Paul's Road, high lying and with a fine outlook over the adjacent country, was purchased, and of this 43 acres were given up to the purpose in view. Plans for laying out the ground as a Park were furnished by Mr. William Henman, of Birmingham. shrubs and trees were planted liberally, ample spaces were set apart for games, the source of the River Rea in the Park itself provided water for an ornamental boating pool, with a swimming bath, a handsome lodge and entrance gates were erected, and in the centre a band-stand and a pavilion for refreshments. Shelters were provided by Robert and John Chance; other presents from various persons included seats, vases, a bell, a greenhouse, flower seeds, fish and a pair of swans for the pool. Before long purchase of more land on the Oldbury side of the Park increased its extent to well over 50 acres. New roads, "Park Road" and "Victoria Road," were made to bound and protect it on the east and west. Moreover, not only did James Chance buy the land and lay it out at his own cost, but also he provided funds for its maintenance, in order that the people of Smethwick should not be burdened therewith; his whole outlay amounting to some £50,000. The property was vested in the first place in his sons and others as Trustees, with power to convey it at a future time to the town of Smethwick, a power exercised by them in 1912. Genner undertook the work of Secretary, and F. Driver was appointed the first park-keeper.

Probably no gift could have given greater satisfaction. Talbot voiced the public sentiment when he wrote in March 1894, on James Chance's intention becoming known:

> Probably nothing is more conducive to the improvement in the tone and manners of people, and to the enlargement of their sympathies with nature and with each other, than frequent visits to well ordered Parks or Gardens, and anyone who confers a permanent benefit of this kind on a population like our own is doing, indeed, as real a service to the course of human progress as one who builds a Church or founds a School. I need not say how we shall all hope that

the Park may carry your own honoured name to remote times, and promote a
constantly recurring recollection, in the memories of thousands, of your long
tried devotion to the highest educational interests of this neighbourhood, and,
it may be hoped, be also an incentive to others to follow so good an example.

On the day appointed for the opening of the Park, September 7, 1895,
holiday was made. Everywhere the streets were gaily decorated, and in the
new Park Road there was a triumphal arch. A procession starting from
the Public Hall, headed by the West Smethwick Park Band, included
members of the Executive Committee, of the Smethwick and Oldbury
Urban District Councils, and of the Smethwick School Board; a company
of the South Staffordshire Volunteers (a force which James Chance had
taken a leading part in raising thirty-five years before)[1] with their band;
nine local Fire Brigades, under command of Superintendent E. J. Forster
; and deputations from a number of Friendly Societies and Trades Unions.
James Chance having declared the Park open to the use of the public for
ever, illuminated addresses were presented to him on behalf of the inhabitants of West Smethwick, the Smethwick and Oldbury District Councils,
the employees of the Oldbury Alkali Company, and the " Old Boys " of
Chance's Schools ; the last represented by Alderman Jabez Jones, James
Robinson, and H. S. Griffiths. Music and various entertainments promoted
enjoyment of the afternoon, and the festivities ended at night with a grand
display of fireworks, concluding with a design portraying the donor.

Together with their Address, the "Old Boys" took opportunity to
present to James Chance a Gold Medal of the value of 100 guineas, bearing his portrait, in commemoration of the jubilee of the opening of the
Schools. The presentation was made by four who had entered them at
their commencement, John Egginton, James Grant, D. Bailey, and William Grigg. The Medal, designed by Hamo Thornycroft, R.A., bore the
inscription :

A.D. 1895. The "Old Boys" of Chance's Smethwick Schools to James T. Chance Esquire in grateful
commemoration of the Schools' Jubilee and of the opening of the Park generously given by him
to the district.

A number of copies were struck in silver and bronze for distribution
to such as desired to possess them. In the evening a reunion of the old

1   "You are the individual in this part of the world who has taken the earliest and most effective
steps in this movement" (Lord Hatherton to James Chance, August 18, 1859).

Boys took place in the Schools, of which Talbot wrote:

> Our O.B.'s put off the Dinner which had been arranged, when they found that you
> would not be able to attend. As they were able to come in to the general scheme of
> presentations on the 7th the change was very happily arranged, and they came in for
> more than their proper share of respect and attention . . . A meeting for refreshment
> and conversation, which was held in the old School in the evening, gave them the
> desired opportunity for seeing each other . . . They were delighted with the meeting,
> and with the pleasant revival of old associations.[1]

Two years later, on October 2, 1897, another occasion was afforded
for festivity, the unveiling, namely, of a fine bronze bust of James Chance
by Hamo Thornycroft, placed in the grounds by members of his family
and backed by a pedimental screen, with an inscription, designed by Mr.
Henman. Again the streets were decorated, and again there was a procession. James Chance was unable to be present himself on this occasion,
but his three sons and many of his relatives were there to meet a large
assemblage. The unveiling ceremony was performed by Mrs. George
Chance, and the leading part on behalf of the inhabitants of Smethwick
was taken by Councillor Smith. A suitable address was presented to the
Trustees. After the proceedings Councillor Smith entertained to a meat
tea, amongst others, 250 poor people over sixty years of age.

After James Chance's death the people of Smethwick determined
upon a memorial of him for the town in the form of a portrait, the commission for which was given to Mr. Joseph Gibbs, before mentioned. The
unveiling ceremony took place in the Public Hall on August 20, 1902,
Alderman Smith, now first Mayor of the new Borough, presiding. The
unveiling was performed by Sir William Chance, James Chance's eldest
son, and other speeches were made by the Rev. George Astbury, Major
Thompson, Aldermen Jones and Cheshire, Messrs. George and Frederick
Chance, Mr. J. E. Mitchell, and the Rev. Henry McKean.

In 1905 a handsome stone fountain, designed and executed as a Memorial of John Chance by Messrs. Henman and Jones, was presented to
the Park by the Company's workpeople, represented for the occasion by
James Field. It is said that every employee in the works contributed to this
spontaneous mark of respect and affection. The presentation ceremony
took place on June 17. K. A. Macaulay represented the Trustees in accept-

---

1   To James Chance, September 11, 1895.

ing the fountain, and the water was turned on by John Chance's daughter, Madame Dracopoli.

Lastly, on July 17, 1912, Sir William Chance, on behalf of the Trustees, formally transferred ownership of the Park to the Borough of Smethwick, again in the presence of a large assemblage of the townsfolk of Smethwick and Oldbury.

CHAPTER XIII

WAR OBLIGATIONS—THE CANTEEN—WELFARE WORK

OUTBREAK of the War cast upon manufacturers heavy anxieties and responsibilities. Anxieties, in view of the financial crash that seemed to impend and of prospective scarcity of labour, fuel and materials; responsibilities, from the necessity of concentrating on satisfaction of the nation's needs. Vastly increased quantities of various commodities were suddenly in demand, at the same time that the customary supply of them from abroad was stopped. Not only had the output of habitual products to be pushed to its practicable limit, but manufactures hitherto strange to be undertaken and, on the other hand, many that were not essential to the War suspended. New plant had to be hurriedly installed and old to be adapted and extended. Chance Brothers & Co. came in for their full share of the general burden.

Their principal duty was to turn out in quantities enormously increased the articles which they alone in Great Britain could supply; optical glass, mirrors and divergers for search-lights and other products of the lighthouse works, and in coloured glass ruby and signal-green and tinted for the soldiers' protective spectacles. Account of these developments has been given in previous chapters. Ordinary glass also was wanted in millions of small pieces for such various purposes as gas-helmets, trench lamps and periscopes, signalling apparatus, bomb-sights, compasses, barometers. Of the circles for the helmets the Company for some months was turning out over 50,000 per day, the eventual supply exceeding eleven millions. In addition were taken in hand for war purposes the alien manufactures of first quality machine tools and 18-pounder shells, the latter entailing provision of entirely new plant and employment of a large number of workpeople day and night for a long period.

All this was in addition to enforced maximum production of the Company's sheet and rolled plate glass. Of the former, before the War, 70 per cent, of British requirements had come from Belgium; now there was call for great additional quantities of it, and yet more of rolled plate, for the Government buildings—military camps, munition works, aerodromes, work and repair shops—that were being established all over the kingdom, while later the extensive damage done by air raids had to be made good. Partly to meet the shortage of sheet glass the Company succeeded, after prolonged experiments, in producing a substitute known as "rolled sheet," which has proved invaluable for factory and horticultural purposes.

A first trouble was deprivation of the services of a great number of the Company's younger men. The first list of those joining up from Spon Lane and Glasgow included nearly 300 names. Of the Managing Directors Walter Chance was detained at his post by special resolution of the Board, but Foster and Wharton went on active service, and so did Williams and Threlfall and others of the staff, to whom the Company looked for its guidance in the future. The older Directors had to give up or to postpone their well-earned retirement and to resume or to continue work, an obligation continued during the whole term of the war and not terminated with the armistice.[1] Foster, invalided, and Wharton and Threlfall, both wounded, have returned to Spon Lane, but Trevor Williams is no more. After serving with distinction as an artillery officer in France, Gallipoli, Egypt and Mesopotamia, he was invalided to India, where he died in hospital at Rawalpindi, as the result of a carriage accident, on April 19, 1918.

On the first news of Great Britain's entry into the war perturbation that was almost panic seized upon manufacturers. It was debated at Spon Lane, as elsewhere, whether, in face of the financial chaos that threatened, it might not be necessary to execute orders only for cash in advance. However, to the great satisfaction of the Company's customers and to the benefit of its reputation, it was decided to carry on business with as little interruption as possible. Accordingly, on August 5, 1914, the following notice was posted:

> After full consideration we have decided to do our best to keep all departments at work (although short time may in some cases be necessary) so long as we can get fuel and materials and dispose of our goods, and provided all employees support us by regular attendance and by doing their best to secure economy in all directions. . . .

---

1   George Chance had ceased to be a Managing Director in 1905, and Macaulay in 1913, the former continuing to be Chairman and the latter a Director. Edward Chance was obliged, in spite of ill-health, to continue at his post.

All those who have gone, or may go, to serve in the Navy, Army, and Auxiliary Forces will have their places kept open for their return.

Payments will be made to dependents of those who so serve as follows: 5*s*. per week for wife or dependent parent, 1*s*. per week for each child under fourteen years of age. We wish to call special attention to the importance of the strictest economy at home both in food and other ways.

In cases where dependents were not in immediate need it was decided to accumulate the allowances for the benefit of the sailor or soldier on his return.

Further resolutions, now or later, were to continue to pay their salaries to members of the staff under agreement, less the military pay that they would receive; to continue for six months the allowances to the dependents of employees killed in action or dying of wounds; to allow 5*s*. a week to clerks joining, or who had joined the forces, with such further sums on their return as would approximately represent what they would have saved if not absent on war service; and others to like effect. Payments on these accounts amounted at the end of 1918 to £4,459, the Company keeping in hand, in addition, the sum of £930 for distribution to the men concerned on their return from service or for allowances to their families in cases of special need.

When the War Loan was issued the employees of the Company were urged to contribute as much as possible. To enable subscriptions to be made in instalments less than the minimum of 5*s*. allowed through the Post Office the Board undertook as follows:

We will advance the total money you undertake to subscribe and deduct a fixed sum from your wages each week until the amount is made up.

We shall credit you with the full allowance of interest at 4 per cent, as received from the Government half-yearly and this will help towards paying for your investment.

When the deductions from wages and allowances of interest reach the total of your subscription we will hand to you a Government Certificate for that amount of War Loan Stock.

If from any unforeseen cause your subscription is not made up by the end of November 1917, we shall pay to you the money then standing to your credit, with interest to date, and the account will be closed.

If you wish to sell the Stock, we shall, at any time while you are in our employment, be prepared to buy it from you at the cost price.

If you leave our employment before the full amount is subscribed you can either pay the balance and receive your Certificate, or withdraw the money standing to your credit.[1]

In 1916, on request from the Ministry of Munitions, a War Savings Association was established, whose object was to enable the 15s. 6d. certificates to be purchased by weekly instalments of any multiple of 3d. Results have been most satisfactory. Figures to March 1919 show 10,762 certificates bought (£8,340), and £790 invested in National War Bonds, withdrawals being under £50. Members of the Association, solely the Company's employees, number about 340. The Company rendered powerful help by the purchase of certificates to the value of nearly £2,000 en bloc, thus encouraging increase both of the membership and of the amount of the weekly receipts, since members were enabled to reserve for themselves a quantity of these securities to be paid for by instalments. At the end of March 1919 only about £370 of these instalments remained outstanding.

Contributions by the Company to charitable organisations coping with war-time necessities amounted by the end of 1918 to nearly £1,600, while a voluntary weekly collection made by the workpeople raised by the same date £1,882. Of this money £500 was given to the Prince of Wales' Fund and the remainder applied to the assistance of local funds and of the Company's employees returned disabled. To the War Loan the Company and its Directors had by March 1917 contributed £105,000, of which £49,000 was "new money."

As from January 1916 the whole of the works were placed under Government control.

One outcome of the War was the establishment of a large Canteen for the use of the workpeople, combining dining and recreation rooms.

To have dining-rooms for the men within the works was an idea of old standing. The subject, as has been noticed, was under consideration as far back as the year 1856. John and Henry Chance were authorised in January 1873 to prepare plans, and four years later it was resolved to build a room on the north bank of the lower canal, abutting on the Spon Lane bridge. Agreement with the Canal Company as to building on to the bridge was reached, the plans were settled, and the Company's counterpart of the agreement was actually sealed, when there came into view in substitution the larger plan of the Institute, which has been mentioned, and the firm's counterpart was left unsigned. In 1887, and again in 1898,

1   Notice issued in 1915.

proposals were put forward for providing dining accommodation in the School buildings, but they also were not carried into effect. At length in 1902 was opened a messroom near the lighthouse works, capable, after later extension, of accommodating 120 men, and on the old side of the works, in 1905, a canteen near the Offices. Other provision was made for the women and girls in various places, but this became wholly inadequate, when war requirements so greatly increased their number.

The new scheme was propounded when at the end of 1915 the School Buildings were no longer occupied. The first idea was to utilise them for social and recreative purposes, perhaps with dining-rooms in connexion. Soon, however, after the first deliberations of a Committee appointed to consider details, the dining-rooms took first place in the scheme. Notes of June 7 on the subject record the Board's opinion that access to the dining-rooms should be from outside the works, that canteens for use at other than meal-times must be inside the works and preferably, for one on the new side, at the existing messroom near the lighthouse department, and generally that plans should be designed with a view to development after a tentative start upon a modest scale. It was set down also that the whole of the work of cooking, serving, &c, might well be done by women under a lady superintendent, who might also undertake "welfare work" among the women and girls, and that the proposed reading and recreation rooms should be open only to subscribing members.

At a meeting on June 26, after consideration of remarks by the Chairman of the Board (George Chance), the Committee arrived at the following conclusions:

> While we agree that a lady social worker is very desirable, we are still of opinion that it will be found necessary to engage a man with some experience to take over the management of the dining and recreation rooms... A lady social worker would find plenty to do in other directions, without attempting to undertake these duties. She would, of course, look after the interests of the girls in the dining and recreation rooms. At all the works we have visited they have found it necessary to have a competent male manager.
>
> As regards closing the Works canteens during meal-times, it could be arranged for the new canteen on the lighthouse side of the Works to be closed for (say) three hours at mid-day, from 11 until 2 o'clock, but we cannot recommend that the present canteen on the old side be closed at this time, as it is just then that it is of most service to the glassmakers. We would point out that to close it at this time when, under the

present licensing arrangements, the public houses are open, would be to defeat one of the main objects for which the canteen was introduced.

On the question of separate dining-rooms for different classes of users the Committee recommended six—namely, for the Directors, the Management Staff and male clerks, foremen, forewomen and lady clerks, workmen, and workwomen. They had considered at one of their earlier meetings, they said, the question of subdivision of these classes, and had come to the conclusion "that it would be very undesirable from every point of view to attempt to classify otherwise than in a very general way."

Having received this report, the Board on July 4 adopted recommendations for continuance of the Committee as constituted, with one of the Directors for its Chairman, and for the establishment and conduct of the institution. In view of the strong opinion expressed they deemed it inadvisable to press, in accordance with their own opinions and especially that of the Chairman, for the appointment of a lady superintendent. They advised again in favour of a start being made on a moderate scale with a view to development, and set down that no meals must be supplied elsewhere than in the authorised dining-room, that the approval of the Canteen Committee of the Central Board should be sought, and that there should be "two canteens inside the Works, one, as now, near the old gatehouse, the other where the messroom is near the lighthouse." Hours of opening to be settled by the Managing Directors in consultation with the Committee.

Establishment of the Canteen having been decided upon, plans for the necessary structural alterations to the School buildings were prepared by Messrs. Wood and Kendrick, of West Bromwich. They were approved by the Central Board Canteen Committee, and outlays were allowed to be charged against excess profits, in respect of such parts of the buildings as were to be used directly for the Canteen, to the amount of £5,000; a sum calculated—insufficiently, as it proved—to cover the expenses of their repurchase from the Local Authority and of their alteration, heating, lighting, and equipment. In November 1916 the contract for the alterations was placed with Messrs. William Lees and Sons, of Smethwick, for the sum of £2,297 10s. The work was put in hand without delay, and on June 11, 1917, the Canteen was opened. Control was vested in a committee partly nominated by the Board and partly elected by the workpeople. Previously to the opening notice was given that as soon as the new premises are in working order, all other Mess Rooms on the Works will be closed, as it is intended that partaking of breakfast or dinner in any other part of the Works shall be strictly forbidden, unless by special permission from the Manager or Foreman of the Department, in consequence of some special circumstances.

Exception, however, for the reason stated above, was made in favour of the men working at the sheet tank or at others near. Also was proposed a portable canteen for the distribution of tea and other temperance drinks, to be taken round the various departments at stated times and "intended to obviate the waste of time caused by sending over to the canteen at all times during the day."

The accommodation at the Canteen includes two large dining halls, a large kitchen, separate dining rooms for the Directors, Staff, foremen, forewomen and lady clerks, a large upstairs hall for social and recreation purposes, and rooms for rest and recreation for the female staff.

Mr. Samuel Nicholls, always interested in the Schools from the fact that, as said in Chapter XI., he had begun his education in them seventy years before, presented nearly 200 volumes as the nucleus of a library. "To the library attached to the Works" he wrote, "I owe an everlasting debt of gratitude."

On February 25, 1918, the following resolutions for governance of the undertaking were adopted.

CANTEEN AND RECREATION SCHEME

There shall be separate Committees for the Canteen and Recreation Scheme.

All furniture, fixtures, fittings, utensils, &c, of whatever kind in the Canteen and Recreation Rooms, are and shall be the property of the Company.

In case any difference arise between the Canteen Committee and the Recreation Committee, or between either of the Committees and those making use of the buildings, the question in dispute may be referred to the Board, whose decision shall be final.

Canteen.

The Canteen shall be under the control of a Committee consisting of not more than four members appointed by those who habitually use the Canteen and an equal number appointed by the Board, who will, in addition, appoint the Chairman and Secretary annually.

The Committee will appoint a Deputy-Chairman.

The Welfare Supervisor will be an ex-officio member of the Committee.

Half-yearly elections shall take place at a General Meeting to be held in April and October of each year, on a date to be fixed by the Committee. At the election in October 1918 two of the elected and two of the nominated members of the Committee shall retire, the selection being decided by lot; and there shall retire at every subsequent election four of those who have held office longest or not less than half the Committee, but retiring members shall be eligible for re-election.

The Committee shall be responsible for the running and control of the Canteen, for the rooms and property therein, and for all arrangements for providing meals and refreshments in the Works.

The Committee shall present to the Board a monthly Statement of receipts and expenditure, and shall not incur any expenditure on fixtures, repairs, &c., in connection with the Canteen without the sanction of the Board.

The Committee shall be responsible for all catering arrangements in connection with the Recreation Scheme.

Recreation Scheme.

The Committee shall consist of 10 members appointed by the Works generally, and an equal number appointed by the Board, who will, in addition, annually appoint the Chairman.

The Committee will appoint a Deputy-Chairman and Secretary. The Welfare Supervisor and Canteen Superintendent will be ex-officio members of the Committee.

Half-yearly elections shall take place at a General Meeting to be held in April and October of each year, on a date to be fixed by the Committee.

At the election in October 1918 five elected and five nominated members of the Committee shall retire, the selection being decided by lot; and there shall retire at every subsequent election ten of those who have held office longest, or not less than half the Committee, but retiring members shall be eligible for re-election.

The Committee may appoint Sub-Committees to look after whist drives, concerts, &c., and may co-opt others, not on the Committee, to assist them in arranging these entertainments; but no entertainment shall be held, except under the direct supervision of the Committee.

The Committee shall be responsible for all financial matters relating to entertainments, including the purchase of prizes, and also for the distribution of any profits which may accrue from such entertainments.

The Committee shall appoint an Auditor (preferably the Secretary of the Company) to whom a balance sheet shall be sent of any entertainment for which a charge for entrance is made; the balance sheet must be handed to the Auditor within a fortnight of each event and, when audited, must be posted up in the Recreation Room for one week, and then filed by the Secretary.

Any person or persons wishing to get up an entertainment may apply to the Secretary of the Recreation Committee, stating the form of the proposed entertainment and the purpose to which the proceeds are to be devoted, and the Secretary shall place such application before the Committee within seven days. If the Committee approve of the entertainment in question they shall place it on the authorised programme.

The Committee shall have sole control of the Recreation Room and its furniture and fittings, the chairs provided for whist drives, and the periodicals, newspapers, &c.

When any of the rooms of the Canteen are required for temporary use, the sanction of the Canteen Committee must be obtained; and ample notice must be given to that Committee, if refreshments are required.

All catering for entertainments must be done by the Canteen Committee at tariff prices, such prices will include the cost of serving, but not of waitresses. The Canteen Superintendent shall be ex-officio a member of the Sub-Committee appointed to arrange an entertainment, where catering is required.

The Committee have not the right of usage of the kitchen or equipment, or of any of the rooms in the Canteen, without the consent of the Canteen Committee.

There shall be a charge of 25s. for the use of the Recreation Room if required for any entertainment, and of 7s. 6d. each extra if any of the three main rooms of the Canteen are also required.[1] This charge includes the cost of preparing the room, or rooms, but not the use of the piano, cards, &c.

The Committee have no right to incur any expenditure on furniture, fixtures, fittings, or repairs, without the sanction of the Board, who have already authorised an expenditure up to £90 for the purchase of seats and tables.

A special account will be opened against which will be debited the salary of a Caretaker (if any), the estimated cost of lighting, heating, and cleaning, &c, and the cost of newspapers, periodicals, &c. The account will be credited with any income arising from hire of rooms, or from other sources.

The Board have undertaken to make up any deficiency incurred during the year 1918, at the end of which period, when the cost of the Recreation Scheme has been ascertained, they will reconsider the question.

The following rules shall apply to both Committees:

(*a*) The dates of ordinary meetings, the time and place of meeting, and the quorum, shall be settled from time to time by the Committee concerned.

(*b*) A special meeting may be summoned at any time by the Chairman, or by the Secretary on request of one-third of the members.

(*c*) Any member leaving the employment of the Company shall vacate his office.

(*d*) The Committees may fill vacancies subject to confirmation at the next General Meeting.

(*e*) The Secretaries shall keep Minutes of the Committee Meetings.

In spite of the contrary opinion at first expressed by the Committee, management of the Canteen has been conducted by ladies, successively Miss Garraway, Miss Cheston, from February 1918 to January 1919, and lastly Mrs. Pinchard, working under control of the "Lady Welfare Supervisor," Mrs. Sleigh.

This social welfare work amongst the women and girls, of whom some 400 were now, under war conditions, employed, was begun in September 1916 by Miss Isabel Campbell Shaw and carried on by her until, unfortunately, her health broke down. Mrs. Sleigh took her place in June 1917.

---

1   Noted, that these charges would probably have to be modified.

It is pleasant to read in her report for the year 1918 of the satisfaction expressed by Miss York, of the Welfare Department of the Ministry of Munitions, on the occasion of a visit of inspection in October. Passages in the report run as follows :

On November 17 Miss York came again, to ascertain if the cessation of hostilities would mean the discharging of women and girls who had been taken on during the war. On leaving, she said how pleasant it was to find a factory so little upset by the change of conditions.

I should like to mention here that, without any exception, the women and girls who have been discharged all say how sorry they are to leave, and ask to be taken on again when opportunity arises. This is very different from the reports one hears from other factories, where girls are deliberately troublesome in order that they may be discharged and so get unemployment benefit. . . .

The first-aid classes are being much better attended than they were last year. The doctor is giving a lecture every other Monday, and I coach the girls in practical work the alternate week.

The room that was formerly used for an ambulance room, by the lighthouse fitting shop, had to be taken for an office, so there has been no ambulance room in the factory. Permission has now been given for a small room in the Old Hall to be used for this purpose, and the necessary alterations are being made in it. There will therefore be a proper ambulance and first-aid room in as central a place as possible. . .

I started a weekly dancing class on November 5, which has proved very popular, being attended by any number up to 40 or so of the girls, and a few young men, chiefly clerks and from the Extra White and Optical Departments.

Two whist drives, and two dances (one of the latter being "fancy dress") have been held on Saturdays under the Recreation Scheme. And on December 30 and 31 I arranged three performances of carols, illustrated by tableaux, to which were invited the staff, the pensioners, and all the women working here.

In her latest Report (July 1919) Mrs. Sleigh notices, amongst other things, the success, both social and financial, of the dancing evenings, and the girls' appreciation of their accommodation in the Canteen.

A development of the welfare work has been its extension to boys, under charge, since April 1919, of Mr. E. D. Burrin, who, after nearly five years spent in the country's naval service, has entered upon his new work with zest, and has schemes in view which, if they mature, should materially advance the well-being of the boys, and indeed of the workpeople in general.

CHAPTER XIV

TRADE

By WALTER LUCAS CHANCE

ALTHOUGH a history of the Glass Trade in this country during the last hundred years does not fall within the scope of this volume, the long-continued and influential part taken by Chance Brothers & Co. in moulding the fortunes of the trade is worth more than a passing mention. In this chapter an attempt will be made to give a brief resume of the general methods adopted for regulating the glass trade in this country, and of the maintenance of the connexions and influence of the firm with those with whom they have been brought into contact in the course of business.

As far back as 1827 there was an "Association of the Crown Glass Manufacturers of England and Scotland," which met twice a year, regulated the prices and general conditions of the sale of crown glass, prohibited the employment of each other's workpeople, established penalties for breaches of rules, and arranged for the adjustment of production to demand. Lucas Chance was invariably present at these meetings, and, though the post of Chairman was apparently taken in rotation, there are many indications that he was regarded as the leader of the trade. Thus, in 1830, as has previously been said, he was appointed Chairman of a special Committee to approach the Government in support of proposals for the commutation of the window duty, in which capacity he wrote an instructive letter to the Chancellor of the Exchequer in that year and signed a Memorial presented on behalf of the whole trade in 1831. So, too, in 1837, a "Fines Fund" of £100 per manufacturer was deposited in the names of Lucas Chance, John Clare, and Richard Pilkington. We also find deputations of representative glass merchants asking Lucas Chance to place their views before his brother manufacturers.

The first regulations for putting into effect "the unanimous opinion of the manufacturers that a restriction of make is expedient," in October 1838, provide interesting reading. The total production of crown glass for the ensuing year was not to exceed one and a half million tables, to be allotted to the manufacturers in proportion to the duty paid by them during the preceding four years, except that no firm's make was to be reduced below 1,200 tables per week. In August 1839 a further reduction of 10 per cent, of make was decreed, and in August 1841 the restriction was extended to include sheet glass also. The following resolutions then passed are worth quoting *in extenso:*—

The Inspector shall deduct 10% from the quantity of Crown glass made by each House from 5th Jan. 1840 to 5th Jan. 1841, and the remainder shall be considered the established quota which each House shall have allowed to it from 5th July 1841.

He shall also deduct 10% from the quantity of Sheet glass made by Messrs. Chance in the year ending 5th Jan. 1841, and that shall be considered their quota of Sheet glass.

Messrs. Cookson & Co., The St. Helens Co., and Messrs. James Hartley & Co. shall be allowed such a quantity of Sheet glass for each of their quotas respectively as shall not exceed half the quantity allowed to Messrs. Chance. Firms were not allowed to transfer portions of their quota to other makers, but in 1842 we find Chance Brothers & Co. permitted to make one-eighth more than their allotted share of sheet glass, and Messrs. Cookson exchanging with them 2,000 cwt. of theirs for 1,250 cwt. of crown glass. The negotiations with the Treasury, which eventually led to the abolition of the glass duties in 1845, naturally demanded constant attention on the part of the Association, Lucas Chance and Mr. Nichol, of the Newcastle Broad & Crown Glass Co., taking the lead in them. The total amount paid in duty on glass of all kinds retained for home consumption averaged £650,000 a year for the ten years prior to 1843, of which sum crown glass contributed more than a half, and Chance Brothers & Co. over £100,000. The firm appear to have made about one-fourth of the total output of crown glass at this time, when there were no fewer than nineteen manufacturers. In 1852 that number had dwindled to ten, and by 1858 to five. Spon Lane sales of crown glass reached their maximum in the four years between 1853 and 1856, during which were sent out an average of about 460,000 tables per annum. By 1866 the output was only half this, and 1876 was the last year in which were sold more than 100,000 tables.[1]

In 1846 the old Association was renamed "The Crown and Sheet Glass Manufacturers' Association," of which the central fund was provided by an assessment of 15s. and 7s. 6d. respectively for each crown and sheet blower employed. In 1849 James Chance supported Lucas Chance at the meetings, and later on Robert Chance regularly attended with his father. Meetings were usually held at the Adelphi Hotel, Liverpool, or at Buxton, Derby, or Newcastle, and occasionally at the London Coffee House. There was a strict rule, with special penalties, against a manufacturer letting any third party know the date or place of the next meeting, and among other rather curious regulations was one that no manufacturer should advertise "patent rough plate or any other kind of window glass," and another that no cullet should be bought from customers above a fixed price.

The principal business, however, of the Association throughout was regulation of prices. At first there appears to have been one discount for "First class London Dealers," and another for everybody else. Later on a most elaborate classification of all glass buyers was instituted, dividing them into six classes lettered A to F, a classification first printed for confidential use in 1858. It is remarkable how very few of the firms then included still exist in the glass trade. Prominent among the survivors are the Chafers, Goslett, and the Farmiloes of London, John Hall of Bristol, the Stocks of Birmingham, Giddings and Dacre of Manchester, and James and Rosewall of Plymouth. Other names, such as Hetley, Chuck, and Cashmore, are still well known in the trade, though no members of those families remain connected therewith.

From 1827 to 1844 the net prices of fourths crown tables to the best buyers fluctuated between 8s. 9d. and 10s. 3d. per table inclusive of duty, which averaged about 6s. 1d. per table. The prices of sheet glass were enormous; coarse 15-ounce crates sold at about is. a foot net, and best quality squares realised up to 4s. 6d. per foot. After the abolition of the duty in 1845 a gradual and continuous fall in prices ensued, so that by 1849 crown tables sold at about 2s. each, and 4/15 crates at under 2½d. per foot. In the early 'sixties the latter price was under 1½d. per foot, and does not seem to have fluctuated materially until the time of the Franco-German war. Meanwhile, rolled plate was becoming increasingly popular. In 1852,

---

1   Cf. pp. 43, 88.

at a meeting of the three allied manufacturers,[1] Chances, Hartleys, and Pilkingtons, prices were fixed for sizes not above 70 inches long. When two other manufacturers had come into the business prices fell steadily, until by 1870 they were in some cases lower than those obtaining in 1914. It is interesting to note that the war of 1870-1 caused an advance much greater than has taken place during the late war, one-eighth rolled plate in 1873 selling for as much as *6d.* a foot.

Despite the fallen prices alike of *c*rown, sheet and rolled plate glass in the 'sixties the firm made substantial profits in those years, due largely to the patent plate and other special departments. The outbreak of war in Europe in 1870 gave an immense impetus to the English glass trade, and for the ensuing four years very high prices were realised, and fine profits obtained. But in the years which followed the boom came a prolonged and disastrous slump. Thus, while in 1879 Chance Brothers & Co. sold two million feet more sheet glass than in 1873, their net sales were over £50,000 lower. It was the adoption of the tank system for making sheet glass by their St. Helens competitors about this time that finally deprived them of the predominant position which they had hitherto held. These years, however, saw the extinction of all other sheet glass manufacturers in this country, and since then there has only been one attempt, and that insignificant and abortive, by any third party to establish the manufacture in Great Britain.

Rolled plate glass has been of high importance among Chance Brothers & Co.'s manufactures for at least a generation past. The introduction of figured rolled in the early nineties proved an immense accession of strength, and though the demand for this glass in the home market has of late years shown a tendency to decline, owing to a change of taste, its sale has gradually extended to every part of the world and is still on the increase in such widely different regions as South America, China, and the East Indies. Its manufacture has been taken up by makers in all European countries and in the United States, and prior to the War was experienced the keenest competition from abroad, to which further reference is made below. Of the many different patterns made it is interesting to see how most of the larger markets have adopted their own special favourites. The firm's trade in this class of ornamental glass was greatly stimulated by the introduction of the "Flemish" type, noticed in Chapter VI. Another development was the manufacture of special prismatic glass, for the better

1   SEE P. 78.

lighting of dark interiors.

For ordinary rolled plate glass for roofing demand increased with great rapidity in the early years of the present century, being stimulated by the growing popularity and efficiency of the new systems of "patent glazing," with metal bars and without putty. The extent of this branch of the trade may be judged from the statement that probably more than one quarter of the rolled plate glass now used in the home market is taken by the patent glaziers.

Of coloured glass, throughout all changes due to fashion or other causes, ruby and signal green have been continuously made, and of recent years in larger quantity and of better quality than ever before, the revival of our supremacy in the manufacture of the first-named being entirely due to the efforts of K. A. Macaulay. The extreme importance of producing these two glasses in this country was realized when continental supplies were cut off by the War, and Spon Lane proved the only source of supply for most urgent necessities of the Army, Navy, and Air Service, as well as for ordinary railway use.

It may be noted that in the years just prior to the War there was quite a revival in the demand for genuine crown glass, and some thousands of tables were sold annually, until the more important war requirements caused suspension of the manufacture of what was purely a luxury. The modern cost of making this glass in small quantities is necessarily high, and the prices of present-day crown are curiously similar to those obtaining nearly a hundred years ago.

Worthy of particular notice in the home trade is the success of Chance Brothers & Co. in securing and maintaining the goodwill of the dealers. The uninterruptedly cordial and in many instances personally friendly relations maintained with the customers have provided one of the most marked features of the commercial history of the concern.

The first London office was at 2 Coleman Street Buildings, Lucas Chance's old place of business as a merchant. In early times no regular commercial travelling was done, but Lucas Chance used periodically to call upon the more important dealers in London, Bristol, Dublin, and other towns. He had many personal friends among them, as James Hetley, of London, Maurice Brooks, of Dublin, and John Hall, of Bristol, whose son, fifty years later, used to recall how Lucas Chance had presented him, as a boy, with a pony. To this day stories are told illustrating his habitual energy and keenness and his effective, if original, methods.

About the year 1848, on expiry of the lease at Coleman Street Buildings, the firm's London office was transferred to 24 Finsbury Circus, where it remained until closed, as said in Chapter V., in 1891. There were also of old time agencies at Dublin, Glasgow, and Liverpool. The first Dublin office was at 124 Abbey Street. For long Maurice Brooks acted as agent there, but, being also a glass merchant, he ultimately gave up the agency in order to develop the sale of foreign glass. After that Robert and John Chance made periodical visits to Ireland, where the former had also religious associations. Later on W. J. Green, who succeeded Samuel Gwilliam as head of the home office at Smethwick, visited Dublin, Belfast, and several other towns quarterly, and was succeeded in turn by George Crowther, Archibald Aitchison, Hugh Gabbey, and J. W. Daughtery. The volume of the firm's Irish trade has revived to a considerable extent in recent years. At Glasgow the first agents were McLaren and Anderson, succeeded by Niven and Pratt, and afterwards by Archibald Aitchison, who carried on the agency until the purchase of the Glasgow works in 1907, when he entered the direct service of the Company. The work is now principally in the hands of his son, John Aitchison, manager of the commercial side at the Glasgow works, under Lindsay Forster. William Stevens, resident agent in London, at one time made a quarterly round of Liverpool, Manchester, Leeds, Bradford and Sheffield, but subsequently J. R. Ross was appointed agent at Liverpool, and from there worked other Lancashire and Yorkshire towns. In 1873 he was succeeded by William Stonier, previously in charge of the office at Nailsea, and he in turn was followed by William Brown, formerly his assistant, who remained in charge until the Liverpool office was closed in 1882.

For some years past the firm has had no agencies in this country, the business being worked direct from the Smethwick and Glasgow offices. George Crowther, who retired in 1915 after over 60 years' loyal service, represented them with conspicuous ability for some 25 years in London, Birmingham, Yorkshire and Lancashire, making periodical visits also to several other districts. He had previously been head of the home office, and his thorough knowledge of the trade, keen judgment, and scrupulous integrity, coupled with extraordinary energy, rendered his services invaluable both to the firm and to its customers, many of whom had good reason to be grateful for his advice and assistance. *At* present the commercial representation of the firm is in the hands of A. E. Bassett, J. W. Daughtery, and John Aitchison.

In this connexion it is not, perhaps, out of place to refer to some of those who owe much of their success in the glass trade to first training at Spon Lane. Chief among these is probably Samuel Nicholls, of whom mention has been made. Leaving the service of the firm in 1862, and walking to London, by the exercise of ability and thrift and by taking proper advantage of a fortunate train of circumstances, he eventually became one of the largest and wealthiest glass merchants in the kingdom. As has been said, he attributes his success in life in principal measure to the training that he received in the Schools under Frederick Talbot, and he marked his appreciation in a practical way by presenting a library of books to the new reading room in 1917. In Birmingham the prosperous business of Ephraim Cutler was founded by Solomon Cutler, father of Ephraim, and formerly a foreman in the sheet warehouse. At Bristol the eminent firm of Samuel Cashmore and Co. takes its name from a former head of the home office at Spon Lane, who was transferred to Bristol, by arrangement between Lucas Chance and John Hall, to look after the latter's business, and who afterwards acquired the old and important concern of John Dix and Co. This business is now virtually in the hands of the Salmond family, who also own the still more important concern of James Hetley and Co., of London. Through it they have long been most intimately connected with Chance Brothers & Co., and they recently acquired a controlling interest in the firm of W. E. Chance and Co. of Oldbury. Enoch Holloway, once a sheet glass cutter at Spon Lane, left to establish a business at Manchester, which is still carried on, though in no large way, by his son Arthur. So, too, Samuel Evans, whose once prosperous business in stained glass has since been taken over by O. C. Hawkes of Birmingham, was formerly employed in the Ornamental department. Many other merchants, while not deriving their experience directly from Spon Lane, are largely indebted for their prosperity to special assistance accorded them by members of the firm.

Hitherto reference has been made only to the home trade, but the export side of the firm's business has always been of special importance. The large interest of Lucas Chance in foreign trade from the first has been noticed, as also have the European tours undertaken on behalf of the firm by John Reynell in 1833-4 and by Bontemps in 1850 and 1852. From the earliest times a considerable trade was done with the United States and later with Canada, and by the early 'forties large consignments of sheet and crown glass were being regularly sent as well to all Australian ports, to India, and to South America, while yet later a considerable trade developed with South Africa and New Zealand. The prices then realized were very similar to those obtaining in the same markets in pre-war times, but the foreigners were not yet in a position to undercut the prices of English manufacturers, although the European tariffs effectually prevented any large exportation of English glass to the Continent.

The firm's first agent at New York was naturally George Chance, resident there from 1816 as partner in the business of William and George Chance, American merchants. On his return to England about 1837 his place was taken by William Chance's second son, William, the firm now trading as W. Chance Son & Co. After this younger William Chance was incapacitated by sunstroke, towards 1860, the business, when set in order by James Chance, was taken over by B. and S. H. Thompson, which firm, though the Thompson interest has long since been eliminated, continue to be Chance Brothers & Co.'s agents for Eastern Canada. At New York James Heroy took over the agency, and after him his son, W. W. Heroy, who, under the style of Heroy and Marrenner, still represents them for certain types of glass. They also have agents in Australasia (the very old connexion of Henry Brooks and Co. still flourishing)., in Egypt, South Africa, India, China, the Malay States and other parts of the Far East, Western Canada, the Argentine, Chili, Bolivia, Brazil, Cuba, and Greece, while regular shipments of rolled glass are made through home shipping houses to many other countries.

It is doubtful if any other class of manufacturer in this country has felt the crushing weight of foreign competition to a greater extent than have those connected with the glass industry. In every branch of the trade it has been the same story, and though the interests of British capital and labour have not, perhaps, been so detrimentally affected in the window glass trade as in the case of flint (or table) glass, of many varieties of which the home manufacture has virtually ceased altogether, yet it is a matter of general knowledge that of the sheet glass used in this country prior to 1914 not less than 70 per cent, came from abroad, almost entirely from Belgium, and strenuous efforts were continually being made by Belgian and German manufacturers to capture both the home and export trade in rolled plate, figured rolled, and other varieties. Of the home trade in coloured and microscopic glass, too, the foreigners had probably more than half, and of the pre-war British demand for optical glass about 70 per cent, was supplied by Germany, and 20 per cent, by France.

The importation of window glass from abroad first began to be troublesome about 1853, a few years after the removal of the duty. Figures compiled by Robert Chance in 1857 set forth that the importation of foreign sheet glass for home consumption jumped from about 20,000 feet per week—a fairly constant average for the previous ten years—to 56,000 feet in 1853, and to 65,000 feet in 1854. From that time onwards, except during the early seventies, when the Belgian output was fully occupied elsewhere, the importation has steadily increased. Foreign competition in rolled plate and figured glass did not attain serious proportions till the beginning of the present century, when a successful attempt to check it by reducing prices was followed, in 1904, by the establishment of a Continental Convention between the British and foreign manufacturers. Although this came to an end in 1910, drastic price reductions by the English makers, coupled with certain private arrangements with the more important of the Continentals, considerably curtailed foreign competition in British markets up to the outbreak of the War. Thereafter, for the first time in history, sheet glass was imported into this country from the United States, while Japan began to develop the manufacture and exportation of both sheet and figured glass so energetically as to constitute a serious menace to British trade in Australian, Indian, and all Eastern markets. Japanese competition, also, is now being met with in microscopic glass, and an endeavour is being made to establish in that country the manufacture of optical glass.

POSTSCRIPT

October 1919

DATING from his sixty-fifth birthday, September 25, 1919, George Chance resigned his Directorship of the Company, and was succeeded in his office of Chairman by Edward Chance. It has not been thought proper in the course of this history to say much about those still living, but exception may be made in the cases of the two mentioned and of K. A. Macaulay, the three who, working together for upwards of thirty years, have steered the Company through the shoals and quicksands that beset its course, and have raised the factory from a state that certainly was backward to one of first-rate efficiency.

George Chance began practical acquaintance with the works already before his school days, frequently going round them with his father. After leaving Harrow he spent a year and a half learning the office work and receiving instruction in chemistry from Dr. George Gore, of Birmingham. He took up regular work at Spon Lane in 1878, on leaving Cambridge, and was soon pronounced to be "the right man in the right place." In 1880 he was made a partner, and then for many years had principal direction of the manufacturing. He succeeded his uncle Henry Chance as Chairman of the Limited Company in 1901, and to his careful guidance as such the sound financial position of the Company is in great measure due.

Macaulay began at the firm's alkali works at Oldbury in 1874, in order to be initiated into office work and laboratory practice before going to King's College, London. Coming to Spon Lane in 1877, he had two years in which to learn and master the manufacturing processes. At the beginning of 1880, however, in consequence of the sad death of Robert Chance's younger son Walter, his services were requisitioned for the offices. Two years later, on John Chance wishing to give up active work, the whole charge of the commercial department, to be conducted with conspicuous success, devolved on him. He was made a partner in 1883, and in due course became in 1889 a Managing Director of the Limited Company. That office he resigned in 1913, remaining a Director. On the outbreak of the War he returned to active work, in the manufacturing departments, to help to fill the place of younger men called away to military service, and to the great benefit of the Company's affairs, continuing until resignation of his Directorship in July 1919.

Edward Chance also began work at Oldbury, in 1884, and also soon transferred the sphere of his activities to Spon Lane, to become likewise a Managing Director of the Company on its formation in 1889. Endowed with extraordinary power of attention to detail, he has been indefatigable in pursuit of improvement. The great development of the rolled plate manufacture, and success with the double-roll machine and the tank-furnaces have been largely due to his initiative and perseverance. Besides which, he has shown special aptitude for and ability in friendly negotiation with other manufacturers, from his success in which mutual interests have greatly benefited.

There has been omission, in Chapter V, to mention one member of the family, John De Peyster, eldest son of John Homer Chance, who, after leaving Harrow in 1870, worked for a time at Spon Lane and at Nailsea. Sent on a voyage to Australia for his health, he died at Melbourne, aged nineteen, on January 22, 1874.

A History of the Firm of Chance Brothers & Co.

Postscript—January 1926.

DURING the years that have passed since 1919, when the "History of the Firm of Chance Brothers & Co." was printed, changes have taken place at Spon Lane.

In March 1921 Edward Chance, who succeeded George Chance as Chairman in 1919, was forced by the breakdown of his health, partly the consequence of over-strain during the War years, to resign his Directorship. Two years later, on May 20th, 1923, he died. In June 1921 George Chance and Macaulay returned to be Chairman and Deputy-Chairman respectively. The former office is still held by George Chance, but Macaulay resigned the latter in April 1924 and his Directorship at the end of 1925. Arthur Chance retired from the Board in July 1924.

Appointed Managing Directors, since 1919, have been: in April 1921, Alfred Lindsay Forster; in August 1922, Albert Edward Bassett; in October 1924, William Hugh Stobart Chance; and from January 1st, 1926, Philip Victor Willingham Gell. Forster's services to the Company have been noticed in the "History." Bassett came to the Works as a lad in 1890, and rose through the offices to be Manager of the Rolled Plate Warehouses in 1898 and Sales Manager in 1915. Hugh Chance, George Chance's second son, joined as Assistant in 1920. He served in the Royal Flying Corps in the Great War, when failure of his engine brought him down within the enemy's lines. He succeeded in burning his machine before capture, but was a prisoner of war in Germany for over two years. Gell also joined as Assistant in 1920 ; in the Great War he served in the Royal Artillery.

Among those mentioned in the "History," who have retired from the service of the Company, are W. P. Fielden and Arthur Stephens (1920), W. A. Jeboult (1921), F. E. Lamplough and R. E. Threlfall (1922), H. Cooper, head clerk in the Lighthouse Office (1924). Three have died: E. D. Burrin in 1921, J. W. Daughtery in 1923, A. L. Simcox in 1924. James Field, senior, principal Manager for so many years, ceased regular attendance as from January 1st, 1925, but his valuable assistance as consultant is retained. Notice should have been taken in the "History" of the great services that he rendered, especially in the successful establishment and extension of the tank furnaces. Stephens has been succeeded as head of the Home Office by Samuel Grice, Cooper by H. P. Sims, Jeboult by Major H. O. Wraith, and Simcox by J. Cameron from the Glasgow works. The Optical Department is now under the charge of W. M. Hampton. Among other newcomers are: Major J. R. Williams as head of the General Works Engineering, Major J. E. Warner in charge of the Internal Works Transport and of the Fire Brigade, and R. L. Butcher as assistant secretary and accountant. Kenneth Mulholland Stobart, H. J. Stobart's second son joined as Assistant in October 1925. In the Great War Williams and Warner served in the Royal Artillery, Wraith in the Royal Engineers.

Valuable work is being done by the Works Scientific Research Committee instituted towards the end of 1923.

Miss Manwaring, after thirty years' service as almoner to the pensioners in succession to her father, was obliged by the burden of increasing years to give up the work in February 1925. Her place was taken by Mrs. Wilkes, a daughter of the late William Rathbone of the Lighthouse Department. On September 6, 1921, the Chairman unveiled a bronze tablet to the memory of the fifty-five employees of the Company who fell in the Great War. They were: —

G. Adams	S. Garfield	J. T. Osborne
E. T. Aldridge	W. Gough	T. Partridge
T. Aspley	C. Griffiths	B. Piper
J. Bird "	R. J. Hill	T. Price
A. R. Brown	W. H. Hillyer	T. Pritchard
A. Caddick	W. Hunt	E. W. Roberts
W. E. Carpenter	G. Iliffe	A. Rowe
F. Cashmore	G. Insley	C. A. Simmonds
A. Clarke	F. H. James	D. Smith
W. Cockbill	B. Kelley	G. Smith
J. H. Corbett	H. Kershaw	G. H. Smith
R. Cox	H. Key	J. Smith
G. Deeley	L. W. Lane	T. Stansfield
E. R. Dunshee	J. Lancaster	J. T. Stokes
W. Eggington	A. E. Lloyd	W. H. Turton
R. Ellis	S. Matthews	I. Waterhouse
H. Fairbrother	W. Mobberley	G. T. Williams
S. Freeth	A. Morris	A. Woodhouse
J. Garbett		

Besides these, 282 others joined the Colours, of whom 242 were able to return to the Works.

At an entertainment given in July 1924 at Great Alne Hall, the residence of Mr. and Mrs. Arthur Chance, to 153 pensioners and medallists in celebration of the firm's centenary, thirty-two more medals for continuous service of fifty years and upwards were presented, raising the number up to that time awarded to 128. One more has been allotted since. Undoubtedly there were many others qualified to receive the medals, who died before the institution of them in 1916. A list of the pensioners of the year 1900, for instance, includes twenty-eight such names. The number of medallists living early in 1925 was sixty-three, of whom thirty-four were still at work.

The date May 18th, 1924, marked the completion of a century since the elder Robert Lucas Chance entered on the manufacture of crown glass at Spon Lane. In connection with the notable display of its productions at the British Empire Exhibition of that year the Company issued, under the title "100 Years of British Glass-making, 1824-1924," a handsome booklet giving a summary account of its activities during the period. Among its newest productions may be noticed "Calorex" glass, for which, in virtue of its power to cut off 80 per cent, of the heat-rays of the sun while transmitting 65 per cent, of the light, there is prospect of a large demand for roof-glazing, especially in hot countries; heat-resisting globes and chimneys and laboratory glass, authoritatively pronounced to be of the highest quality, this an important new British industry; an entirely new orange glass for railway distant signals; rolled ruby cathedral and figured; pressed dioptric lenses and other ship lights; and flashed sheet glass of various new tints. The manufacture of wired glass has been successfully undertaken, and that of flashed opal revived; all the outdoor globular lamps at the British Empire Exhibition were made with this opal glass. There have been many developments and improvements in optical glass, and a large crown disc of 27 inches diameter has been successfully produced for the Johannesburg Observatory.

New exhibits at the British Empire Exhibition of 1925 included a particularly fine object-glass of 15½ inches diameter for a telescope; specimens of Mr. Lamplough's "Vitaglass," specially manufactured by the Company for the inventor; and a second-order lighthouse apparatus destined for Bird Rock in the Bahamas. This is of a special type, single-flashing and three-sided, revolving once every 15 seconds to give every 5 seconds a flash of 0·17 second duration. The apparatus, illuminated by an incandescent oil-burner, will give an intensity of flash of over half-a-million British candles. "Vitaglass" is a window-glass which has the property of transmitting a much greater proportion of the ultra-violet rays of sunlight than has any previously known. In view of recent appreciation in the medical world of the curative power of the ultra-violet rays, this glass is of special importance.

In the Lighthouse Department further developments include advances in connection with unattended acetylene lights. As the result of considerable experimental work a satisfactory system has been evolved, and complete unattended acetylene lights, made by the department, are now working in various parts of the world. Mr. Lamplough's light-valve ("History",167 page 167) has now come into regular use for beacon lights, and has been installed also in many lighthouses at home and abroad.

In the autumn of 1922 was taken up the manufacture of the diaphone system of fog signals, and this was followed in 1924 by an important series of trials at St. Catherine's Lighthouse of different types of diaphone, lent to the Trinity House. The tests were carried out in the presence of the Deputy-Master and other Elder Brethren, and the advantages of the diaphone over the standard syren instrument were demonstrated very strikingly.

The Glasgow works have been enlarged by the purchase of adjoining property, and considerably re-organised.

Social work among the women and girls has been continued, and its scope extended, under the supervision of Mrs. Sleigh, helped by Mrs. Howe, the forewoman in the Globe Department. In 1921 a Works Company of Girl Guides, the first to be formed in the neighbourhood, was started. This company has steadily progressed, and there are now over sixty members. Dancing classes have been a special feature of the work, and in each year since the company was formed its girls have been chosen to take part in the annual display at the Birmingham Town Hall, arranged by the Union of Girls' Clubs. In 1924 they gained one shield for the dramatic section, one shield for sewing and one certificate for dancing, and in 1925 both senior and junior certificates for knitting, plain-sewing, and toy-making, as well as for folk-dancing. Further, in the Girl Guides'

Division competition at West Bromwich in November 1925 they won the senior and junior shields for handiwork, and were awarded highest marks in the display of national dancing and club swinging.

The work of the Ambulance Committee has also been considerably extended, and a well-equipped ambulance room has been set up in the Old Hall. All cases of cuts and injuries are now treated by a properly qualified member of the Ambulance Brigade, and the value of this work is shown in the reduction in the number of cases of septic poisoning resulting from slight injuries.